D1363291

A2-Level
Biology

The Revision Guide

Contents

Editors:
Ellen Bowness, Kate Houghton, Simon Little, Kate Redmond, Rachel Selway, Jennifer Underwood.

Contributors:
Ben Aldiss, Gloria Barnett, Wendy Butler, Martin Chester, James Foster, Julian Hardwick, Derek Harvey, Katherine Reed, Ed Robinson, Adrian Schmit, Emma Singleton.

Proofreaders:
Ben Aldiss, Sue Hocking, Gloria Steward.

Published by Coordination Group Publications Ltd.

This book is suitable for:

OCR, AQA A, AQA B and Edexcel.

There are notes at the tops of double pages to tell you if there's a bit you can ignore for your syllabus.

ISBN: 978 1 84146 365 0

Groovy website: www.cgpbooks.co.uk
Jolly bits of clipart from CorelDRAW®
Printed by Elanders Hindson Ltd, Newcastle upon Tyne.

Text, design, layout and original illustrations © Coordination Group Publications Ltd 2005.
All rights reserved.

What to revise...

*This book covers **OCR, EDEXCEL, AQA A** and **AQA B**.
These two pages show you **exactly which pages to revise**
for each examined unit of these syllabuses.*

EDEXCEL

Unit 4 — Respiration and Coordination and Options	
Core material	Section 1 — p2-9 Section 5 — p62-75, 82-87 Section 10 — p134
Option A *Microbiology and Biotechnology*	Section 13 — p176-185, 190-192, 196-199
Option B *Food Science*	Section 15 — all pages
Option C *Human Health and Fitness*	Section 14 — all pages Section 5 — p76-77 Section 10 — p136-137 Section 13 — p194-195 Section 12 — p170, 172

Unit 5B — Genetics, Evolution and Biodiversity	
5B.1 Photosynthesis	Section 1 — p12-19 Section 6 — p90-91
5B.2 Control of Growth in Plants	Section 5 — p88-89
5B.3 Biodiversity	Section 2 — p20, p24-25, p28-37 Section 4 — p54-59
5B.4 Genetics and Evolution	Section 3 — All pages Section 4 — p56-59 Section 7 — All pages

AQA B

Core Modules	
Module 4 *Energy, Control and Continuity*	Section 1 — p2-7, 12, 16-19 Section 3 — p40-49 Section 4 — p54-59 Section 5 — p62-87
Module 5 (a) *Environment*	Section 2 — p20-25, 27-31, 36-39

Option Modules	
Option Module 6 *Applied Ecology*	Section 2 — p26-27, 32-33, 37 Section 4 — p60 Section 8 — p108-109, 111-115, 118-121
Option Module 7 *Microbes and Disease*	Section 12 — p170-171 Section 13 — p176-183, 188, 190-199
Option Module 8 *Behaviour and Populations*	Section 10 — p140-141 Section 11 — 142-143, 154, 156, 160-163 Section 12 — all pages Section 15 — p212-217

OCR

Module 2804 — Central Concepts	
Section 5.4.1 *Energy and Respiration*	Section 1 — p2-11
Section 5.4.2 *Photosynthesis*	Section 1 — p12, p14-19
Section 5.4.3 *Populations and Interactions*	Section 2 — p24-25, p28-31, p34, p36
Section 5.4.4 *Meiosis, Genetics and Gene Control*	Section 3 — p40-41, p43-44, p46-51
Section 5.4.5 *Classification, Selection and Evolution*	Section 3 — p53 Section 4 — all pages
Section 5.4.6 *Control, Coordination and Homeostasis*	Section 5 — p62-64, p66-70, p72-75, p86-89

Module 2805 — Optional components	
Component 01 *Growth, Development and Reproduction*	Section 11 — all pages
Component 02 *Applications of Genetics*	Section 2 — p39, Section 3 — p42, p44-45, p48 Section 7 — p102-103, p104-107, Section 9 — all pages Section 12 — p175
Component 03 *Environmental Biology*	Section 2 — p27, p30-33, p36-39 Section 8 — p109-112, p114-119
Component 04 *Microbiology and Biotechnology*	Section 13 — p176-179, 180-189, p192
Component 05 *Mammalian Physiology and Behaviour*	Section 5 — p64-65, p76-83 Section 6 — p96-99 Section 10 — all pages

AQA A

Unit 4 — Inheritance, Evolution and Ecosystems	
Inheritance and Evolution	Section 2 — p38-39 Section 3 — all pages Section 4 — p54-57
Ecosystems	Section 2 — p20-23, 26-29, 34-37
Energy Processes	Section 1 — p2-10, 12, 16-19

Unit 5 — Physiology and the Environment	
Physiological Control	Section 4 — p60 Section 5 — 62-69 86-87 Section 6 — p90, 92-95
Digestion and Absorption	Section 6 — p96-101
The Nervous System	Section 5 — p70, 72-75, 77-81 Section 10 — p140 Section 14 — p200-201

There are also notes throughout the book telling you where to skip bits that aren't in your syllabus.

Energy and the Role of ATP

All animals and plants need energy for life processes and also for reading books like this.
This stuff is pretty tricky and we're diving in at the deep end, so hang on...

Biological processes need **Energy**

Cells need **chemical energy** for biological processes to occur. Without this energy, these processes would stop and the animal or plant would just **die**... not good.

Energy is needed for **biological processes** like:
- active transport
- muscle contraction
- maintenance of body temperature
- reproduction and growth

Plants need energy for **metabolic reactions**, like:
- photosynthesis
- taking in minerals through their roots

* ATP lowers activation energy needed for metabolic reactions

ATP carries **Energy** around

It only **gets worse** from here on in for the rest of the section. But **don't worry**, ATP didn't make any sense to me at first — it just clicked after many **painful hours** of reading dull books and listening to my teacher going on and on.

Here goes...

1)   ATP (**adenosine triphosphate**) is a **small water-soluble** molecule that is easily transported around cells.

2)   It's made from the nucleotide base **adenine**, combined with a **ribose sugar** and **three phosphate groups**.

3)   ATP is a **phosphorylated nucleotide** — this means it's a nucleotide with extra phosphate groups added.

4)   ATP **carries energy** from **energy-releasing** reactions to **energy-consuming** reactions.

How ATP carries energy:

* ATPsynthase catalyses resynthesis of ATP
* ATPase catalyses breakdown of ATP → ADP

1) ATP is **synthesised** from **adenosine diphosphate** (ADP) and an **inorganic phosphate** group using the energy produced by the **breakdown of glucose**. The enzyme **ATPsynthase** catalyses this reaction.

2) ATP **moves** to the part of the cell that requires energy.

3) It is then **broken down** to ADP and **inorganic phosphate** and **releases chemical energy** for the process to use. **ATPase** catalyses this reaction.

4) The ADP and phosphate are **recycled** and the process starts again.

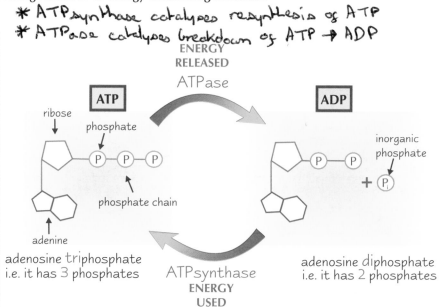

Cells Release Energy *(to make ATP)* by Respiration

Cellular respiration is the process where cells **break down glucose**, it produces carbon dioxide and water and releases **energy**. The energy is used to **produce ATP** from ADP and P_i. There are two types of respiration:

1) **Aerobic respiration** — respiration **using oxygen**.
2) **Anaerobic respiration** — respiration **without oxygen** (see page 8 for more). Both types produce ATP.

You need to learn the summary equation for **aerobic respiration**.

$$C_6H_{12}O_6 \text{ (glucose)} \ + \ 6O_2 \ \longrightarrow \ 6CO_2 \ + \ 6H_2O \ + \ \text{Energy}$$

Energy and the Role of ATP

Respiration *takes place in the* **Mitochondria** *of the* **Cell**

1) **Mitochondria** are present in all **eukaryotic** (i.e. plant, animal, fungi and protoctist) cells. They're 1.5 to 10 µm long.

2) Cells that use lots of energy, e.g. **muscle cells**, **liver cells** and the middle section of **sperm**, have lots of mitochondria.

3) The **inner membrane** of each mitochondrion is folded into **cristae** — structures that increase surface area.

4) **ATP** is produced via the **stalked particles** on the cristae of the inner mitochondrial membrane, in a stage called the **electron transport chain** (see page 7).

5) The **Krebs Cycle** (page 6) takes place in the **matrix** of mitochondria.

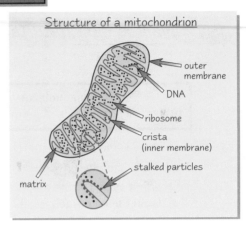

Structure of a mitochondrion

outer membrane
DNA
ribosome
crista (inner membrane)
stalked particles
matrix

Respiration *is a* **Metabolic Pathway**

There are some pretty confusing technical terms about reactions in this section. If you do chemistry you'll be laughing — if not, you'd better concentrate:

*   **Metabolic pathway** — a **series** of **small reactions**, e.g. respiration or photosynthesis, controlled by enzymes.

*   **Catabolic reactions** — breaking **large molecules** into **smaller ones** using enzymes, e.g. breaking down glucose in respiration.

*   **Anabolic reactions** — **combining smaller molecules** to make **bigger ones** using enzymes.

*   **Phosphorylation** — **adding phosphate** to a molecule, e.g. ADP is phosphorylated to ATP.

*   **Hydrolysis** — the **splitting** of a molecule using **water**.

*   **Photolysis** — the **splitting** of a molecule using **light** energy.

Redox reactions — reactions that involve **oxidation** and **reduction**.

1) If something is **reduced** it has **gained electrons**, and may also have lost oxygen or gained **hydrogen**.
    If something is **oxidised** it has **lost electrons**, and may also have gained oxygen or lost **hydrogen**.

2) Oxidation of one thing always involves reduction of something else.

3) The enzymes that catalyse redox reactions are called **oxidoreductases**.

4) Respiration and photosynthesis are riddled with redox reactions.

Oxidation	Reduction
electrons are lost	electrons are gained
oxygen is added	oxygen is lost
hydrogen is lost	hydrogen is gained

One way to remember electron movement is "OILRIG" = Oxidation Is Loss of e⁻ and H, Reduction Is Gain of e⁻ and H.

Practice Questions

Q1 How is energy released from ATP?

Q2 Write down five metabolic processes in animals which require energy.

Q3 What is the purpose of the cristae in mitochondria?

Q4 What are oxidoreductases?

Exam Questions

Q1 What is the connection between phosphate and the energy needs of a cell?   [2 marks]

Q2 ATP is a small, water-soluble molecule which can be rapidly and easily converted back into ADP if ATPsynthase is present. Explain how these features make ATP suitable for its function.   [3 marks]

I've run out of energy after that little lot...

You really need to understand what ATP is, because once you start getting bogged down in the complicated details of respiration and photosynthesis, at least you'll understand why they're important and what they're producing. It does get more complicated on the next few pages, so take your time to understand the basics before you turn the page.

Glycolysis

You can split the process of respiration into four parts — that way you don't have to swallow too many facts at once. The first bit, glycolysis, is pretty straightforward.

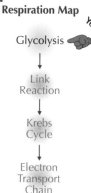

*Glucose: $C_6H_{12}O_6$
*Pyruvate: $C_3H_6O_3$

Respiration Map

Glycolysis 👉 *You are here*

↓

Link Reaction

↓

Krebs Cycle

↓

Electron Transport Chain

Glycolysis is the First Stage of Respiration

So, to recap... most cells use carbohydrates, usually glucose, for respiration.

> Glycolysis splits **one molecule** of glucose into **two** smaller molecules of **pyruvate**.

Glucose is a hexose (6-carbon) molecule.
Pyruvate is a triose (3-carbon) molecule.
Pyruvate is also known as pyruvic acid.

1) Glycolysis is the first stage of respiration (see the map to the right).
2) It takes place in the **cytoplasm** of cells.
3) It's the **first stage** of both aerobic and anaerobic respiration, and **doesn't need oxygen** to take place — so it's **anaerobic**.

There are Two Stages of Glycolysis — Phosphorylation and Oxidation

1 Stage One — Phosphorylation

1) Glucose is split using water (**hydrolysis**).
2) Glucose is **phosphorylated** by adding 2 **phosphates** from 2 molecules of ATP.
3) 2 molecules of **triose phosphate** and 2 molecules of ADP are created.

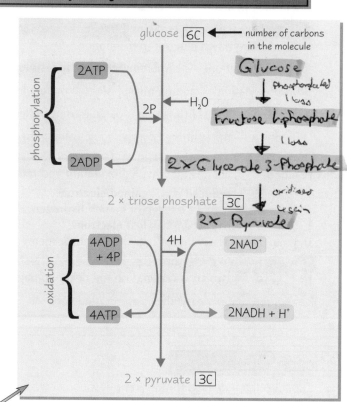

glucose 6C ← number of carbons in the molecule

Glucose
↓ Phosphorylated
1 loss
Fructose biphosphate
↓ 1 loss
2 × Glycerate 3-Phosphate

phosphorylation:
2ATP
2P ← H_2O
2ADP

2 × triose phosphate 3C ↓ oxidised & again

2 × Pyruvate

2 Stage Two — Oxidation

1) The triose phosphate is **oxidised** (loses hydrogen), forming **two** molecules of **pyruvate**.
2) **Coenzyme NAD⁺** collects the hydrogen ions, forming **2 reduced NAD (NADH + H⁺)**.
3) **4 ATP** are produced, but 2 were used up at the beginning, so there's a **net gain of 2 ATP**.

oxidation:
4ADP + 4P
4H
2NAD⁺
4ATP
2NADH + H⁺

2 × pyruvate 3C

> A coenzyme is a <u>helper</u> molecule that carries chemical groups or ions about, e.g. NAD removes H⁺ and carries it to other molecules.

A → B
1 ↓ 2
These arrows in diagrams just mean that A goes into the main reaction and is converted to B. A will normally release or collect something from molecule 1, e.g. hydrogen or phosphate.

A triose phosphate is just a simple 3-carbon sugar with a phosphate group attached. Different books use different names, but this is the easiest to remember.

Next in Aerobic Respiration...

*Reduced NAD = NADH + H⁺

1) The **2 molecules** of **reduced NAD** go to the **electron transport chain** (see page 7).
2) The **two pyruvate** molecules go in to the matrix of the **mitochondria** (by active transport) for the **link reaction** (a small reaction that **links** glycolysis to the second stage, the **Krebs cycle**). It's so exciting I bet you can't wait...

Glycolysis

The Link Reaction converts Pyruvate to Acetyl Coenzyme A

The link reaction is fairly simple and goes like this:

1) One **carbon atom** is removed from pyruvate in the form of CO_2.

2) The remaining **2-carbon molecule** combines with **coenzyme A** to produce **acetyl coenzyme A** (**acetyl CoA**).

3) Another oxidation reaction occurs when **NAD⁺** collects more **hydrogen ions**. This forms **reduced NAD** (**NADH + H⁺**).

4) **No ATP** is produced in this reaction.

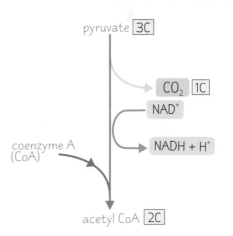

pyruvate 3C

CO_2 1C

NAD⁺

coenzyme A (CoA)

NADH + H⁺

acetyl CoA 2C

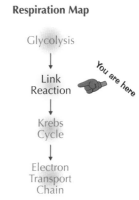

Respiration Map

Glycolysis

↓

Link Reaction

You are here

↓

Krebs Cycle

↓

Electron Transport Chain

The Link Reaction occurs Twice for every Glucose Molecule

1) For each **glucose molecule** used in glycolysis, **two pyruvate** molecules are made.

2) But the **link reaction** uses only **one pyruvate** molecule, so the **link reaction** and the **Krebs cycle** happen **twice** for every glucose molecule which goes through glycolysis.

The Products of the Link Reaction go to the Krebs Cycle and the ETC

So for each glucose molecule:

- Two molecules of **acetyl coenzyme A** go into the Krebs cycle (see next page).
- Two **carbon dioxide molecules** are released as a waste product of respiration.
- Two molecules of **reduced NAD** are formed and go into the **electron transport chain** (which is covered on the next two pages).

Practice Questions

Q1 What do the terms hydrolysis and phosphorylation mean?
Q2 Why is there only a net gain of 2 ATP during glycolysis?
Q3 How many molecules of ATP are produced in the link reaction?
Q4 Where is acetyl CoA formed?

Exam Questions

Q1 Describe simply how a 6-carbon molecule of glucose can be changed to pyruvate. [6 marks]

Q2 Describe what happens in the link reaction. [4 marks]

Acetyl Co-what?

It's all a bit confusing, but you need to know it, so it's worth taking a bit of time to break it down into really simple chunks. Don't worry too much if you can't remember all the little details straight away. If you can remember how it starts and what the products are, you're getting there. You'll get the hang of it all eventually, even if it seems hard right now.

Krebs Cycle and Electron Transport Chain

And now we have the third and fourth stages of the respiration pathway. Keep it up — you're nearly there.

The **Krebs Cycle** is the **Third Stage** of *Aerobic Respiration*

The Krebs cycle takes place in the **matrix** of the mitochondria. It happens once for each pyruvate molecule made in glycolysis, and it goes round twice for every glucose molecule that enters the respiration pathway.

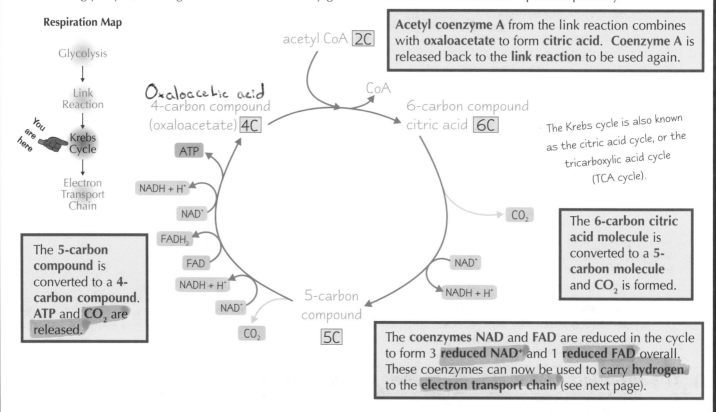

Respiration Map

Glycolysis

Link Reaction

You are here → Krebs Cycle

Electron Transport Chain

Oxaloacetic acid
4-carbon compound (oxaloacetate) 4C

acetyl CoA 2C

CoA

6-carbon compound citric acid 6C

Acetyl coenzyme A from the link reaction combines with **oxaloacetate** to form **citric acid**. Coenzyme A is released back to the **link reaction** to be used again.

The Krebs cycle is also known as the citric acid cycle, or the tricarboxylic acid cycle (TCA cycle).

ATP

NADH + H⁺

NAD⁺

FADH₂

FAD

NADH + H⁺

NAD⁺

CO₂

5-carbon compound 5C

NAD⁺

NADH + H⁺

CO₂

The 5-carbon compound is converted to a **4-carbon compound**. ATP and CO₂ are released.

The 6-carbon citric acid molecule is converted to a **5-carbon molecule** and CO₂ is formed.

The **coenzymes NAD and FAD** are reduced in the cycle to form 3 **reduced NAD⁺** and 1 **reduced FAD** overall. These coenzymes can now be used to carry **hydrogen** to the **electron transport chain** (see next page).

Products of the **Krebs Cycle** are used in the **Electron Transport Chain**

Some products are **reused**, some are **released** and others are used for the **next stage** of respiration:

- One **coA** is **reused** in the next **link reaction**.
- **Oxaloacetate** is **regenerated** so it can be **reused** in the next **Krebs cycle**.

- Two **carbon dioxide** molecules are released as a **waste product** of respiration.
- One molecule of **ATP** is made per turn of the cycle — by **substrate level phosphorylation**.

- Three reduced NAD and **one reduced FAD** co-enzymes are made and carried forward to the **electron transport chain**.

The **Electron Transport Chain** is the **Final Stage** of *Aerobic Respiration*

Before we get too bogged down in all the details, here's what the electron transport chain is all about:

All the products from the previous stages are used in this final stage. Its purpose is to **transfer** the **energy** from molecules made in glycolysis, the link reaction and the Krebs cycle to ADP. This forms **ATP**, which can then deliver the energy to parts of the cell that need it.

The electron transport chain is where **most of the ATP** from respiration is produced. In the whole process of aerobic respiration, **32 ATP molecules** are produced from one molecule of glucose: 2 ATP in glycolysis, 2 ATP in the Krebs cycle and 28 ATP in the electron transport chain.

The electron transport chain also **reoxidises NAD and FAD** so they can be reused in the previous steps.

Krebs Cycle and Electron Transport Chain

The *Electron Transport Chain* produces *lots of ATP*

The **electron transport chain** uses the molecules of **reduced NAD** and **reduced FAD** from the previous three stages to produce **28 molecules of ATP** for every molecule of glucose.

1) **Hydrogen atoms** are released from **NADH + H⁺** and **FADH$_2$** (as they are oxidised to NAD⁺ and FAD). The H atoms **split** to produce **protons (H⁺)**, and **electrons (e⁻)** for the chain.

2) The **electrons** move along the electron chain (made up of three **electron carriers**), losing energy at each level. This energy is used to **pump** the **protons (H⁺)** into the space **between** the inner and outer **mitochondrial membranes** (the **intermembrane space**).

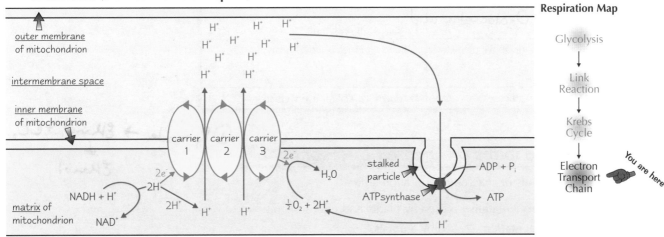

3) The **concentration** of protons is higher in the intermembrane space than in the mitochrondrial matrix, so an **electrochemical gradient** exists.

4) The **protons** then move back through the inner membrane **down** the **electrochemical gradient**, through specific channels on the **stalked particles** of the **cristae** — this drives the enzyme **ATPsynthase.** By 'spinning like a motor', this enzyme supplies **electrical potential energy** to make **ATP** from ADP and inorganic phosphate.

5) The **protons** and **electrons** recombine to form **hydrogen**, and this combines with **molecular oxygen** (from the blood) at the end of the transport chain to form **water**. Oxygen is said to be the final **electron acceptor**.

The **synthesis of ATP** as a result of the energy released by the electron transport chain is called **oxidative phosphorylation**.

This is how the **electron transport chain** produces 28 molecules of **ATP** from **1** molecule of **glucose**:

- 1 turn of the Krebs cycle produces **4** molecules of **reduced NAD** (including **1** from the **link reaction**) and **1** of **reduced FAD**.
- **2** molecules of **pyruvate** enter the Krebs cycle for **each** molecule of **glucose**, so overall **8 NAD⁺** and **2 FAD** are reduced.
- **2 reduced NAD** are also produced from the first part of respiration, **glycolysis** (see p. 4).
- Each **reduced NAD** can produce **2.5 ATP**, and each **reduced FAD** can produce **1.5 ATP**.
- So: 8 reduced NAD + 2 reduced NAD = **10** reduced NAD. **10 × 2.5 = 25 ATP.** **2 reduced FAD × 1.5 = 3 ATP.**
  In total, 25 + 3 = **28 molecules of ATP.**

(There are also **2 ATP** produced by **glycolysis**, and 2 for each molecule of glucose in the **Krebs cycle** = 32 ATP produced in total by **respiration**.)

Practice Questions

Q1 How many molecules of CO$_2$ are made in one turn of the Krebs cycle?

Q2 Name the two coenzymes that are reduced in the Krebs cycle.

Q3 Which molecule finally accepts the electrons passed down through the electron transport chain?

Exam Question

Q1 Calculate the number of ATP molecules that are produced by aerobic respiration from one molecule of glucose. Show your working in detail.   [14 marks]

Cheers for that Mr Krebs...

...you can keep your cycles to yourself, in future. Phew, this biochemistry stuff is tough going. The key to learning this stuff is to learn the big facts first — glycolysis, link reaction, Krebs cycle, electron transport chain. Once you know what the main parts are and roughly what happens at each stage, you stand some chance of learning the more detailed stuff.

Anaerobic Respiration

Anaerobic respiration is respiration without oxygen — so read on — and don't hold your breath.
You can skip these two pages if you're doing AQA B.

Anaerobic Respiration is Different from Aerobic Respiration

Anaerobic Respiration	Aerobic Respiration
Does not need oxygen	Needs oxygen
Takes place in cytoplasm	Takes place in cytoplasm and mitochondria
Uses glycolysis and alcoholic or lactate fermentation	Uses glycolysis, Krebs cycle, link reaction and the electron transport chain
Pyruvate not completely oxidised	Pyruvate oxidised in link reaction
Can follow either of two metabolic pathways 1. alcoholic fermentation 2. lactic acid fermentation	Follows one metabolic pathway
Produces 2 ATP for every glucose	Produces 32 ATP for every glucose

① Pyruvate → Ethanal + CO_2
↓
Ethanol

There are Two forms of Anaerobic Respiration

The **two different kinds** of anaerobic respiration are:

1) **Alcoholic fermentation** — used by **plants** and **microorganisms** like yeast.

2) **Lactate fermentation** — used by **animals**.

They both occur in the **cytoplasm**, and both start with the **glycolysis reaction**. *② Pyruvate → Lactic acid*

Alcoholic Fermentation makes Alcohol

Alcoholic fermentation's a bit simpler than aerobic respiration — thank goodness.

1) It starts with **glucose (6C)**. The process of glycolysis turns this into **pyruvate (3C)** (see page 4).

2) CO_2 is removed from pyruvate to form **ethanal (2C)**.

3) The **reduced NAD** made in glycolysis is **reoxidised** and transfers **two hydrogen atoms** to **ethanal** to form **ethanol**.

4) The **reoxidised NAD** can then be reused in **glycolysis**.

<u>Alcoholic Fermentation</u>

2ADP 2ATP NAD⁺ NADH + H⁺ CO_2 NADH + H⁺ NAD⁺

glucose ——————————→ pyruvate ——→ ethanal ——→ ethanol
              Glycolysis

Alcoholic fermentation is used in Industrial Processes

Some industrial processes use fermentation reactions to help produce their products.
The most useful products that **microorganisms** produce by fermentation are **alcohol** and **carbon dioxide**.

Two examples of industries that use fermentation are:

1) **Brewing** — **Yeasts** ferment fruit or grain to produce **alcoholic drinks** like wine, cider and beer.

2) **Breadmaking** — **Yeasts** produce CO_2 which helps the bread to rise. The alcohol **evaporates** during baking.

Fermentation reactions are normally carried out on an industrial scale in vessels called **fermenters** or **bioreactors** (see page 182 for more).

Anaerobic Respiration

Lactate Fermentation Occurs in Animal Cells

Some animal cells can also respire without oxygen when they need to, for a short time:

1) It starts with **glycolysis**, producing **pyruvate** (see page 4).

2) **Pyruvate** is **reduced** to **lactic acid** (lactate) by adding **two H atoms** from **reduced NAD**.

3) **NAD** is returned to glycolysis to be **used again**.

Pretty simple, really.

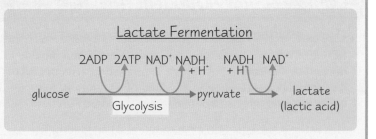

Lactate Fermentation

Anaerobic respiration often occurs in Muscle Cells during Vigorous Exercise

When the muscles have been working hard, the supply of pyruvate from glycolysis is **more than** the oxygen supply available for the later stages of **aerobic respiration**. This is called an 'oxygen debt' and is when cells start respiring anaerobically by **lactate fermentation**.

As lactate builds up in the muscles, you can get a painful, **burning sensation** (stitch). The lactate is removed from the muscle cells by the **blood**. When oxygen is available again, the **liver** converts the lactate back to **glucose** and then **glycogen** for storage, or back to pyruvate.

Olympic sprinters run out of oxygen in their muscle cells after about 10 metres of a 100 metre race. From then on they're using anaerobic respiration to complete the race.

Anaerobic Respiration is Not very Efficient

In **aerobic** respiration, about **32 ATP** molecules are made for every glucose molecule (see p. 7).

In **anaerobic** respiration, only **2 ATP** molecules are made (during glycolysis).

So **aerobic** respiration releases **loads more energy** than **anaerobic** respiration.

Practice Questions

Q1 Name the two types of anaerobic respiration.

Q2 What molecule is made when CO_2 is removed from pyruvate during alcoholic fermentation?

Q3 What is the final product of alcoholic fermentation?

Q4 Which type of anaerobic respiration occurs in animals?

Q5 Why is anaerobic respiration so inefficient compared to aerobic respiration?

Exam Question

Q1 a) Compare and contrast the two forms of anaerobic respiration. [10 marks]

b) In terms of the amount of ATP produced, how much more efficient is aerobic respiration at releasing energy than anaerobic respiration? [3 marks]

c) Olympic sprinters use anaerobic respiration during most of a 100 m race, even though it is less efficient at releasing energy than aerobic respiration. Suggest why this might be. [2 marks]

Did somebody mention beer...

Anaerobic respiration isn't too bad. It's not overly-complicated, and fermentation even has some rather tasty products. That makes it a lot more interesting than most of this section, I reckon. And everyone knows what lactate fermentation feels like. This is the stuff that makes biology slightly less tedious, so you've got no excuse for not learning this.

Measuring Respiration

Skip these two pages if you're doing AQA B or Edexcel. AQA A can skip page 11.
You're probably wondering when this biochemical torture will ever end — well, hang in there a bit longer, and learn how you can measure respiration. Then you can go and eat a whole tub of chocolate ice-cream.

The **Respiratory Quotient (RQ)** tells you what **Substrate** is being **Respired**

During respiration, oxygen is consumed and carbon dioxide is produced.

RQ (Respiratory Quotient) is the amount of **carbon dioxide** produced,
divided by the amount of **oxygen consumed** in a set period of time.
It's useful because it shows what **kind** of substrate is being used for
oxidation, and what sort of **metabolism** an organism has.

$$RQ = \frac{\text{Molecules of } CO_2 \text{ produced}}{\text{Molecules of } O_2 \text{ used}}$$

This is the basic equation for respiration using glucose:

$$C_6H_{12}O_6 \; + \; 6O_2 \longrightarrow 6CO_2 \; + \; 6H_2O \; + \; \textbf{Energy}$$

RQ = molecules of CO_2 produced ÷ molecules of O_2 used
    = 6 ÷ 6 = 1.

So if cells only used pure glucose, the RQ would always just be 1. But **most cells don't just use glucose**...

Different **Substrates** have **Different RQ Values**

Under normal conditions, the **usual RQ** for humans is
between **0.7** and **1.0**. An RQ in this range shows that some
fats (**lipids**) are being used for respiration as well as
carbohydrates like glucose. Protein isn't normally used for
respiration by the body unless there's nothing else.

Respiratory Substrate	RQ
Lipids (triglycerides)	0.7
Proteins or amino acids	0.9
Carbohydrates (e.g. glucose)	1
Anaerobic respiration of carbohydrate	> 1

1) **High RQs** (greater than 1) often mean that an organism is short of oxygen,
and having to respire **anaerobically** as well as aerobically.

2) If the organism is respiring **lipids** the RQ will be **lower than 1**.
This is because more oxygen is needed to oxidise fat than to oxidise carbohydrate.

3) **Plants** sometimes have a **low RQ**. This may be because the CO_2 released in
respiration is used for **photosynthesis** (so it's not measured).

The **Rate of Respiration** can be measured with a **Respirometer**

The amount of **oxygen taken up** and the amount of **carbon dioxide produced** are used as **indicators** of rate of
respiration. **Respirometers** measure the rate of **oxygen** being taken up. The **more** oxygen used in a certain
time, the **faster** the rate of respiration.

It's fairly easy to measure rates of respiration in **small invertebrates**, like woodlice, using this **simple apparatus**.

1) The **carbon dioxide** produced is
absorbed by the **sodium hydroxide
solution**, so a **decrease** in the volume
of the air in the tube will be due to
oxygen consumption by the woodlice.

2) The decrease in the volume of the air
reduces the pressure in the tube.
This causes the coloured liquid in the
manometer to move towards the tube.

3) So, the movement of the coloured
liquid in the manometer shows the
oxygen consumption over time.

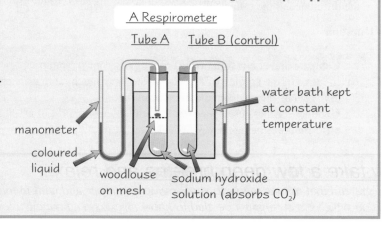

A Respirometer

Tube A   Tube B (control)

water bath kept at constant temperature

manometer

coloured liquid

woodlouse on mesh

sodium hydroxide solution (absorbs CO_2)

Measuring Respiration

Temperature *Affects* Respiration Rate

The graph on the right shows how temperature would typically affect rate of respiration in an organism. The rate of respiration **doubles** for every rise of **10°C**. Above **45°C** the rate of respiration starts to decrease because important proteins (e.g. enzymes) are becoming **denatured**.

For organisms that depend on environmental temperature (like **bacteria** and ectothermic organisms), temperature will greatly affect the rate of respiration. For endothermic organisms this principle is less important because their core temperature is kept constant.

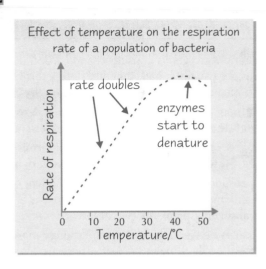

Effect of temperature on the respiration rate of a population of bacteria

You can measure the effect of **temperature** on the rate of respiration using a **respirometer**, by placing the apparatus inside a water bath and varying the temperature of the water.

Practice Questions

Q1   Give the expression used to calculate RQ.

Q2   Under normal conditions what is the usual RQ for humans?

Q3   When is protein used as a respiratory substrate?

Q4   An organism has an RQ of 1.2. Suggest what could be causing this high RQ.

Q5   What apparatus is used to measure rate of respiration?

Q6   Describe what typically happens to the rate of respiration in an organism as its temperature increases.

Exam Questions

Q1   The equation for the respiration of the fat tripalmitin is $2C_{51}H_{98}O_6 + 145O_2 \longrightarrow 102CO_2 + 98H_2O$. What is the RQ for tripalmitin?
    [2 marks]

Q2   The graph shows the respiratory quotients of maize seeds and sunflower seeds during germination. Account for the difference in the shapes of the curves on the graph.

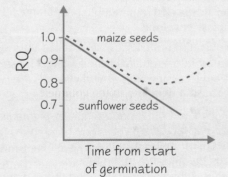

Now take a few deep breaths and relax...

Just be grateful that no one wants to shut you up in a room and measure your BMR. Anyway, that was the last page on respiration which I'm sure you'll be glad to know. Next, we've got eight lovely pages on photosynthesis. Well actually, the first four pages are fairly lovely. The last four are... well, not so lovely. You'll find out. I wouldn't want to spoil the surprise.

Photosynthesis

You remember photosynthesis — carbon dioxide and water plus light energy makes tasty glucose, and the rather useful by-product of oxygen. It's very popular in plants, where it mostly tends to go on in the leaves. Sound familiar yet?

Leaves are Adapted for Photosynthesis

Leaves are **organs** — they contain groups of **tissues** that work together to carry out **photosynthesis**.

1) Leaves are usually **broad**, **thin** and **flat** — this helps them absorb as much of the available **light** as possible. It also means that **carbon dioxide** entering the leaf through the **stomata** can reach inner cells easily.

2) **Veins** branching through the leaf contain **xylem vessels** to bring **water** from the roots, and **phloem vessels** to carry away the **sugars** made in photosynthesis to the rest of the plant.

3) The waxy **cuticle** covering the **upper epidermis** protects the leaf from damaging **UV rays** in the sunlight. It's also **waterproof**, which helps to prevent **dehydration**, and clear so light can penetrate it.

4) The **palisade cells** in the upper part of the leaf contain the most **chloroplasts**, because they're close to the light striking the top of the leaf. Chloroplasts are where photosynthesis happens — they contain the **light-absorbing pigments** (see below).

5) **Air spaces** between the cells in the **spongy mesophyll** layer allow gases like **carbon dioxide** and **oxygen** to diffuse easily through the leaf. There are also spaces between the palisade cells for the same reason.

6) **Stomata** (pores) in the lower surface of the leaf allow exchange of carbon dioxide and oxygen between the leaf and the **atmosphere**.

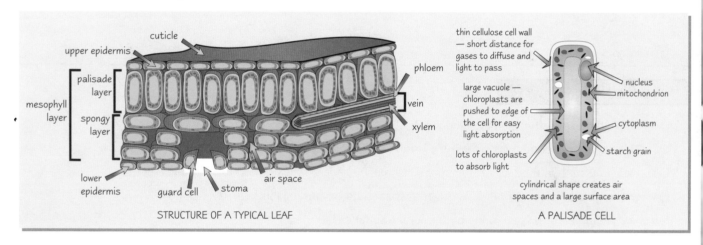

STRUCTURE OF A TYPICAL LEAF   A PALISADE CELL

Chloroplasts are the Site of Photosynthetic Reactions

1) Chloroplasts are usually about **5 µm** in diameter.

2) They're surrounded by a double membrane called the **chloroplast envelope**.

3) **Thylakoids** (fluid-filled sacs) are stacked up inside the chloroplast into structures called **grana**. These structures have a large surface area. The thylakoids are where the **light-dependent reaction** of photosynthesis occurs.

4) **Chlorophyll** and other photosynthetic pigments are found on the **thylakoid membranes.** They form a complex called **photosystem II** (see page 16). Some thylakoids have **extensions** that join them to thylakoids in other grana. These are called **inter-granal lamellae**, and they're the sites of **photosystem I** (see page 16).

5) The thylakoids are embedded in a gel-like substance called the **stroma.** The stroma is where the **light-independent reaction** of photosynthesis (called the **Calvin cycle**) happens. It contains **enzymes** (for the Calvin cycle), **sugars** and **organic acids**.

6) Carbohydrates produced by photosynthesis and not used straight away are stored as **starch grains** in the **stroma**.

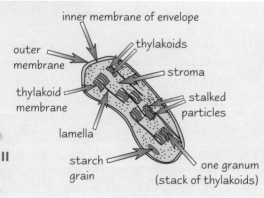

Photosynthesis

Photosynthetic Pigments absorb Visible Light

This page is just for Edexcel.

The **thylakoid membranes** inside chloroplasts contain **photosystems**, which are groups of **pigment molecules**.

Three of the most important pigments found in plants are **chlorophyll a**, **chlorophyll b** and **carotene** (a carotenoid).

Experiments have shown that each of these pigments absorbs a **different wavelength** (colour) of light:

1) The different pigments can be separated using **chromatography**.

2) A **spectrometer** is used to measure the absorption of light of different wavelengths by the different pigments.

3) This gives an **absorption spectrum**, showing the **absorption** of each pigment over the wavelength range of **visible light** (see diagram).

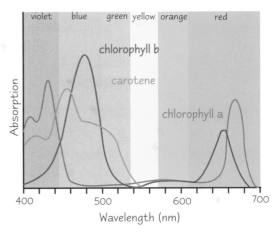

Having three different pigments allows the plant to absorb as many of the **different wavelengths** in white light as possible, as **efficiently** as possible. As the graph shows, the three pigments together are able to absorb most of the **red** and **blue** light in sunlight for photosynthesis. There aren't any pigments that absorb well in the **green** part of the spectrum though, so plants **reflect** green light instead of absorbing it. That's why they look green.

Action Spectra show the Rate of Photosynthesis at Different Wavelengths

Absorption spectra show which wavelengths are **absorbed** by plants. Further experiments have confirmed that these same wavelengths of light are then used in **photosynthesis** (pretty obvious I'd have thought, but hey).

A typical experiment involves:

1) Exposing leaves to light of **different wavelengths** (colours) for a **set period**.

2) Measuring the **volume of oxygen** or **amount of carbohydrate** formed in each case (which corresponds to the amount of **photosynthesis** happening).

3) This is used to produce an **action spectrum** — a graph showing the **rate of photosynthesis** at **different wavelengths**.

There's a pretty **close correlation** between the **absorption** and **action spectra** for plants, as shown in the graph. They absorb mostly **blue** and **red** light, and these are also the wavelengths at which the most **photosynthesis** happens.

Practice Questions

Q1 What reaction occurs in the thylakoid membranes?

Q2 Name three pigments used in photosynthesis.

Q3 What does an absorption spectrum show?

Q4 Which pigment has a high absorption spectrum at about 680nm?

Q5 What does an action spectrum show?

Exam Questions

Q1 Describe how a leaf is adapted for photosynthesis. [5 marks]

Q2 Describe how a chloroplast is adapted for photosynthesis. [5 marks]

I'm sorry, chlorophyll, but we're just not on the same wavelength...

If you're not much of a physicist, don't start panicking just because you see the word 'wavelength' on this page. Light just travels in waves, and each colour of light has a different length of wave. But don't worry about any of that — just remember that plants have different pigments so they can absorb different colours and get as much light as possible for photosynthesis.

Limiting Factors in Photosynthesis

Okay, at first glance this might seem quite a boring page — but I'm telling you now, appreciate it while it lasts...
These two pages are for OCR and Edexcel only.

Plants also need a supply of magnesium for photosynthesis, as it's an important part of chlorophyll.

There are **Optimum Conditions** for Photosynthesis

The **ideal conditions** for photosynthesis vary from one plant species to another, but the conditions below would be ideal for most plant species in temperate climates like the UK.

High light intensity of a certain **wavelength**
- Light is needed to provide the **energy** for the **light-dependent stage** of photosynthesis — the higher the **intensity** of the light, the **more** energy it provides.
- Only certain **wavelengths** of light are used for photosynthesis (see the **absorption spectrum** on page 13).

Temperature around **25°C**
- **Less than 10 °C** means that enzymes are **inactive** — they need some energy from heat to function.
- **More than 45 °C** and enzymes may start to **denature**. **Stomata** close at high temperatures to avoid losing too much water. This causes photosynthesis to stop because CO_2 can't enter the leaf when the stomata are closed.

Water — a constant supply is needed
- Too **little** and photosynthesis obviously has to stop, because it's one of the **reactants**.
- Too **much** waterlogs the soil and reduces uptake of minerals such as **magnesium**, which is needed to make **chlorophyll a**.

Carbon dioxide at **0.4%**
- Carbon dioxide makes up **0.04%** of the gases in the atmosphere.
- Increasing this to **0.4%** gives a **higher rate** of photosynthesis, but any higher and the stomata start to **close**.

Light, **Temperature**, **Water** and **CO₂** can all **Limit Photosynthesis**

All four of these things need to be at the right level to allow a plant to photosynthesise as quickly as it can. If any **one** of them is too low, it will be **limiting photosynthesis**. The other three factors could be at the perfect level, and it wouldn't make **any difference** to the speed of photosynthesis. On a warm, sunny, windless day, it's usually **carbon dioxide** that is the limiting factor in photosynthesis. At night, the limiting factor will obviously be **light intensity**. But **any** of those four factors could become the limiting factor, depending on the **environmental conditions**.

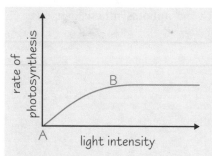

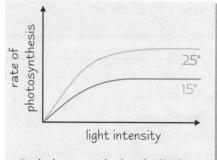

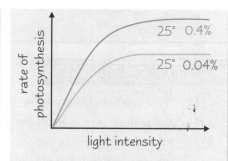

Between points A and B, the rate of photosynthesis is limited by the **light intensity**. So as the light intensity **increases**, so can the rate of photosynthesis. Point B is the **saturation point** — increasing light levels after this point makes no difference, because **something else** has become the limiting factor. The graph now **levels off**.

Both these graphs level off when **light intensity** is no longer the limiting factor. The graph at **25 °C** levels off at a **higher point** than the one at **15 °C**, showing that **temperature** must have been a limiting factor at **15 °C**.

This graph shows that **carbon dioxide concentration** must have been a limiting factor when the previous graph levelled off at **25 °C**. This can be seen by comparing the two lines above, which are at the **same** light intensity and temperature.

Water is essential for photosynthesis and can also be a limiting factor, but lack of water causes lots of **other damage** to the plant — by the time it becomes a limiting factor in photosynthesis, that'll be the least of the plant's troubles.

EDEXCEL ONLY

Limiting Factors in Photosynthesis

Plants can get CO₂ from Respiration

Plants **respire** as well as photosynthesise. Respiration **produces** carbon dioxide, and photosynthesis **uses it up**. Photosynthesis starts at dawn and the rate **increases** as the light gets **brighter**. At one point, **all** the carbon dioxide produced by respiration is **re-used** in photosynthesis, and this is known as the **compensation point**.

After this, **light** and **temperature** increase and respiration can't produce enough CO_2 for the amount of photosynthesis happening. The plant has to use CO_2 diffusing in from the **atmosphere** too. Then when dusk arrives and photosynthesis starts to slow down, the CO_2 from respiration is **all re-used** again. The **compensation point** has been reached again.

Limiting Factors can be Investigated using Pondweed

Canadian pondweed (*Elodea*) is often used to measure the **rate of photosynthesis** under different conditions. The rate at which **oxygen** is produced by this plant can easily be measured, and this **corresponds** to the rate of photosynthesis. You can vary the **light intensity** (as shown below) or another factor, and measure the **effect** this has on the rate of photosynthesis. You can do several different experiments, varying a **different limiting factor** each time, and so find the **optimum conditions** for photosynthesis:

1) Optimum **light intensity** — a source of white light is placed at **different distances** from the plant. The **volume of oxygen** is measured to discover the **optimum distance** (i.e. the one that produces the most oxygen).

2) Optimum light **wavelength** — the light is kept at a **fixed distance** from the pondweed, and the wavelength altered by **adding coloured filters** between the light source and the plant. The volume of oxygen produced is measured to find the **best wavelength** of light for photosynthesis.

3) **Temperature** can be varied by, for example, placing the test tube of pond weed in a beaker containing water of **different temperatures**.

4) **Carbon dioxide** levels can be altered by using different quantities of **sodium hydrogen carbonate** to supply the CO_2.

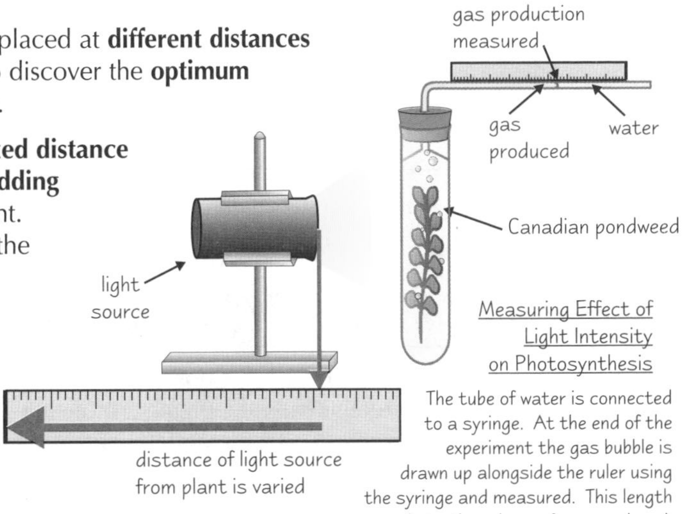

Some factors aren't as easy to measure. Supplying the plant with different amounts of **water** affects the plant in all sorts of ways, so it's not directly testing the effect on **photosynthesis**.

gas production measured

gas produced — water

— Canadian pondweed

Measuring Effect of Light Intensity on Photosynthesis

The tube of water is connected to a syringe. At the end of the experiment the gas bubble is drawn up alongside the ruler using the syringe and measured. This length corresponds to the volume of gas produced.

light source

distance of light source from plant is varied

> Plant growers use **greenhouses** to keep the temperature, water, light intensity and CO_2 concentration **optimal**. But the **cost** of controlling the variables that limit photosynthesis, i.e. **heating**, **lighting** and CO_2 **supply**, mustn't exceed the profit from the **extra yield**, or the growers' efforts won't be worthwhile.

Practice Questions

Q1 Name three factors that can limit photosynthesis.

Q2 Name two sources of CO_2 for plants.

Q3 What is the compensation point?

Exam Questions

Q1 Explain whether a lack of nutrients in the soil can affect photosynthesis. [2 marks]

Q2 Would measuring the uptake of CO_2 by a plant give a true measurement of the rate of photosynthesis? Explain your answer. [2 marks]

Aah, Canadian pondweed — a biology student's best friend...

Or else a piece of damp weed that you're forced to stare at endlessly while it produces small bubbles in a very boring way. It all depends on your point of view, really. Anyway I hope you appreciated this nice but dull page, because the next four pages are a wee bit trickier. Nothing to panic about, I hasten to add, but just... trickier. You'll see. Brace yourselves...

The Light-Dependent Reaction

Don't worry if this seems hard at first. Read it through carefully a couple of times, and it'll start to make sense.

Photosynthesis can be Split into Two Stages

Here's the overall equation for photosynthesis. Hopefully it'll look pretty familiar. When you were doing your GCSEs this little equation was all you had to worry about, but those days are long gone, my friend.

$$6CO_2 + 6H_2O + Energy \xrightarrow{\text{chlorophyll}} C_6H_{12}O_6 + 6O_2$$

Photosynthesis happens in the **chloroplasts** (see page 12), and nowadays you need to know that it consists of two **stages**:

1) The **light-dependent reaction** (which, as the name suggests, needs **light energy**) takes place in the **thylakoid membranes** of the chloroplasts. Light energy is absorbed by pigments in the **photosystems** (see below), and used to provide the energy for the next stage — the light-independent reaction. There are **two different reactions** going on in this next stage — **cyclic photophosphorylation** and **non-cyclic photophosphorylation**. The plant can **switch between** the two, depending on whether it needs **reduced NADP** or just **ATP** (see below).

2) The **light-independent reaction** or **Calvin cycle** (which, as the name suggests, doesn't use light energy) happens in the **stroma** of the chloroplast. The **ATP** and the **reduced NADP** molecules that were made in the light-dependent reaction supply the **energy** to make **glucose**. See pages 18-19 for more on the Calvin cycle.

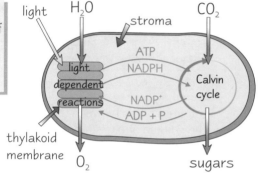

The diagram shows how the two different reactions, light-dependent and light-independent, fit together in the chloroplast.

Photosystems I and II capture Light Energy

1) **Photosystems** are made up of **chlorophyll a**, **accessory pigments** (like **chlorophyll b** and **carotenoids**) and **proteins**. The proteins hold the pigment molecules in the best positions for **absorbing** light energy and **transferring** this energy to the **reaction centre** of the photosystem.

2) The **reaction centre** is a particular **chlorophyll molecule** called a **primary pigment**. The energy from absorbing light is passed from one **accessory pigment** to another until it reaches this **primary pigment**.

3) The energy is then used to **excite** pairs of **electrons** in the reaction centre pigment. The electrons move up to a higher **energy level**, ready to be used in the **light dependent reactions**.

4) There are **two** different photosystems used by plants to capture light energy. **Photosystem I** (or PSI) uses a chlorophyll molecule that absorbs light at wavelength **700 nm** in its reaction centre. PSI is found mostly in the **lamellae** in a chloroplast. **Photosystem II** (PSII) uses a chlorophyll molecule that absorbs light best at around **680 nm** in its reaction centre. It's found mostly in the **thylakoids** of the chloroplast.

The Light-Dependent Reaction makes ATP in Photophosphorylation

The **energy** captured by the photosystems is used for **two** main things:

1) Making **ATP** from **ADP** and **inorganic phosphate** (**phosphorylation**). It's called **photo**phosphorylation here, as it uses **light**.

2) Splitting **water** into H^+ ions and OH^- ions. This is called **photolysis**, because the splitting (lysis) is caused by light energy (photo). Photolysis is covered on the next page.

Photophosphorylation involves the **excited electrons** in the reaction centre of a photosystem being passed to a special molecule called an **electron acceptor**. These electrons are then passed along a **chain** of other electron carriers, each at a slightly **lower energy level** than the one before, so that the electrons **lose energy** at every stage in the chain. The energy given out is used to add a phosphate molecule to a molecule of ADP — and this is photophosphorylation.

You don't need to know the actual mechanism in detail, but it's very similar to how ATP is produced in **respiration**, involving the flow of **hydrogen ions** (H^+) through **stalked particles** (which you can read all about on page 7). The H^+ ions used in photosynthesis come from the **photolysis of water**, which is covered on the next page.

The difference between **cyclic** and **non-cyclic photophosphorylation** is in what happens to those electrons that have been moving through the chain of carriers. This is also explained on the next page.

The Light-Dependent Reaction

Cyclic Photophosphorylation just produces ATP

Cyclic photophosphorylation only uses **photosystem I**. It's called cyclic photophosphorylation because the electrons from the chlorophyll molecule are simply **passed back** to it after they've been through the chain of carriers — i.e. they're **recycled** and can be used repeatedly by the same molecule. This **doesn't** produce any **reduced NADP** (**NADPH + H⁺**), but there **is** enough energy to make **ATP**. This can then be used in the **light-independent reaction**.

Non-cyclic Photophosphorylation produces ATP, NADPH and Oxygen

Non-cyclic photophosphorylation uses both **PSI** and **PSII**. It involves **photolysis**, which is the splitting of **water** using light energy. Photolysis only happens in **PSII**, because only PSII has the right **enzymes**.

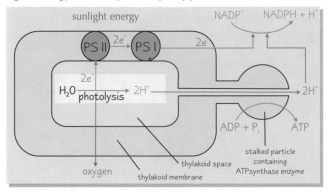

1) Light energy enters **PSII** and is used to move electrons to a **higher energy level**.

2) The electrons are passed along the chain of electron carriers to **photosystem I**. Most of the energy lost by the electrons during this process is used to make **ATP** (like in cyclic photophosphorylation).

3) Light energy is absorbed by PSI, which excites the electrons **again** to an **even higher energy level**.

4) The electrons are passed to a **different electron acceptor**, and **don't** return to the chlorophyll.

5) For the chlorophyll to keep working, the electrons have to be replaced from **somewhere else** — so they're taken from a molecule of **water** (water is the electron donor). This makes the water molecule split up into **protons** (H⁺) and **oxygen**.

6) The **protons** (H⁺) from the water molecule combine with the **electrons** currently with the second electron acceptor to give **hydrogen atoms**. These are used to react with a substance called **NADP** to produce **NADPH** and **H⁺**. These are needed for the **light-independent reaction** (see pages 18-19).

So when electrons move back from the **second electron acceptor** to the chlorophyll molecule, that's **cyclic** photophosphorylation. If they don't, and the replacement electrons come from **water** instead, that's **non-cyclic**.

	cyclic photophosphorylation	non-cyclic photophosphorylation
photosystem	I	I and II
what's needed	light, ADP, inorganic phosphate	light, water, NADP, ADP, inorganic phosphate
what's produced	ATP	ATP, NADPH + H⁺, O₂

Practice Questions

Q1 What is the full equation for photosynthesis?

Q2 Where in the chloroplast does the light-independent reaction of photosynthesis happen?

Q3 What is the reaction centre of a photosystem?

Q4 What two main things is the light energy captured by photosystems used for?

Q5 What is the difference between cyclic and non-cyclic photophosphorylation?

Q6 What useful waste product of photosynthesis is produced during non-cyclic photophosphorylation?

Exam Question

Q1
  a) Where precisely in the plant does the light-dependent stage occur? [1 mark]
  b) Which two compounds produced in the light-dependent stage are used in the light-independent stage? [2 marks]
  c) Which of the light-dependent reactions of photosynthesis are involved in producing these compounds? [3 marks]

Photophosphorylate that, if you can...

*If you're feeling filled with despair as you read this tip, well, don't. You **will** understand this, don't give up. I guarantee it'll seem clearer every time you go through it, until at last you're left wondering what all the fuss was about. By the way, don't be put off when it says protons instead of hydrogen ions. That always confused me, but they mean the same thing.*

The Light-Independent Reaction

The second stage of photosynthesis doesn't need light energy, but that doesn't mean it can happily carry on in the dark. It still needs the NADPH and ATP from the light-dependent reaction, so if that stops, so does the light-independent stage.

The **Light-Independent** Reaction is also called the **Calvin Cycle**

The Calvin cycle makes **hexose sugars** (sugars with **6 carbons**, like **glucose** and **fructose**) from **carbon dioxide** and a **5-carbon** compound called **ribulose bisphosphate**. It happens in the **stroma** of the chloroplasts. There are a few steps in the reaction, and it needs **energy** and **H+ ions** to keep the cycle going. These are provided by the products of the **light-dependent reaction**, **ATP** and **reduced NADP (NADPH + H+)**.

The diagram shows what happens at each stage in the cycle. The numbers in brackets (5C, 3C etc.) show how many **carbon atoms** there are in each molecule — the cleverest bit of the cycle is how it turns a **5-carbon** compound into a **6-carbon** one.

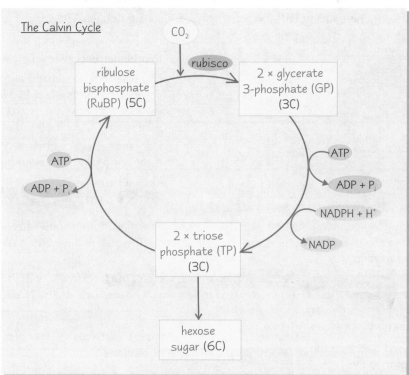

The Calvin Cycle

1) **CO$_2$** enters the leaf through the **stomata** and diffuses into the **stroma** of the chloroplast.
2) There it's taken up by **ribulose bisphosphate (RuBP)**, a **5-carbon** compound. This gives an **unstable 6-carbon** compound, which quickly breaks down into **two** molecules of a **3-carbon** compound called **glycerate 3-phosphate (GP)**.
3) This reaction is catalysed by the enzyme **ribulose bisphosphate carboxylase (rubisco)**.
4) **ATP** from the **light-dependent stage** of photosynthesis is now used to provide the energy to turn the **3-carbon** compound, **GP**, into a **different** 3-carbon compound called **triose phosphate**.
5) This reaction also needs **H+ ions**, which are provided by the **reduced NADP (NADPH + H+)** made in the **light-dependent reaction**.
6) **Two** triose phosphate molecules then **join together** to give **one hexose sugar** (e.g. glucose).

Five out of every **six** molecules of **triose phosphate** produced in the Calvin cycle are **not** used to make hexose sugars, but to **regenerate** RuBP. Making RuBP from triose phosphate molecules uses the rest of the ATP produced by the light-dependent reaction.

The Calvin cycle is the starting point for making **all** the substances a plant needs — plants can't take in **proteins** and **lipids** like animals can.

- Other **carbohydrates**, like **starch**, **sucrose** and **cellulose**, can be easily made by joining the simple hexose sugars together in different ways.
- **Lipids** are made using **glycerol** synthesised from **triose phosphate**, and **fatty acids** from **glycerate 3-phosphate**.
- **Proteins** are made up of **amino acids**, which are also synthesised from **glycerate 3-phosphate**.

The Light-Independent Reaction

The Calvin Cycle needs to turn 6 times to make 1 Glucose Molecule

On the last page, you saw that **five** out of every **six triose phosphate molecules** go back into the cycle to regenerate **RuBP**, rather than going to make new hexose sugars. This means that the cycle has to happen **six times** just to make one new sugar. The box below shows why.

Remember, the photosynthesis equation uses **6 CO₂**.

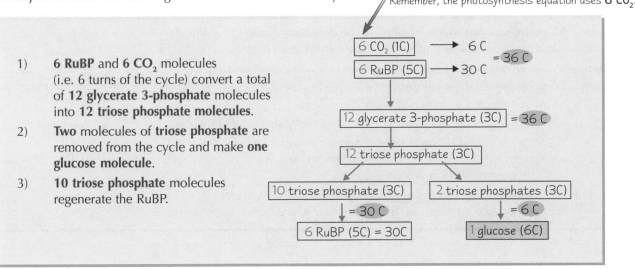

1) **6 RuBP** and **6 CO₂** molecules (i.e. 6 turns of the cycle) convert a total of **12 glycerate 3-phosphate** molecules into **12 triose phosphate molecules**.

2) **Two** molecules of **triose phosphate** are removed from the cycle and make **one glucose molecule**.

3) **10 triose phosphate** molecules regenerate the RuBP.

This might seem a bit inefficient, but it keeps the cycle going and makes sure that there's always **enough RuBP** there ready to combine with CO_2 taken in from the atmosphere.

Practice Questions

Q1   Give two examples of a hexose sugar.

Q2   How many carbon atoms are there in a molecule of triose phosphate?

Q3   Name the enzyme that catalyses the reaction between carbon dioxide and ribulose bisphosphate.

Q4   How is the Calvin cycle involved in making lipids?

Q5   How many CO_2 molecules need to enter the Calvin cycle to make one glucose?

Exam Questions

Q1   Which molecule in photosynthesis:
a) is the carbon dioxide acceptor?   [1 mark]
b) provides the hydrogen ions to reduce glycerate 3-phosphate?   [1 mark]
c) is regenerated in the Calvin cycle?   [1 mark]
d) is known as rubisco?   [1 mark]
e) is made of 5 carbon atoms?   [1 mark]

Q2   Look at the diagram on the right and describe what is happening:
a) between points a and b,   [1 mark]
b) between points b and c,   [1 mark]
c) at point c.   [1 mark]

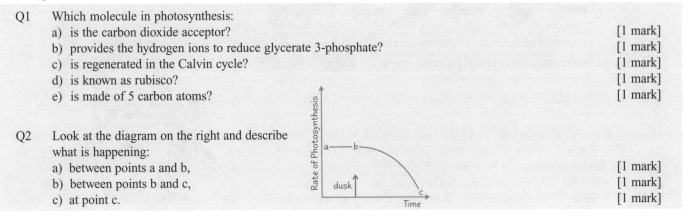

Don't worry — I promise the next section will be a lot easier...

Now don't start sinking into despair again. I know it's a lot to take coming after that last page, but actually this one is probably a bit easier. Learn that cycle on the last page and you're three quarters of the way there. Don't worry too much about learning the maths bit in the box on this page either — as long as you've got the general gist, that's enough.

Ecosystems and Energy Transfers

If you're doing OCR then you don't need to learn these pages. Edexcel just need to learn the definitions on page 20.
These two pages deal with loads of words that you need to know. They also deal with the way energy flows through ecosystems. Not too difficult, but you need to understand the basic principles.

You need to learn some **Definitions** to get you started

Ecosystem	An **ecological unit** which includes all the **organisms** living in a particular area and all the **abiotic** (non-living) features of the local environment.
Population	All the **individuals** of a particular **species** living in a given area.
Community	All the **living organisms** in an ecosystem. These organisms are all **interconnected** by food chains and food webs.
Habitat	The **place** where the communities live, e.g. a rocky shore, a field, etc.
Niche	The **'role'** an organism has in its environment — where it lives, what it eats, where and when it feeds, when it is active etc. Every species has its own **unique** niche.
Environment	The **conditions** surrounding an organism, including both abiotic factors (e.g. temperature, rainfall) and biotic factors (e.g. predation, competition).

The opposite of abiotic is biotic (to do with living things).

Don't get confused between population size (how many in total), and density (how many in a given area).

Energy **Flows Through** ecosystems

Energy comes into the ecosystem from sunlight and is fixed into the ecosystem by plants during **photosynthesis**. The energy stored in the plants can then be passed onto other organisms in the ecosystem along **food chains** — each link in a food chain is called a **trophic level**. During this process a lot of the energy is gradually lost from the food chain — this is **dissipation**. **All** the food chains in an **ecosystem** are linked together in **food webs**.

Scientists call food chains and webs <u>dynamic feeding relationships</u>.

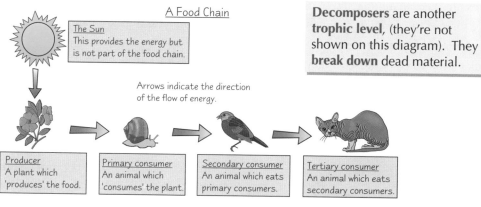

A Food Chain

The Sun
This provides the energy but is not part of the food chain.

Arrows indicate the direction of the flow of energy.

Producer
A plant which 'produces' the food.

Primary consumer
An animal which 'consumes' the plant.

Secondary consumer
An animal which eats primary consumers.

Tertiary consumer
An animal which eats secondary consumers.

Decomposers are another **trophic level**, (they're not shown on this diagram). They **break down** dead material.

1) At each trophic level about **10%** of the energy is used for **growth** and **storage** — that's the energy that can be passed onto the **next** level when the organism is **consumed**.

2) So about **90%** of energy is **wasted** between one trophic level and the next. The cow diagram shows what typically happens to this energy.

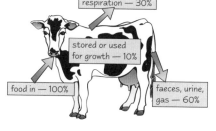

respiration — 30%

stored or used for growth — 10%

food in — 100%

faeces, urine, gas — 60%

All this wastage means that food chains rarely get longer than five links.

Ecosystems and Energy Transfers

Food chains can be shown as *Three Types* of *Pyramid*

In all food pyramids the **area** of each block tells you about the **size** of the trophic level.
The food chain can be shown in terms of number, biomass or energy:

Pyramids of Number

These are the **easiest to produce** — they show the **numbers** of the different organisms so it's just a question of counting.

They're **sometimes misleading**, though — the nice pyramid shape is often messed up by the presence of small numbers of big organisms (like trees) or large numbers of small organisms (like parasites).

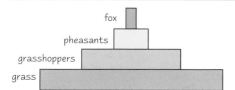

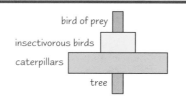

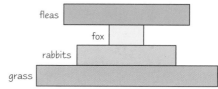

Pyramids of Biomass

These are produced by measuring biomass (the **dry mass** of the organisms in kg/m²).
It's **difficult to get the raw data** for them (you'd have to kill the organisms) but they're pretty accurate — they nearly always come out pyramid shaped.

Pyramids like this are always symmetrical and they're always drawn to scale.

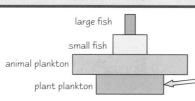

The amount of plant plankton is quite small at any given instant, but, because it has a short life and reproduces very quickly, there is a lot of it around over a period of time. This is why the plant plankton level is smaller than the animal plankton level.

Pyramids of Energy

These measure the **amount of energy** in the organisms in **kilojoules** per **square metre** per **year** ($kJm^{-2}yr^{-1}$).

This data is **very difficult to measure** but these pyramids give the **best picture of the food chain** and are **always proper pyramids**.

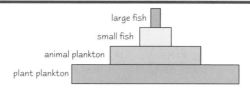

Practice Questions

Q1 What is a community?

Q2 What is a trophic level?

Q3 Why are there very rarely more than five links in a food chain?

Q4 What are the three types of pyramid?

Exam Questions

Q1 Explain the differences between a habitat and an ecosystem. [4 marks]

Q2 Pyramids of number and pyramids of biomass are not always pyramid-shaped, but pyramids of energy are.
Explain why this is so. [4 marks]

Ah... pretty coloured pyramids — after section 1, this is pure heaven...

This stuff is fairly straightforward but there are quite a few definitions you need to get in your head before the rest of the section. Otherwise, come the harder stuff, you'll be struggling to remember what ecosystems, populations, communities and habitats are. Make sure you know how energy flows through ecosystems and the different types of pyramid as well.

Nutrient Cycles

Skip these pages if you are doing OCR or Edexcel.
The amount of carbon and nitrogen on Earth is fixed — they can exist in different forms but no more can be made. The good news is they are constantly cycled around so they won't run out. The bad news is that you have to learn how that happens — just like you did for GCSE — only this time it's a bit more complicated. Great.

The **Carbon Cycle** is fairly straightforward

The carbon cycle involves four basic processes – **photosynthesis**, **respiration**, **death and decay** and **combustion**.

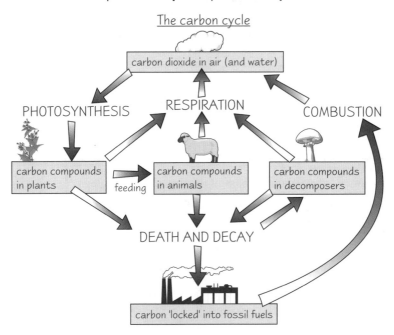

The carbon cycle

These are the seven things that you need to **remember** about the carbon cycle:

1) The only way that carbon gets into ecosystems is through **photosynthesis**.

2) Herbivores get their carbon by eating **plants**, carnivores get theirs by eating **other animals** and omnivores eat a **mixture** of plants and animals.

3) **Decomposers** get their carbon by digesting dead organisms.
Feeding on dead material is **saprobiontic** or **saprotrophic nutrition**.

4) All living organisms return carbon to the air in the form of **carbon dioxide** through **respiration**.

5) If plants or animals **die** in situations where there are no decomposers (e.g. deep oceans) the carbon in them can get turned into **fossil fuels** over millions of years.

6) The carbon in fossil fuels is released when they are burned — **combustion**.

7) **Microorganisms** are important in the cycle because they can quickly get the carbon in dead material **back into the atmosphere**.

The **Carbon Cycle** keeps atmospheric carbon dioxide levels **Constant**

In a totally **natural** situation the carbon cycle would keep atmospheric levels of carbon dioxide **more or less the same**.
Nowadays people are affecting the **global carbon balance** in two key ways:

- We burn huge quantities of **wood** and **fossil fuels** each year, which **adds** loads of carbon dioxide to the air.

- We are **clearing large areas of forest**, which would normally help to absorb some of the carbon dioxide in the atmosphere.

- In combination these two activities mean that the total amount of carbon in the atmosphere is much **higher** than it naturally would be.

Nutrient Cycles

The **Nitrogen Cycle** is a bit more **Complicated**

You need to be familiar with all the stages in the nitrogen cycle:

Don't worry if you get a diagram that looks different to this in the exam - all the information will be basically the same.

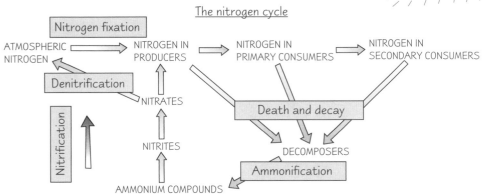

The nitrogen cycle

Plants and animals need nitrogen for **proteins** and for nucleic acids, but despite the atmosphere being 78% nitrogen, neither of them can use nitrogen gas. The key thing you need to remember is how important **bacteria** are (see box below). Without bacteria to produce the **nitrates** that plant roots can absorb, plants and animals couldn't exist.

1) Atmospheric nitrogen is **fixed** by **bacteria**. Some **live free** in the soil (e.g. *Azotobacter*), others, like *Rhizobium*, are found inside root nodules of **leguminous** plants (that's peas, beans and clover to you and me). Atmospheric nitrogen is changed into **ammonia**, then nitrites, then nitrates, which all plants can absorb and use to make protein.

2) The nitrogen in the plant proteins is passed onto animals through **food chains**.

3) When living organisms **die** their nitrogen is **returned** to the soil in the form of **ammonium compounds** by **microorganisms**. Animals get rid of excess amino acids via **deamination** in their livers — the nitrogen gets back into the soil via their **urine**.

4) Ammonium compounds are changed into nitrates by **nitrifying bacteria**. Firstly *Nitrosomonas* changes ammonium compounds into nitrites, then *Nitrobacter* changes the compounds into nitrates.

5) Nitrates are **converted back** into atmospheric nitrogen by **denitrifying bacteria** like *Pseudomonas* and *Thiobacillus*.

You don't need to learn the names of the microorganisms for AQA.

Sometimes you'll see a couple of extra things on diagrams of the nitrogen cycle:
1) **Industrial processes** like the Haber Process produce ammonia and nitrate fertilisers directly from atmospheric nitrogen.
2) **Lightning** naturally converts nitrogen into nitrates.

Practice Questions

Q1 What process in living things extracts carbon dioxide from the air?

Q2 What is 'saprobiontic nutrition'?

Q3 What types of plants have root nodules?

Q4 In the nitrogen cycle, what chemical changes occur during 'nitrification'?

Exam Questions

Q1 Explain how the carbon cycle has maintained the level of carbon dioxide in the atmosphere and how human activity has disrupted this balance. [10 marks]

Q2 Describe the role and significance of microbes in the nitrogen cycle. [6 marks]

The Carbon and Nitrogen Cycles — surely, not again...

Here we are in A2 Biology and the Carbon and Nitrogen cycles are back again. When you realise that without nitrogen recycling bacteria, plants and animals couldn't exist, then you can see how important the cycles are. Perhaps that's why they keep cropping up, or perhaps the examiners are just torturing you — either way it's got to be learnt.

Population Sizes

Skip these pages if you are doing AQA A.

Some terms to know and some graphs to interpret, but these two pages are more common sense than rocket science.

OCR ONLY

Population Growth follows a Standard Pattern

The graph below shows the way in which a **population of microbes grows** in a laboratory culture. Outside a lab, lots of **other factors** complicate things, but this is how any population would grow if it was left completely alone.

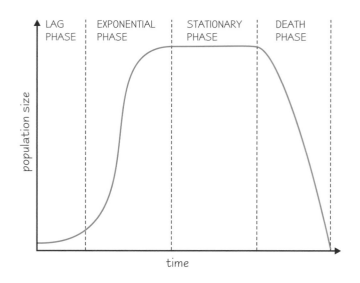

Typical bacterial growth curves are always <u>sigmoid</u> (S-shaped).

You need to be able to identify and explain the **four phases**:

> **THE LAG PHASE** happens when the microbes are starting to **produce** enzymes to digest the new food supply. They are "getting used to their surroundings" and they are reproducing quite slowly.
>
> **THE EXPONENTIAL PHASE** is a phase of **maximum population growth**.
> There is an **ample** supply of food and little build up of toxic wastes. It is also called the **LOG PHASE**.
>
> **THE STATIONARY PHASE** is when the number of **new** microbes **equals** the number **dying** (due to lack of food or build-up of toxins). The population has reached **carrying capacity** — the number of microbes that the area can sustain.
>
> **THE DEATH PHASE** is when the population **dies off**. This is due to things like the **exhaustion of food** or the build up of **toxic wastes**. This always happens in laboratory populations but it's rare in natural ones.

Population Growth is stopped by Limiting Factors

Limiting factors are things that put an **upper limit** on a population's size.

Many limiting factors are **abiotic** (non-living) — examples include: **temperature**, **soil** or **water pH**, **oxygen availability**, and the supply of **mineral nutrients**.

Some of these things (e.g. temperature and oxygen availability) can affect the **rate** of **population growth** as well as limiting the total **size**.

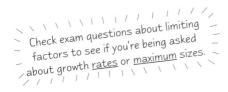

Check exam questions about limiting factors to see if you're being asked about growth <u>rates</u> or <u>maximum</u> sizes.

Interactions between Organisms can Limit population size

There are two important interactions that limit population size — **competition** and **predation**.

1) Competition occurs when a lot of organisms are competing for some sort of limited 'resource' — very often it's food, but it can be other things like **shelter**, **nesting sites** and **mates**.

2) If the organisms competing are of the **same** species, it is called **intraspecific** competition.
If they are from **different** species, it is called **interspecific** competition.

Population Sizes

Interspecific Competition affects population Distribution

Sometimes interspecific competition doesn't just affect the population **size** of a species, it totally **prevents** the species living in an area at all.

Since the introduction of the **grey squirrel** into Britain, the native **red squirrel** has **disappeared** from large areas because of interspecific competition. In the few areas where both species still live, both populations are **smaller** than they would be if there was only one kind of squirrel there.

Distribution of red squirrels, 1998

Distribution of grey squirrels, 1998

maps courtesy of Forest Research

Predator and Prey populations are interlinked

The presence of **predators** affects the **size** of populations. The **graph** shows **population fluctuations** of the **lynx** and its prey, the **snowshoe hare**, in Canada — it's a good example of how closely related predator and prey numbers can be.

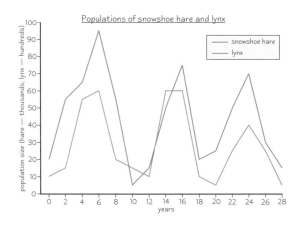

Populations of snowshoe hare and lynx

1) The theory is that as the **population** of **prey grows**, there is **more food for predators**, and so the predator **population grows**.
2) As the **predator** population gets **larger**, **more prey** is **eaten** so the **prey** population **falls**.
3) Then, there is **less food** for the **predators** so their **numbers go down**, and so on.
4) In nature you rarely get this **exact pattern** because it's unusual to get a situation where a species is eaten by just **one** predator, or a predator just eats **one** prey species.

Some regulating factors are 'Density Independent'

Factors which limit population size can be **density dependent** or **density independent**. **Density dependent** factors have **more impact** as a population gets more **dense** — e.g. disease, food availability, competition for mates or shelter.

Density independent factors affect populations of **any** density in the same way — like flooding, forest fires or severe weather. A crowded population and a sparse population will be **equally affected**. Density independent factors are often **abiotic**, while density dependent factors tend to be **biotic**.

Practice Questions

Q1 What are the four growth phases seen in laboratory populations?
Q2 What does the term 'carrying capacity' mean?
Q3 State three abiotic factors that might limit the size of a population.
Q4 Explain the difference between interspecific and intraspecific competition.

Exam Questions

Q1 Growth of a laboratory population usually follows a 'sigmoid' (S-shaped) growth curve. Populations in nature, however, rarely follow such a regular pattern. Suggest reasons for this. [8 marks]

Q2 Factors that regulate population size can be density dependent or density independent. Explain the difference between these two types. [4 marks]

Don't let your revision enthusiasm run into the Death Phase...

Again there's some stuff you've seen lots before here — like predator-prey relationships, but it's jumbled up with new stuff too. So don't be fooled into thinking, "yeah yeah, been there, done that, got the T-shirt — in fact I invented the T-shirt me, this A2 biology lark's a cinch, reckon I might just sack this revision off"... you get the point, this stuff needs learning.

Diversity

You can skip these pages if you're doing Edexcel. If you're doing OCR, you just need page 27.
If you're doing AQA B you need page 27, and page 26 covers topics from optional module 6, Applied Ecology.

Diversity is all about how many different species there are in an ecosystem. In total, about 1.5 million species have been described but scientists reckon that the total number of species on Earth may be as high as 10 million.

There is a **Link** between **Diversity** and **Stability** in an ecosystem

Ecosystems with a **large diversity** of species tend to be **more stable** than those that are less diverse.
There are two ways to tell how **stable** an ecosystem is:

> 1 Stable ecosystems are **resistant** to **change**.
> 2 If **disrupted** in some way, stable ecosystems return to their **original state** quite quickly.

If you think about it it makes sense — **low diversity** means that **predator** species don't have much **choice** of prey.
If the population of a prey species is **reduced** or **wiped out**, then the predator species will be **at risk**.

When diversity is **higher**, the **predator** species will have a **large selection** of possible prey species.
If one of the prey species is wiped out, there will still be plenty of other species that predators can eat.

1) **Extreme** environments like **tundra**, **deserts**, **salt marshes** and **estuaries** are all ecosystems with **low diversities**. **Monocultures** are agricultural areas where only one crop is grown (e.g. wheat fields) — they have **artificially** low diversities. In areas of low diversity, plant and animal populations are mainly affected by **abiotic** factors.

2) **Ecosystems** with **high diversities** are usually mature (i.e. old), and natural (e.g. oak woodlands), with environmental conditions that aren't too hostile. In these ecosystems populations are mostly affected by **biotic** factors.

Diversity is measured by the 'Diversity Index'

The simplest way to measure diversity is just to count up the number of species. But that takes no account of the **population size** of each species. Species that are in an ecosystem in very **small** numbers shouldn't be treated the same as those with **bigger** populations.

The **diversity index** is an equation for diversity that takes different population sizes into account. You calculate the diversity index (**d**) of an ecosystem like this:

$$d = \frac{N(N-1)}{\sum n(n-1)}$$

Where...
N = **Total number** of organisms of **all** species
n = **Total number** of **one** species
Σ = '**Sum of**' (i.e. added together)

Pete wasn't sure that the company's new increased diversity policy would be good for productivity.

The **higher** the number the **more diverse** the area is. If all the individuals are of the same species (i.e. no diversity) the diversity index is 1.

Here's a simple example of the diversity index of a field:

There are 3 different species of flower in this field, a red species, a white and a blue.
There are 11 organisms altogether, so N = 11.
There are 3 of the red species, 5 of the white and 3 of the blue.
So the species diversity index of this field is:

$$d = \frac{11(11-1)}{3(3-1) + 5(5-1) + 3(3-1)} = \frac{110}{6 + 20 + 6} = 3.44$$

When calculating the bottom half of the equation you need to work out the n(n-1) bit for each different species then add them all together.

A variety of **Microclimates** leads to **Higher Diversity**

Microclimates are **small areas** where the **abiotic** factors are **different** from the surrounding area. For example, the underneath of a rock has a different microclimate than the top surface — it's cooler and more humid.

Each microclimate provides a slightly different **habitat** that will suit **certain species**. So, ecosystems that have a variety of microclimates can support a **high diversity**. Basically: **more microclimates = more species = higher diversity**.

Diversity

You need to know how to **Use Quadrats**

Ecologists look at three key **factors** when they're working out diversities:

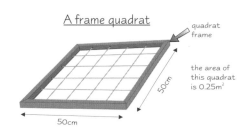

SPECIES FREQUENCY This is how **abundant** a species is in an area.

SPECIES RICHNESS This is the **total number** of **different species** in an area.

PERCENTAGE COVER This is how much of the surface is covered by a particular plant species (you can't use it for **animals** because they move around too much).

1) To measure all of these you use a piece of equipment called a **frame quadrat** — a square frame made from metal or wood. The area inside this square is known as a **quadrat**.

2) **Quadrat frames** are **laid on the ground** (or the river / sea / pond bed if it's an aquatic environment). The **total number** of **species** in the quadrat frame is recorded as well as the number of **individuals** of each species.

A frame quadrat

quadrat frame

the area of this quadrat is 0.25m²

50cm

50cm

3) Generally it's not practical to collect data for a whole area (it would take you ages) so **samples** are taken instead. This involves measuring lots of quadrats from different parts of the area. The data from the samples is then used to **calculate** the figures for the **entire area** being studied. **Random sampling** (see p.38) is used to make sure that there isn't any **bias** in the data.

4) **Species frequency** is measured by counting **how many quadrats** each species appears in and is given as a percentage (e.g. if a species was found in 5 out of 20 quadrat samples, the frequency would be 25%).

5) **Species richness** is measured by counting up the **total number of species** found in all the samples. You assume that the number of different species in your sample is the same as the number in the whole area that you are studying.

6) **Percentage cover** is measured by dividing the area inside the quadrat frame into a **10 × 10 grid** and counting **how many squares** each species takes up. Sometimes plants **overlap** so the total percentage cover ends up being **more** than **100%**.

There's lots more information about ecological field techniques on p36-37.

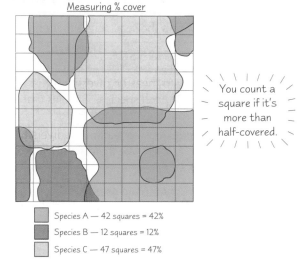

Measuring % cover

You count a square if it's more than half-covered.

Species A — 42 squares = 42%

Species B — 12 squares = 12%

Species C — 47 squares = 47%

Practice Questions

Q1 What is meant by a 'stable' ecosystem?

Q2 Give three examples of ecosystems that are likely to be unstable.

Q3 What is the difference between the diversity index and species richness?

Q4 Why do percentage cover measurements sometimes add up to more than 100%?

Exam Questions

Q1 Explain the link between the stability of an ecosystem and the diversity of species within it. [5 marks]

Q2 Describe in detail how you would measure the percentage cover of clover on a school field. [8 marks]

It's just all a bit random...

Well actually, that's the point, it's supposed to be random, or it wouldn't be a fair test. If you've used a quadrat, you've probably already realised it's just a fancy name for four bits of wood nailed together. They're probably expensive to buy, too. If you're a business studies student, you might see a money-making opportunity here, but get lost, because I saw it first.

Succession

Succession is all about how ecosystems change over time. Apart from a few fancy words I think that it's one of the easiest things in A2 biology — it's a lot more straightforward than the Krebs cycle and photosynthesis.

There are **Two** different types of **Succession**

Succession is sometimes called ecological succession in exams.

Succession is the process where **plant communities** gradually develop on **bare land**. Eventually a **stable climax community** develops and after that big changes don't tend to happen.

There are **two** different types of succession...

Primary succession

Happens on land where there is no proper soil and **no living organisms**. New land created by a **volcanic eruption** is a good example of a place where primary succession will occur.

Secondary succession

Happens when most of the living organisms in an area are **destroyed**, but the **soil** and **some** living organisms remain. Examples include: woodland that has been burned by a **forest fire**, areas subject to severe **pollution**, or land that is cleared by **people** for things like **housing** or **new roads**.

Each stage in the succession of an area is called a **seral stage**. In every seral stage the plants change the environmental conditions, making them suitable for the next plants to move in.

You need to **Learn** an **Example** of succession

This example shows the seral stages that change **bare** sand dunes into **mature woodland**.

You need to learn this example, including all the names of the species.

The first plants to colonise an area have to be specialised so they can deal with the harsh abiotic conditions. These plants are known as pioneer species. They are usually herbaceous (non-woody).

climatic climax — tree community

shrub community

grass community

marram grass

sea

embryo dune

1) The first **'pioneer'** species to colonise the area need to be able to cope with the **harsh abiotic conditions** on the **sand dunes** — there is **little fresh water** available, there are **high salt levels**, the **winds** are **strong** and there is **no proper soil**. Marram grass has good **xerophytic** adaptations (see pages 60-61) so it is usually the first to start growing.

2) As the pioneer species begin to **die** they are broken down by microorganisms. The dead marram grass adds **organic material** to the sand creating a very basic 'soil' which can hold more water than plain sand.

3) This soil means that the abiotic conditions are **less hostile** and so other, less specialised grasses begin to grow.

4) These new grasses will eventually **out-compete** the original colonisers via **interspecific** competition.

5) As each new species moves in, more **niches** are created making the area suitable for even more species.

6) After the grass communities have all been out-competed, the area will be colonised by **shrubs** like **brambles**.

7) Eventually the area becomes dominated by **trees** — in Europe the trees will usually be things like **birch** and **oak**. The trees dominate because they prevent light from reaching the herbaceous plants below the leaf canopy.

Succession

Diversity Increases and Species Change as succession progresses

When succession happens in any environment the general pattern of change is always the same:

- The species present become **more complex** e.g. a forest starts with simple mosses and finishes with trees.
- The **total number** of organisms **increases**.
- The **number of species increases**.
- **Larger species** of plants arrive.
- **Animals** begin to move into the area — with each seral stage **larger** animals move in.
- **Food webs** become more **complex**.
- Overall, these changes mean that the ecosystem becomes more **stable.**

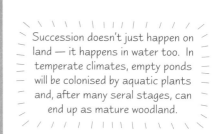

Succession doesn't just happen on land — it happens in water too. In temperate climates, empty ponds will be colonised by aquatic plants and, after many seral stages, can end up as mature woodland.

There are different Types of Climax Community

Various factors can **stop succession** going any further and lead to a **climax** community. The climax is **classified** according to **what** has prevented the succession from going any further...

1) In a **climatic climax**, the succession has gone as far as the **climate** in the area will allow. E.g. Trees can't grow at high altitudes, so high up on alpine mountains the largest plants are **shrubs**.

2) **Human activities** can stop succession by felling trees, ploughing fields or grazing animals on farmland. Some ecosystems are deliberately '**managed**' to keep them in a particular state, for example, the heather on moorland is burned every 5-7 years to prevent woodland from developing. When succession is stopped **artificially** like this the climax community is called a **plagioclimax**.

Farmland is an example of a Plagioclimax

Succession is stopped by regular **ploughing** or by the **grazing** of stock. In a grazed field, **grass** can survive because it is fast-growing, but slow growing plants get eaten before they can get established.

If the **grazing stops**, then slower growing plants can gradually begin to establish themselves. As they do, the **grass** will become **less dominant**. The new plant species will attract a wider range of **insects** and so the area will **increase** its **diversity**. Eventually, the field will become replaced by **woodland**.

Practice Questions

Q1 What is secondary succession?
Q2 What are the stages in succession called?
Q3 What is a pioneer species?
Q4 What name is given to a climax community brought about by human intervention?

Exam Questions

Q1 Define the term 'ecological succession', explaining how it occurs and the different types of climax communities that can be produced. [8 marks]

Q2 a) Suggest three features of a plant species that might make it a successful pioneer species. [3 marks]
   b) Suggest two reasons why such a pioneer species may disappear during the early stages of succession. [2 marks]

There are many different types of climax...

If your enthusiasm for all these Biology facts is waning, why not try reading some ICT... "Remote management supports users on a network. In the event of a significant problem or recurring error, the network administrator takes temporary control. Afterwards, the user can be advised on how to avoid similar problems in the future..." You see what I'm saying?

Agriculture and Ecosystems

Skip these pages if you are doing AQA A.
Agriculture has always changed ecosystems.
You need to know the facts and have a balanced view on modern and organic farming methods.

Farming can be **Intensive** or **Extensive**

The **traditional** methods of farming and those used by **organic** farmers are examples of
extensive food production. In the later part of the twentieth century, these methods
were replaced by **intensive farming**. The table summarises the differences:

Intensive	Extensive
Requires a lot of capital investment	Less capital investment needed
Requires less space	Requires a lot of space to grow the same amount of produce
Not labour intensive — much is done by machines	Labour intensive
Production concentrated on single crop (monoculture) or animal species	Farm usually grows several crops and/or produces livestock
Large amounts of pesticides used Biological control sometimes used	Fewer or no pesticides used Biological control more common
Large scale use of inorganic fertilisers	Emphasis on organic fertilisers

Intensive Farming methods can **Affect** the environment

Intensive farming methods have **boosted food production**, stopping people starving in some parts of
the world and making sure there is always plenty of good quality food on supermarket shelves.

But these methods can cause **environmental problems**:

- **Hedgerows** are **removed** to allow access for large machines or to
  enlarge fields. This **destroys** the **habitat** of many species and reduces
  species diversity.
- Large amounts of **inorganic fertiliser** are used. This can **leach** through the soil
  and **pollute** nearby streams and rivers (see below).
- '**Factory farming**' methods of keeping many animals in a small space produce
  large amounts of **waste**, which can cause pollution if not disposed of carefully.
- Large scale use of **pesticides** can result in water **pollution** and may also kill
  non-pest species. Food can become contaminated.

The use of **Fertilisers** can lead to **Eutrophication**

Fertilisers come in **two types** — **inorganic** (man-made chemicals) and **organic** (manure).

1) **Organic** fertiliser is better for the **environment**, but is **more difficult** to **store** and **apply**.
   Its bulk improves the soil structure.

2) Fertilisers often get **washed** into nearby streams and rivers, which is a **waste**, and can also cause **eutrophication**.

Eutrophication occurs when **fertilisers** stimulate the **growth** of **algae**
in ponds and lakes. These algae prevent light from reaching the water
plants below.
Eventually the plants **die** because they are unable to photosynthesise
enough. The dead plants are digested by **bacteria**. The bacteria
drastically **reduce** the **oxygen** levels in the water, which kills fish and
other aquatic organisms.
Phosphates entering water from **sewage** also have this effect.

Digging up <u>farmland</u> that hasn't been
used for a while releases nitrates which
can cause eutrophication. Nitrates from
soil can also <u>contaminate drinking water</u>.

Agriculture and Ecosystems

Farmers can be Environmentally Friendly

Organic farmers set out to be 'green' and avoid all use of chemicals. However, non-organic farmers often manage their land well, encouraging species diversity and avoiding pollution. Below are listed some 'good practices' for farmers.

Planting hedgerows

This increases **species diversity** and some of the species attracted might well feed on potential pests, helping the farmer. Habitat variety can be further increased if the farmer leaves small areas of woodland on his land.

Preventing Soil Erosion

Top-soil blowing away reduces the fertility of the soil. Planting trees as **wind shields** can prevent this. This is particularly important in dry climates. Planting **hedgerows** also helps reduce soil erosion.

Using **organic fertilisers** can improve the soil structure, reduce pollution and help dispose of animal waste in a useful way.

Intercropping

This is the practice of growing two or more crops in the same field at the **same time**. It can produce a greater yield on a given piece of land, by using space that would otherwise be wasted with a single crop. Careful **planning** is required, taking into account the soil, climate, crops and varieties. An example is planting a deep-rooted crop with a shallow-rooted crop to make maximum use of soil nutrients. Intercropping encourages **biodiversity**, and it can also reduce pests — each crop may contain a chemical that repels the pest species of the other.

Planting Legumes

Legumes (peas, beans and clover) naturally **restore nitrates** to the soil. If planted alternately with other crops, there's less need to **fertilise** the soil. Peas and beans can be sold as a food crop, too.

Using Biological Control

This is an **alternative** to the use of chemical pesticides, and will be dealt with in the next section.

Reducing Eutrophication

Farmers can reduce the risk of eutrophication by adding fertiliser only when plants are growing and not to **bare soil**. It shouldn't be added when it is **raining**, and a strip of **unfertilised** soil should be left around the outside of the field. The soil should be **tested** before it is fertilised to check the nitrogen levels, and the amount of fertiliser carefully measured so that excess nitrogen is not put into the soil.

Practice Questions

Q1 Give four differences between intensive and extensive farming methods.
Q2 Why do intensive farming methods often result in the destruction of hedgerows?
Q3 What is the advantage of planting legumes in between two plantings of other crops?

Exam Questions

Q1 Describe the benefits and problems of intensive food production. [6 marks]

Q2 Explain the advantages and disadvantages of the use of organic fertilisers as an alternative to inorganic ones. [5 marks]

Q3 Explain what eutrophication is and how it may be caused as a result of farming practices. [8 marks]

Phew, that was a bit intensive...

Actually, that's not true. In fact, I'd say the page was pretty darn straightforward. Isn't legume a funny word though? You expect some kind of alien plant with legs... but no... they're just plain old peas and beans. Anyway, I feel we're losing track a bit, so what was I going to say... oh yes, the most important thing to make sure you understand on this topic is Sorry — I've haven't got space to tell you.

Controlling Pests

Skip these pages if you are doing AQA A.

We had a pest in our cupboard once. He was a mouse and he ate all my rice so we called him Paddy.

There are **Three** methods of **Pest Control**

Pests are species that negatively affect human activities. They can be **weeds** that out-compete crops, **insects** that carry diseases or **fungi** that reduce crop yields. They could also be animals such as **rabbits** or **deer** that eat crops. There are three main approaches to dealing with pests:

1) **Chemical control** uses chemicals like pesticides to kill the pest species.

2) **Biological control** uses living organisms to control the pest. These include **predators, parasites** and **pathogens** which kill the pest or prevent it from reproducing.

3) **Integrated pest management** mixes both chemical and biological control together to combat pests.

Chemical Pesticides are pretty **Effective**

1) **Herbicides** are used to combat weeds. There are two kinds:

   • **Contact herbicides** kill weeds when they are sprayed onto their **surface**.

   • **Systemic herbicides** have to be **absorbed** into the weed to kill it.
   High doses of synthetic auxins (plant hormones) are used as systemic herbicides.

2) **Fungicides** are used to fight fungal infections. Like herbicides they can be either contact or systemic.

3) **Insecticides** are used against... guess what... insect pests. There are three types this time:

   • **Contact insecticides** have to come into direct contact with the insect.

   • **Systemic insecticides** are absorbed by the plant and carried in its **phloem**.
   They kill the insects that feed on the plant's sap.

   • **Stomach ingestion insecticides** are sprayed over the crop and are
   consumed when pests eat the plants.

Chemical Pesticides can cause **Problems**

Some pesticides are **persistent** — that means that they don't break down in the environment or within the tissues of living organisms. This causes big problems for species that aren't supposed to be damaged by the pesticide:

Bioaccumulation happens when persistent pesticides stay in an organism's living tissue. As the organism consumes more and more of the pesticide the levels in its tissues become **increasingly toxic**. The pesticides are passed from one **trophic level** to the next, and each time they become increasingly **concentrated**. The species that are highest up in the food chain (e.g. birds of prey) can end up with **lethal** concentrations in their bodies (see page 110 for more).

Pesticides that aren't specific cause lots of problems because they kill other species, as well as the ones that they're supposed to attack.

Biological Control can be Better for the **Environment**

Biological control avoids the need for putting extra **chemicals** into the environment.
Organisms that are used in biological control include:

1) **Insect parasites** which are specific to the pest. Most lay their **eggs** on or in the host and then when the eggs hatch the larvae eat the host from the **inside**. Nice...

2) **Predators** are **carnivorous** species used to control insects and other animal pests.

3) **Pathogens** are **bacteria** and **viruses** used to kill pests. For example, the bacterium *Bacillus thuringiensis* produces a **toxin** which kills a wide range of **caterpillars**.

Controlling Pests

Biological control has **Advantages** and **Disadvantages**

Advantages

1) The control organism is usually **specific** and only the pest species will be affected.
2) **No chemicals** are used, so problems of bioaccumulation and pollution are avoided.
3) Control organisms usually establish a population so there is **no need** for **re-application**.
4) Pests don't usually develop genetic resistance to the biological control agent (this is a big problem with chemical fertilisers).

Disadvantages

1) Control is **slower** than using chemical pesticides because you have to wait for the biological control agent to establish a large enough **population** to control the pests.
2) Generally biological control **doesn't permanently exterminate** the pest — instead it's reduced it to a level where it is no longer a big problem.
3) It can be **unpredictable** — it's really hard to work out what all the **knock-on effects** of the introduction of the biological control species will be.
4) A lot of **scientific research** into the relationship between the pest species and the biological control agent is needed. This research takes **time** and costs **money**.
5) If the control species has several sources of food it's population levels might **grow** and it may **become a pest** itself.

There are **Other Ways** of dealing with **Pests**

In recent years farmers have started to use **integrated pest management** where chemical and biological methods are combined.

Biological control is used to keep the pest down so that it doesn't affect the **profitability** of the farm. If a pest **outbreak** occurs then the farmer uses a specific pesticide for a **short** period of time.

Crop rotation can help control pests — changing the crop each year makes it difficult for pests that can only feed on one species to become established.

Genetic modification can be used to produce crops which are **resistant** to pests, so no pest control is required. (Although it is possible for pests to evolve so they can feed on the genetically modified organisms.)

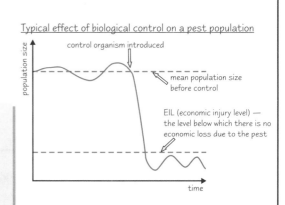

Typical effect of biological control on a pest population

Practice Questions

Q1 What is the difference between a contact pesticide and a systemic pesticide?
Q2 What problem is caused to food chains by persistent pesticides?
Q3 What is integrated pest management?

Exam Questions

Q1 A chemical company is developing a new weed killer. State three features that the company should aim to include in their herbicide to ensure that it causes as little environmental damage as possible. [3 marks]

Q2 Explain why some farmers prefer to use pesticides rather than biological control to deal with pests. [5 marks]

There's no need to be pestimistic, this stuff's easy...

When you realise how difficult it is to control pests you can understand why some people think that genetic modification is the way forward. The thing that worries people is that, like biological control, it's hard for scientists to know exactly what the consequences of growing GM foods will be in the long term. It could be a great thing, or it could be a scary mistake.

Managing Ecosystems

Skip these two pages if you are doing AQA B.

If you want to maintain ecosystems, you can't just leave them alone, because succession is likely to alter them.
These pages look at how ecosystems can be managed for the good of both humans and the natural environment.

There *Isn't much* Woodland Left

Britain used to be 85% covered in forest, now only about 8% is forest.
Deforestation has taken place for a number of reasons:

1) Trees have been removed for **timber**.
2) Trees have been used to make products like **paper** and **chipboard**.
3) Woodland has been cleared to make more room for **agriculture**.
4) Woodland has been cleared to allow for the building of **roads and houses**.

Deforestation Changes *Abiotic Conditions*

1) In a forest, most of the mineral nutrients are stored as **living biomass** in the trees. In autumn the trees drop their leaves and the minerals are **returned** to the **soil** where they are used again by plants growing in the spring. If the trees are removed from the habitat, the **minerals** are **lost** and the cycle is **interrupted**.

2) Removing trees makes the land more exposed to **wind** erosion. Tree roots usually bind the soil together, so when they are removed the soil structure becomes less stable.

3) Woodlands absorb a lot of **rainfall** — removing them creates greater risk of flooding and soil loss.

> <u>Large scale</u> deforestation has large scale effects because it affects the global balances of carbon dioxide (which is used in photosynthesis) and oxygen (which is given out by trees).

Woodland can be Sustained *by* Good Management

Today, most woodland in the UK is used for either timber production or leisure.
Woodland is managed so that it can continue to provide habitats for wildlife.

Timber Production

Wood that is grown for timber is **deliberately** planted — species such as **pine** are used because they **grow quickly**. As soon as one crop of trees has been felled, more are planted.

Coppicing is a **traditional** method of timber production. Trees like hazel and sweet chestnut are cut down to their base so that they begin to sprout — each shoot is harvested when it matures. Coppicing produces lots of long poles without the trees themselves having to be felled.

Pollarding is similar — trees have their tops cut off to encourage the formation of a crown of branches.

a tree that has been coppiced

Leisure

Management for leisure use involves the creation of **mixed woodland** with paths and open spaces. Planting native shrubs like blackthorn, hawthorn and holly in shady parts of the wood can provide excellent habitats for birds such as wrens.

Dead wood is left where it is as it provides food and shelter for **fungi** and **invertebrates** like woodlice and wood-boring beetles. These are the foundation of many **food chains**, attracting many birds e.g. woodpeckers and nuthatches, bats and other small mammals. The management techniques needed to maintain mixed woodland are shown in the diagram...

grassy open (mow every Autumn) / taller ground flora (mow every 2 years) / edge shrubs (cut every 5 years) / under canopy (leave) / woodland

Managing Ecosystems

Grassland needs Management too

In Britain, grassland is mainly used for livestock or parkland. You might think that a field of grass grazed by cows or sheep just sorts itself out, but it actually needs fairly careful management.

Grazing needs to be carefully controlled. Sheep and dairy cattle prefer short grass, but beef cattle do better when it is slightly longer. The farmer needs to rotate his stock at intervals to ensure that pastures are evenly grazed. Grazing also stops shrubs from growing which can shade the grass and stop it growing so well.

Mowing is used to maintain areas of parkland. Like grazing, it stops the development of larger plant species that would eventually displace the grass.

If mowing or grazing is inadequate, **manual shrub clearance** of the grassland is sometimes needed.

Amazingly, even **fire** can be used for the management of grassland. Perennial grasses have growing points which are below the soil, so they can survive fire. Fire can improve grass quality, quantity and palatability (how tasty it is) — it removes dead plant material, increases new growth, and controls competing weeds.

The European Union has set up the Natura 2000 Project

In 1992, the EU Habitats Directive set up **Natura 2000** to ensure that sensitive **habitats** and **endangered species** across Europe are **conserved**. Each member state was asked to provide a list of sensitive sites and areas where threatened species were found, and to agree a plan of action to manage them.

EDEXCEL ONLY

Practice Questions

Q1 State 3 ways in which deforestation can damage the environment.

Q2 Why do perennial grasses survive when fire sweeps the land?

Exam Questions

Q1 Describe ways in which woodland can be managed for sustained timber production. [4 marks]

Q2 The graph shows the pattern of grass growth and the requirements of livestock living off it, throughout a year.

(a) Suggest what measures would be needed to maintain the grassland during May to September. [1 mark]

(b) Suggest how the stock might be properly fed between September and the following May. [1 mark]

(c) Land managers are sometimes asked to leave a strip around 2m wide round the edge of their fields, where the grass is left to grow long. Suggest a possible environmental benefit of this. [1 mark]

Natura 2000 — isn't that a nudist colony...

There's something so appealing about the idea of running free in the great outdoors, feeling the wind in your hair, and the sun on your back. How I long to be out there with the little bunny rabbits and the birds and the little fishies, communing with nature. With my clothes on, of course. Now enough of this nonsense — there's biology to learn here, y'know.

Investigating Ecosystems

If you've been on an ecology field trip you'll be familiar with this stuff. You'll be relieved to know that you can revise this in the comfort of your own bedroom — you won't be asked to stand in a river catching horrible squirmy things.

You need to know how to take **Abiotic Measurements**

Temperature is easy enough — just use a **thermometer**.

pH measurements are only taken for soil or water. **Indicator** paper / liquid or an electronic **pH monitor** are used to get the data.

Light intensity is difficult to measure because it varies a lot over short periods of time. You get the most accurate results if you connect a **light sensor** to a data logger and take readings over a period of time.

Oxygen level only needs to be measured in aquatic habitats. An **oxygen electrode** is used to take readings.

Air humidity is measured with a hygrometer.

Moisture content of soil is calculated by finding the mass of a soil sample and putting it in an oven to dry out. The amount of mass that has been lost is worked out as a % of the original mass.

Quadrat Frames are a Basic Tool for Ecological Sampling

The method for using quadrats is described on page 27. Remember that samples must be taken **randomly**. It's also important to consider the **size** of the quadrat — smaller quadrats give more accurate results, but it takes longer to collect the data and they're not appropriate for large plants and trees.

Plotting a graph of cumulative number of species found against number of quadrats sampled should show you how many quadrats you'd need to sample in further studies of the same type of habitat.

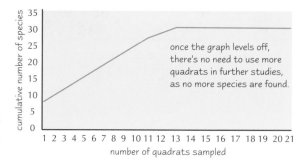

once the graph levels off, there's no need to use more quadrats in further studies, as no more species are found.

Point Quadrats are an alternative to quadrat frames

Pins are dropped through holes in the frame and every plant that each pin hits is recorded. If a pin hits several **overlapping** plants, **all** of them are recorded. A tape measure is laid along the area you want to study and the quadrat is placed at regular intervals (e.g. every 2 metres) at a right angle to the tape.

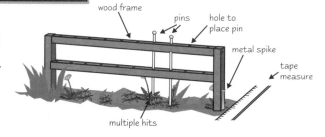

wood frame · pins · hole to place pin · metal spike · tape measure · multiple hits

Line and *Belt Transects* are used to *Survey* an area

The line you select to sample across the area is called a **transect**. Transects are useful when you want to look for **trends** in an area e.g. the **distribution of species** from low tide to the top of a rocky shore.

A **line transect** is when you place a tape measure along the transect and record what species are touching the tape measure.

A **belt transect** is when data is collected between two transects a short distance apart. This is done by placing frame quadrats next to each other along the transect.

If it would take ages to count all the species along the transect, you can take measurements at set intervals, e.g. 1 m apart. This is called an **interrupted transect**.

The data collected from belt or line transects is plotted on a **kite diagram** (that's just a fancy kind of graph) and trends across the area can be observed.

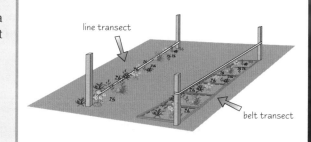

line transect · belt transect

Investigating Ecosystems

To sample **Animals**, you've got to **Catch** them

Most animals are **mobile** so they can't be sampled using quadrats or transects. There's various methods for catching animals depending on their **size** and the **kind of habitat** being investigated.

Nets can be used to trap flying insects and aquatic animals.

Pitfall traps can be used to catch walking insects on land. The insects fall into the trap and are... well, trapped.

Pooters are used to catch individual insects which are chosen by the user.

Tullgren funnels are used to extract small animals from soil samples. The animals move away from the light and heat produced by the bulb and eventually fall through the barrier into the alcohol below the funnel.

Pitfall trap   Tullgren funnel   Pooter

The **Mark-Release-Recapture** Technique is for estimating **Population Sizes**

The mark-release-recapture method is basically this:

1) **Capture** a sample of the population.
2) **Mark** them in a harmless way.
3) **Release** them back into their habitat.
4) Take a **second sample** from the population.
5) **Count** how many of the second sample are marked.
6) Estimate the **size** of the whole population using the **Lincoln index**.

$$\text{Population size (S)} = \frac{n_1 \times n_2}{n_m}$$

n_1 = number of individuals in first sample
n_2 = number of individuals in second sample
n_m = number of marked individuals in the second sample

The accuracy of this method depends upon these **assumptions**:

1) The marked sample has had enough **time** and **opportunity** to **mix** back with the population.
2) The marking has not affected the individuals' **chances of survival**.
3) **Changes** in population size due to **births**, **deaths** and **migration** are **small**.
4) The marking has **remained visible** in all cases — so it needs to be waterproof.

Good ways of marking animals include using a UV pen or cutting a little bit of the animal's fur off.

Practice Questions

Q1 How would you calculate the moisture content of soil?
Q2 What is the difference between a line transect and a belt transect?
Q3 What piece of apparatus is used to extract small organisms from a soil sample?

Exam Questions

Q1 When measuring light intensity in an ecosystem, why is it not sufficient to take a single light-meter reading? [3 marks]

Q2 Under what circumstances would you use a transect rather than random sampling of an ecosystem? Give an example in your answer. [2 marks]

Q3 A population of woodlice were sampled using pitfall traps. 80 individuals were caught. The sample was marked and released. Three days later, a second sample was taken and 100 individuals were captured. Of these, 10 had marks. Use the Lincoln index to estimate the size of the woodlouse population. [2 marks]

What do you collect in a poo-ter again?

Aren't you glad that we don't use the mark-release-recapture technique to measure our population size. I don't fancy falling in a pitfall trap and then getting a chunk of my hair cut off. Seems kind of barbaric, now I think about it. When we did this experiment at school, we never caught any of the woodlice we'd marked again. They'd all disappeared...

Data Analysis

Skip these two pages if you are doing Edexcel. *Sheets and sheets of muddy paper with your field results don't tell you much about an ecosystem. To work out what they really tell you, you need to do some sums…*

Sampling has to be *Random*

If you're sampling a small section of an area and then drawing conclusions about the whole ecosystem, it's important that the sample **accurately** represents the ecosystem **as a whole**.

One way to avoid bias in your answer is to pick the sample sites **randomly**, e.g. you could divide the whole area you're studying into a **grid** and then use a random number generator or table to select each coordinate.

Once you have your data you've got to *Analyse It*

In a **normal distribution** of data, most of the samples are **close to the mean** (the average value), with relatively **few** samples at the **extremes**. On a graph, a normal distribution produces a **bell-shaped curve**, like the one below.

Standard deviation is often used to analyse data sets — it tells you how much a set of data is **spread out** around the **mean**.

E.g. you might use standard deviation to find the **variation** in the **number** of apples produced by trees in an apple orchard.

The formula for standard deviation (s) is:

$$s = \sqrt{\dfrac{\sum x^2 - \dfrac{\left(\sum x\right)^2}{n}}{n-1}}$$

s is the standard deviation
$\sum$ means 'sum of'
x is an individual result
n is the total no. of results

When all the samples have **similar** values then the distribution curve is **steep** and the standard deviation is **small**.

When the samples show a **lot** of **variation**, the distribution curve is relatively **flat** and there is a **large** standard deviation.

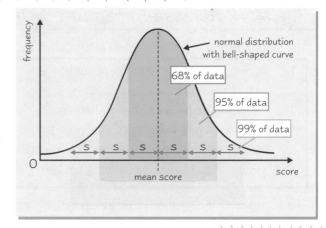

normal distribution with bell-shaped curve
68% of data
95% of data
99% of data

frequency

s s s s s s

mean score

score

You can use the standard deviation to split any normal distribution graph into the areas shown:

These numbers are *always the same*.

- The middle area (within **1 standard deviation** either side of the mean) shows you where **68%** of the data lies.

- The next area (within **2 standard deviations** either side of the mean) shows you where **95%** of the data lies.

- The biggest area (within **3 standard deviations** either side of the mean) shows you where **99%** of the data lies.

You need to work out *Degrees of Freedom* to do *Statistical Tests*

To use statistical tests, you have to work out the **degrees of freedom** (df) for your data. The degrees of freedom for a **t-test** are just the **total number of results** in both sets **– 2**. For a **chi square test**, it's the number of **categories** of data **– 1**.

The *t-test* has nothing to do with *Hot Beverages*

You don't need to know about the t-test if you're doing AQA A or AQA B.

Sometimes two sets of data look different. You can just say they are different, but to be absolutely sure you need to **prove** they are using statistics. If they have been proven different they are said to be **significantly different**. Your **null hypothesis** would be that the means **aren't** significantly different, and you're seeing if you can prove this wrong. You use the **t-test** to calculate whether the **mean** of two sets of data are **significantly different**.

The formula for t is:

$$t = \dfrac{\text{mean of set 1 results} - \text{mean of set 2 results}}{\text{standard error}}$$

Work out the standard error like this:

$$\sqrt{\dfrac{s_1^{\,2}}{n_1} + \dfrac{s_2^{\,2}}{n_2}}$$

where s is the standard deviation, and n is the number of results, for each of the 2 sets of data.

Data Analysis

When you've **calculated t** and the **degrees of freedom** you just need to **compare** your values to those in a **t-table**. This tells you if your means are **significantly different**, **how confident** you can be that your results actually mean something, and that the difference isn't just **due to chance**.

The table gives numbers for three different **P values** (P = **probability**). If your t value is bigger than the number in the P = 0.05 column, that means there's only a 1 in 20 chance that the pattern in your results has happened by chance.

You can now be 95% confident that the difference you see in your results is **significant**. Biologists are pretty careful characters, so unless you're **95% certain**, your results shouldn't be counted as meaningful. If your t number is bigger than P = 0.01 or P = 0.001, even better — you can be even more confident in them.

Confused? Thought so, but don't worry, it will all make sense once you've been through a nice worked example...

degrees of freedom → ↓ different probabilities ↘

df	P = 0.05	P = 0.01	P = 0.001
1	12.75	63.66	636.61
2	4.30	9.92	31.60
3	3.18	5.84	12.92
4	2.78	4.60	8.61
5	2.57	4.03	6.87
6	2.45	3.71	5.96
7	2.36	3.50	5.41
8	2.31	3.36	5.04
9	2.26	3.25	4.78
10	2.23	3.17	4.59
11	2.20	3.11	4.44
12	2.18	3.05	4.32
13	2.16	3.01	4.22
14	2.14	2.98	4.14
15	2.13	2.95	4.07

WORKED EXAMPLE

The heights of six trees at two different sites, one at sea level and one at an altitude of 700m are measured. The t value is calculated to be 4.36 and the degrees of freedom is 10 (12-2).

We look up the values in the table for **10 df**. The calculated t (4.36) is bigger than the one in the table for P=0.05 (2.23) and P=0.01 (3.17) so there is a **significant difference** between the mean height of the two populations of trees. So we are up to **99% confident** there is a difference, but not 99.9% confident (because the t value is smaller that the one in the table for P=0.001 (4.59)).

Remember if **t** > the number in table (P = 0.05) the means are **significantly different**.

The **Chi-squared** test is really **Useful** in ecology

The **chi-squared** value [χ^2] tells you if a set of results differs **significantly** from an **expected** result. For example, you could see what effect a **fertiliser** had on the yield of a **turnip** crop.

You'd do the experiment by growing two crops of turnips, one **with** the fertiliser and one **without**. The null hypothesis is that the **yield** will be the same in both fields (I know you wouldn't *really* expect that, but it's an assumption of the test). The formula for chi squared is:

$$\chi^2 = \sum \frac{(O-E)^2}{E}$$

Where:
O = observed result
E = expected result

Don't panic about these tests. You'll probably need them for your coursework, but you don't have to remember the formulae for the exams. You may be tested on how to look up values in the statistical tables, though.

When you've got your chi-squared value, you look it up in a **chi-squared table of results** — just the same as you would in a t-table (remember you need to work out the degrees of freedom too). As with the t-test, if the number in the table is bigger than your calculated one, the result is significant.

Practice Questions

Q1 What does standard deviation give an indication of?

Q2 In a statistical test, if you have one set of data with 20 results, how many 'degrees of freedom' does this represent?

Q3 Which statistical test tests whether the means of two sets of results are significantly different?

Exam Questions

Q1 Explain how you would test to see if the use of an automatic irrigation system increased the yield of cucumbers in a greenhouse. [3 marks]

Q2 A scientist investigated whether adjusting the diet of cattle improved their milk yield. She compared two groups of cattle, with 8 cows in each. One group were fed the new diet and the other remained on the previous diet. She found the mean yields for the groups on the old and new diets, did a t-test and found that t=1.77. Does the new diet make a significant difference? Explain your answer [use the t-table on this page]. [3 marks]

Difficulty with statistics is a standard error...

The important thing is being able to interpret results. That means that you need to be able to tell the examiner if things are significant or not, and you can tell that a large standard deviation means there's lots of variation. Make sure that you're confident with the differences between chi square and the t-test too. Then run away and hide under a bush somewhere.

Meiosis

You might remember meiosis from AS — it's the one needed for sexual reproduction (no sniggering at the back, please).

Meiosis is a Special type of Cell Division

Meiosis **halves** the chromosome number. It's used for sexual reproduction in plants and animals.

In animals, meiosis produces the **gametes** (sperm and egg cells) and takes place in the **testes** and **ovaries**. When the **gametes** fuse at **fertilisation** they combine their chromosomes, so the chromosome number is **restored**. These two processes make sure that **chromosome numbers stay constant** overall from generation to generation.

Meiosis creates Haploid Cells

Meiosis has **two divisions**, and each one is made up of stages called **interphase**, **prophase**, **metaphase**, **anaphase** and **telophase**. The **first division** (parts 1-5 in the diagram) **halves** the chromosome number, and the **second division** (parts 6-11) **separates** the pairs of chromatids that make up each chromosome, like in **mitosis**.

1. INTERPHASE

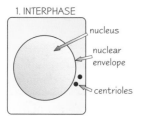

The cell is diploid
(it has two sets of chromosomes.)
It replicates its DNA, ready to divide.

2. PROPHASE I

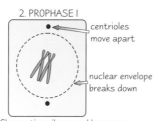

Chromatin coils up and becomes visible as chromosomes, each chromosome consisting of 2 chromatids. Homologous chromosomes pair up — the pairs are known as bivalents.

3. METAPHASE I

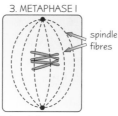

Pairs of chromosomes go to the centre of the cell.

4. ANAPHASE I

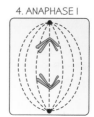

Members of each homologous pair of chromosomes separate and are pulled apart along the spindle fibres.

5. TELOPHASE I

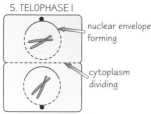

Two haploid cells are forming.

6. INTERPHASE

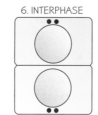

This is the beginning of meiosis II. The cells prepare to divide again.

7. PROPHASE II

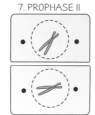

Chromosomes coil up once more. Each chromosome consists of 2 sister chromatids.

8. METAPHASE II

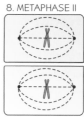

Chromosomes line up in centre of cells.

9. ANAPHASE II

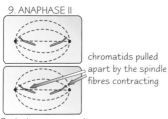

chromatids pulled apart by the spindle fibres contracting

Each chromosome splits into its chromatids.

10. TELOPHASE II

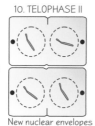

New nuclear envelopes form. Cells divide.

11. FOUR HAPLOID CELLS (e.g. sperm cells)

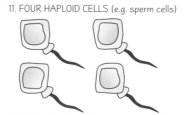

Haploid cells have only one set of chromosomes.

The diagram only shows what happens to one pair of chromosomes. In human cells, there are 23 pairs of chromosomes in total, all doing the same thing.

Cells divide Twice in Meiosis

First division (Meiosis I)

The chromosome pairs come together. The two chromosomes of a pair are called **homologous** chromosomes (see the next page). When they pair up, the pair is called a **bivalent**. Then, these homologous chromosomes move to opposite ends of the cell, and the cell divides. Now, there are **two haploid cells** instead of one diploid cell.

Second division (Meiosis II)

This is **similar** to mitosis (check in your AS notes if you can't remember).
Each new **haploid cell** divides, and each chromosome splits into its **chromatids**.

Meiosis

Crossing Over happens between Chromatids during Prophase I

During **prophase I**, the homologous chromosomes **exchange** pieces of their chromatids. This is called **crossing over.**

Crossing over happens randomly between the homologous chromosomes at any place along them.

The place where crossing over occurs is called a **chiasma** (plural: **chiasmata**). Crossing over helps to mix up alleles in new combinations and creates **variation**.

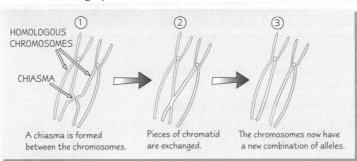

HOMOLOGOUS CHROMOSOMES

CHIASMA

A chiasma is formed between the chromosomes.

Pieces of chromatid are exchanged.

The chromosomes now have a new combination of alleles.

You need to Learn some Key Terms

There's a lot of **fancy words** used about genetics and inheritance. Make sure you know these **important ones**:

chromosome	A strand of genetic material (DNA) found in the nucleus of a cell. Each chromosome consists of one molecule of DNA and histone proteins.
chromatid	One of the two identical strands of genetic material that make up a chromosome during cell division.
homologous	Homologous chromosomes are a pair of equivalent chromosomes with the same structure and arrangement of genes — usually one is inherited from the mother and one from the father.
bivalent	A pair of homologous chromosomes.
haploid	A cell with only half the number of chromosomes of the parent organism (only one copy of each chromosome), e.g. sperm and egg cells.
diploid	A cell with the full number of chromosomes — in pairs of homologous chromosomes.
gene	A section of DNA on a chromosome which controls a characteristic in an organism. It carries the genetic code to make one or more polypeptide or protein, or to make RNA.
locus	The position on a chromosome where a particular gene is located.
allele	An alternative form of a gene. E.g. in pea plants, the gene for height has two forms — one allele for tall plants and one allele for short plants.
genotype	The alleles a particular individual has.
phenotype	An individual's characteristics, e.g. eye colour, blood group.
homozygous	An individual with two copies of the same allele for a particular gene.
heterozygous	An individual with two different alleles for a particular gene.
dominant	The condition in which the effect of only one allele is apparent in the phenotype, even in the presence of an alternative allele.
codominance	The phenomenon in a heterozygote in which the effects of both alleles are apparent in the phenotype.
recessive	The condition in which the effect of an allele is apparent in the phenotype of a diploid organism only in the presence of another identical allele.
linked	Genes located on the same chromosome that are often inherited together.

Practice Questions

Q1   At which stage in meiosis: a) do the cells become haploid? b) does crossing-over occur?

Q2   Place these events in meiosis in the correct order: A. chromatids separate; B. homologous chromosomes pair up; C. two haploid cells are produced; D. homologous chromosomes separate; E. four haploid cells are produced.

Exam Questions

Q1	In which organs of the human body does meiosis occur?	[2 marks]
Q2	Explain the difference between: a) a gene and an allele; b) haploid and diploid.	[4 marks]
Q3	Explain the importance of meiosis in the life-cycles of sexually reproducing organisms.	[3 marks]

How do you tell the sex of a chromosome? Pull down its genes...

Remember that genes are carried on the chromosomes, so whatever the chromosomes do (like separating and re-combining), the genes will do too. It's a huge diagram, but just break it down into meiosis I and II, and learn the names of each phase, and what happens in each one. Use your AS notes to help you if you're still not sure.

Variation

Ever wondered why no two people are exactly alike? No, well nor have I, actually, but it's time to start thinking about it. This variation is partly genetic and partly due to differences in the environment.

Variation can be Continuous or Discontinuous

Discontinuous variation

This is when there are two or more **distinct types**, and each individual is one of these types, for example:

Sex — you're either male or female

Blood group — you can be group A, group B, group AB or group O, but no intermediates

Continuous variation

This is when the individuals in a population vary along a **range**, with **no distinct types**, for example:

Height — you could be any height over a range

Weight — you could be any weight over a range

Skin colour — any shade from very dark to very pale

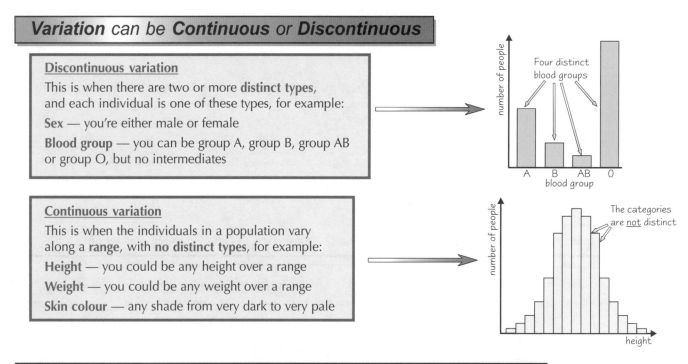

Discontinuous Variation is influenced by One or a few Genes

Discontinuous variation has clear-cut categories because it depends on only one or a few genes (it is monogenic). So, there's a **limited number** of possible phenotypes. Discontinuous variation isn't so strongly influenced by the environment.

Because it's pretty straightforward, discontinuous variation is the kind of variation usually studied in genetic diagrams (see pages 46-47), but most variation in real life is actually continuous variation.

Continuous variation is Polygenic

Polygenic means that several genes affect the same characteristic.
Continuous variation can be more strongly affected by the **environment**. Because of the interaction of loads of different genes plus the effect of the environment, there's lots of possible **phenotypes**.

Example:
Human body mass shows **continuous variation**. Your mass is **partly genetic** (big parents often have big children), but body mass is also strongly affected by **environmental factors** like diet and exercise.

Sexual Reproduction helps create Variety

Sexual reproduction mixes up the alleles in new combinations, creating more **variety** in a species. This means that **survival** (of at least some individuals in a population) is more likely, so there's less chance of the species becoming extinct. If all individuals were the same, one set of environmental conditions or one disease could easily wipe them **all** out.

Sexual reproduction isn't the same as "sexual intercourse", or mating. Plants, fungi and protoctists can reproduce sexually, as well as animals. It just means reproduction where two gametes (sex cells) from two different individuals fuse together to produce a new individual.

Variation

Meiosis creates new Chromosome Combinations

Meiosis (see pages 40-41) does more than just halve the chromosome number. It also helps create **genetic variety**, by producing new combinations of alleles. Here's how it happens:

Independent assortment of chromosomes

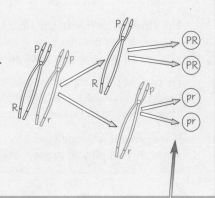

During **meiosis I**, the pairs of homologous chromosomes **separate** (at anaphase). The chromosomes from each pair end up **randomly** in one of the new cells, so you can get **different chromosome combinations**. In **meiosis II**, there's also random assortment of **chromatids**.

One pair of chromosomes would give **2** different types of haploid cell.

- Two pairs would give 2^2 possible haploid cells = **4 possibilities**.
- 23 pairs, like in humans, give 2^{23} possible haploid cells = over **8 million possibilities**. (Your parents would have to have millions of children before they stood any chance of having two genetically the same — unless they have twins.)

Crossing Over

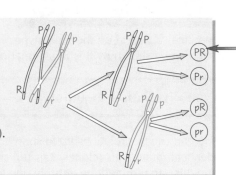

Chromosomes often swap parts of their chromatids during **prophase I** (see page 41). This creates **new combinations** of alleles on those chromosomes, separating alleles that are normally inherited together (**linked**).

The letters in the diagrams represent the alleles of a particular gene found on these homologous chromosomes.

Mutation

Mistakes sometimes happen during cell division, producing a **completely new** characteristic (see page 50).

Meiosis also produces variation because it lets **fertilisation** take place.
The **random fusion** of gametes from two individuals at fertilisation creates unique **combinations** of alleles.

Practice Questions

Q1 Give two examples of characteristics that show continuous variation, and two that show discontinuous variation.

Q2 Why it is important to have variety within a species?

Q3 State three ways in which meiosis helps to create variety within a species.

Exam Questions

Q1 If the body cells of an organism contain three pairs of chromosomes, how many different chromosome combinations can be produced in the gametes of this organism as a result of independent assortment? [2 marks]

Q2 Compare characteristics showing continuous and discontinuous variation with reference to:
a) the extent to which they are affected by the environment. [1 mark]
b) the number of genes that control them. [1 mark]

Variety is the spice of... meiosis...

By now you should have a pretty good idea of how meiosis creates variety in species. It's amazing to think of how many things influence the way that we look and behave. It's the reason we're all so lovely and unique... my parents often said they were glad they'd never have another child quite like me — I can't imagine why.

Environment and Phenotype

The fact that you're such a fantastic person is partly because of the genes you got from your parents, but it's also because of the way they brought you up. Either way, you can thank them for it.

Environment affects Phenotype too

A lot of variation in characteristics (phenotype) is due to differences in genotype, but **environment** also has an effect:

1) The **Himalayan rabbit** is mainly white, but some parts of its fur (at the ears, feet and tail) are black. The growth of the black fur is caused by environmental temperature — these parts of the body are cooler, and the black colour only develops when the skin temperature is below about 25 °C.

2) People are on average much **taller** today than they were 200 years ago (if you're a strapping six-footer, you'll probably bump your head on the ceilings of an old house). This is thought to be because our diet is much better.

3) **Plant growth** is strongly affected by the environment — plants show better, healthier growth when there are more **nitrates** and other minerals available in the soil.

OCR ONLY

Learn this Relationship to do with Variation

The effects of genetic and environmental variation on the phenotype (the characteristics of an individual) can be shown by this **fancy relationship:**

> **Variation in phenotype = variation in genotype + variation in the environment**

Or: $$V_P = V_G + V_E$$

You can use this relationship to show that genetically identical individuals (**clones**) can still show variation, and this variation will be **entirely** due to **environment**:

$$V_G = 0, \text{ so } V_P = V_E$$

OCR ONLY

Environmental Changes can Switch On Bacterial Genes

An example of how the environment can affect genes is seen in the bacterium ***E. coli***.

1) The bacterium feeds on **glucose** but can use other food sources, like **lactose** (milk sugar) if it's available.

2) If lactose is present, *E. coli* makes enzymes to digest it, including an enzyme called **ß-galactosidase**. If there's no lactose around, it doesn't waste energy making enzymes it doesn't need.

3) The production of **ß-galactosidase** is controlled by an **operon**. An operon is a group of genes that work together — an **operator gene** and several **structural genes**. When there's no lactose around, the **operator gene** is **inhibited** by a **repressor** molecule, produced by a **regulator gene**. The repressor binds to the operator gene and in this case stops it producing **ß-galactosidase**.

4) When lactose is present, it binds to the repressor molecule, **inhibiting** it. This lets the operon switch on **enzyme synthesis** — the **operator gene** 'instructs' the **structural genes** to make the enzyme.

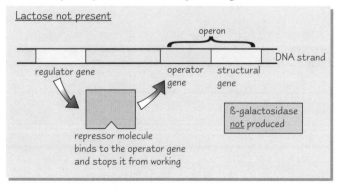

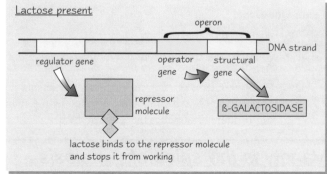

The operon is also **inhibited** by the presence of **glucose**. This means that if there's **both** glucose and lactose available, the bacterium will use up the glucose before it starts on the lactose. It therefore shows **diauxic growth** (see page 180).

Environment and Phenotype

Genes and Environment interact in the Phenotype

Pea plants provide a clear example of the **interaction** between genes and environment that produces a **phenotype**.

Pea plants come in tall and dwarf forms.
This characteristic (tall or dwarf) is passed on from one generation to the next, so we can tell that it is **genetic**.

However, the tall plants vary in height, and so do the dwarf plants, so **environment** is involved too.

> Tall or dwarf is discontinuous variation. Height variation among the plants of each type is continuous variation.

You need these graphs for the exam questions on this page.

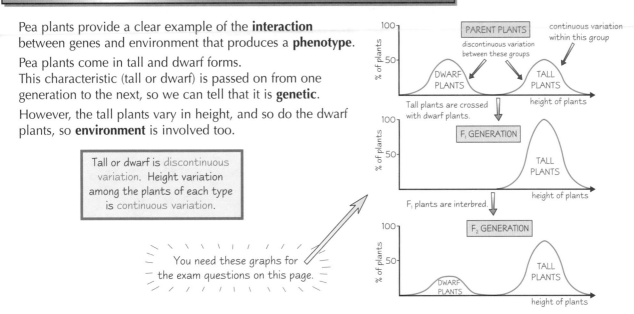

Twins can show the effects of Genes and Environment

Studies of **twins** have been used to find out if a human characteristic is mainly influenced by **genes** or by **environment**.

> **Monozygotic** twins ("identical twins") have **identical genes** and **alleles**, because they both developed from the **same fertilised egg**. This means that if there are any differences in their characteristics, they may be due to the **environment**. Occasionally, monozygotic twins are raised **separately**, and comparing differences between them (compared with twins raised together) could show how important these environmental influences are.

Practice Questions

Q1   Describe the relationship between the effect of genetic and environmental variation on phenotype.

Q2   What is the name of the enzyme, produced by *E. coli*, that digests lactose?

Q3   Explain the meaning of these terms: a) regulator gene; b) operator gene; c) operon; d) gene repressor

Q4   Why don't monozygotic twins always have identical characteristics?

Exam Questions

To answer these questions, look at the data for the heights of pea plants in the graphs on this page.

Q1   How does the information about the parent plants suggest that:
a) height is partly genetically determined?   b) height partly depends on the environment?   [2 marks]

Q2   Do the heights of the F_1 generation show continuous variation, or discontinuous variation, or both?
Explain your answer.   [2 marks]

Q3   Compare the heights of the parental generation with the F_2 generation.
How does this comparison support the idea that height is partly genetically determined?   [1 mark]

Having an identical twin is like having a clone...

That's quite a weird thought... if you've got same-sex twins in your class, it's interesting to investigate how similar they are. How can you tell them apart? If they're identical twins, it must be an environmental difference, e.g. one's dyed their hair pink, or has a tattoo. It's often hard to tell if they're really identical though, or just fraternal twins who look very similar.

Inheritance

Brace yourself for two pages of genetic diagrams. You need to get comfortable with these, because in the exam you'll not only have to interpret them, you might have to draw some of your own. It's probably a smart idea to learn some of the most common patterns and ratios, then you'll be able to apply them to new examples in the exam.

Monohybrid Inheritance *Involves* One Characteristic

Each individual has **two copies** of a gene. But they **segregate** when the sex cells are formed in meiosis, so each **gamete** contains only **one copy** of **every** gene. Monohybrid inheritance is the **simplest** form of inheritance — it's just inheritance where a single gene is being considered. A **monohybrid cross** is a genetic cross for only one gene:

<u>Example</u>

In fruit flies, the allele for **normal wings** is **dominant** (N), and the allele for **vestigial** (short) wings is **recessive** (n).

A normal-winged fruit fly is crossed with a fruit fly that has vestigial wings. **All** the offspring are normal-winged. These flies then **interbreed**, and the next generation shows a **3:1 ratio** of normal wings to vestigial wings, i.e. a 75% chance of normal wings and a 25% chance of vestigial wings.

> The first set of offspring from an experiment like this, where the two parents are true-breeding (homozygous), is called the F_1 generation. If you then breed these offspring together, you produce the F_2 generation.

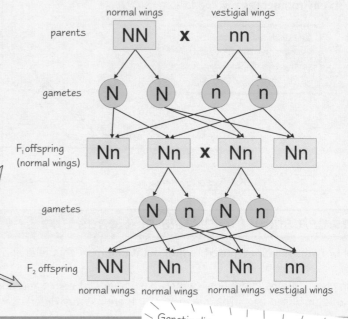

> *Genetic diagrams like this show all the possible combinations of gametes from the parents. Remember to use a capital letter for a dominant allele, and a small letter for a recessive allele.*

A Test Cross *helps you find out an Individual's* Genotype

Sometimes, you might cross a normal-winged fly with a vestigial-winged fly and get a **1:1 ratio** of normal wings to vestigial wings in the offspring, instead of all normal. This happens if the normal-winged fly is **heterozygous** — it has **one allele** for **normal** wings, and **one allele** for **vestigial** wings. Because the allele for vestigial wings is recessive, it doesn't show up in the phenotype of heterozygous flies — vestigial wings is a recessive condition.

Compare this with the first diagram on this page. In each case, the normal fly **looks** the same (they have the same **phenotype**). The only way of telling its genotype is by a **breeding experiment** where you mate it with a recessive individual — in this case that's a fly with **two alleles** for **vestigial** wings (remember this because it'll crop up in the exam). This is called a **test cross**.

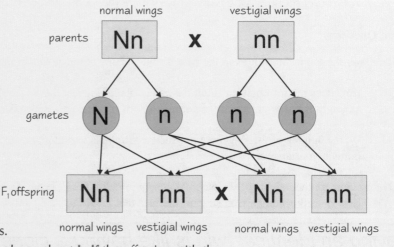

- a **homozygous** normal winged fly produces **all** offspring with the **dominant** characteristic when it's crossed with a fly with vestigial wings.

- a **heterozygous** normal winged fly produces about **half** the offspring with the **recessive** characteristic when it's crossed with a fly with vestigial wings.

Inheritance

Alleles can be Codominant

Occasionally, alleles show **codominance**.
One example in humans is the allele for **sickle-cell anaemia**:

- Normal people have two alleles for normal haemoglobin ($H_N H_N$).

- People with **sickle-cell anaemia** have two alleles for the disease ($H_S H_S$). Their red blood cells are sickle shaped, and can't carry oxygen properly. They usually die quite young.

- Heterozygous people ($H_N H_S$) have an in-between phenotype, called the **sickle-cell trait**. Some of their blood cells are normal, and some are sickle-shaped. The two alleles are **codominant**, because they're **both** expressed in the **phenotype**.

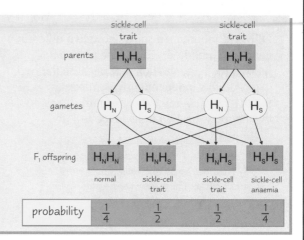

Some Genes have Multiple Alleles

Inheritance is more complicated when there are more than two alleles of the same gene — **multiple alleles**.
E.g. in the **ABO blood group system** there are **three alleles** for blood type:

I^O is the **recessive** allele for blood group **O**. I^A is the allele for blood group **A**. I^B is the allele for blood group **B**.

Alleles I^A and I^B are **codominant** — people with copies of **both** these alleles will have a **phenotype** that expresses **both** alleles, i.e. blood group **AB**. In the diagram below, if members of a couple who are both **heterozygous** for blood groups A and B have children, those children could have one of **four** different blood groups — A, B, O or AB.

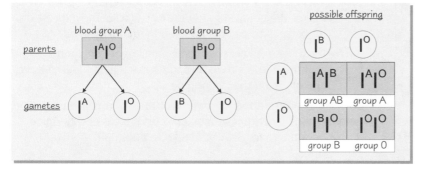

Recessive blood groups are normally really rare, but it just so happens that loads of people in Britain descended from people who were $I^O I^O$, so it's really common.

Practice Questions

Q1   What genetic ratios do you expect from each of these crosses?
   a) Aa × Aa   b) AA × Aa   c) Aa × aa

Q2   Do the genetics of the ABO blood group system show multiple alleles, codominance, or both?

Q3   What is sickle-cell anaemia? What type of inheritance pattern do sickle-cell alleles show?

Exam Questions

Q1   List the six possible genotypes for the human ABO blood groups.   [3 marks]

Q2   In pea plants, the allele for purple flowers is dominant over the allele for white flowers.
       How would you find out if a purple-flowered plant is homozygous or heterozygous?   [3 marks]

It's hard to do test crosses on humans...

If you're wondering whether you're heterozygous for a particular trait, it's probably not an option to breed with a recessive person, and then have lots of babies and see what they look like, unless you take your science homework very seriously.

Inheritance

There's so much to say about inheritance that we've generously stuck in another two pages for you to enjoy.

Genes on **Different Chromosomes** Segregate **Independently**

Dihybrid inheritance shows how **two** different genes are inherited. Each gene gives a 3:1 ratio in the F_2 generation, but because the two genes do this **independently**, it makes a **9:3:3:1 ratio** overall. This diagram shows how this happens for two traits in the fruit fly.

Crossing an F_1 fly with a double recessive fly (vestigial wings and ebony body) gives a **1:1:1:1 ratio**. Check your understanding by working this out yourself.

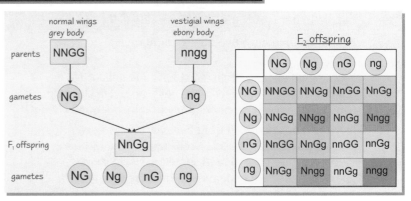

An **Epistatic** gene **Controls** another Gene

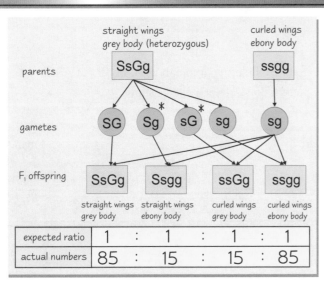

Sometimes, **two different genes** control the **same** characteristic, and they **interact** in the phenotype. This is called **polygenic** inheritance.

One example of this is when one gene can prevent the other one from being expressed — **epistasis**.

E.g. in the genes controlling fur colour in mice:

1) **One gene** (C/c) decides whether the fur is **coloured** (C) or **albino** (c).

2) This gene is **epistatic** over a **second** gene (G/g) which makes the colour (if any) **grey** or **black**. The **expression** of the second gene is affected by the first gene.

3) If a mouse is **recessive** for the first gene (cc) then the mouse will be **albino** and the second gene doesn't have any effect on the phenotype.

Epistasis can produce some **weird ratios** when you start crossing heterozygous individuals together — in the case of the mice, it produces the ratio **grey: 9, black: 3, albino: 4**.

Genes on the **Same Chromosome** are **Linked**   *Skip this bit if you're doing AQA A or B.*

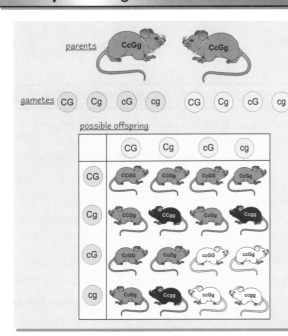

expected ratio	1 :	1 :	1 :	1
actual numbers	85 :	15 :	15 :	85

The crosses so far all assume that two different genes assort (**segregate**) independently. This is usually true, **but** if the genes are on the same chromosome, they're inherited together. This is called **linkage**. If genes are **linked**, when you do a test cross, an **expected** 9:3:3:1 ratio will turn out more like a 3:1 ratio, and a 1:1:1:1 ratio turns out more like 1:1. It's not quite that clear-cut though, due to **crossing over** (see page 43). This means the results might turn out like in this diagram of a fruit fly cross — the gametes **Sg** and **sG** (marked with a star) are **only** produced as a result of **crossing over**, so they occur much **less often** than you'd expect. In this example, it happens just **15%** of the time.

Genes on the same chromosome tend to be inherited together:

Inheritance

In *Mammals* Sex is Determined by the **X** and **Y** Chromosomes

The genetic information for your **gender** is carried on two **specific** chromosomes:

1) In mammals, **females** have **two X** chromosomes, and **males** have **one X** and **one Y**. The probability of having male or female offspring is **50%**.

2) The Y chromosome is **smaller** than the X chromosome and carries **fewer genes**. So most genes carried on the sex chromosomes are only carried on the X chromosome. These genes are sex-linked. Males only have **one copy** of the genes on the X chromosome. This makes them more likely than females to show **recessive phenotypes**.

3) Genetic disorders inherited this way include **colour-blindness** and **haemophilia**. The pattern of inheritance can show that the characteristic is **sex-linked**. In the example below, females would need **two copies** of the recessive allele to be colour blind, while males only need one copy. This means colour blindness is **much rarer** in **women** than **men**.

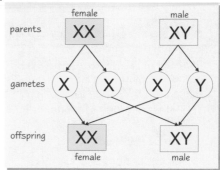

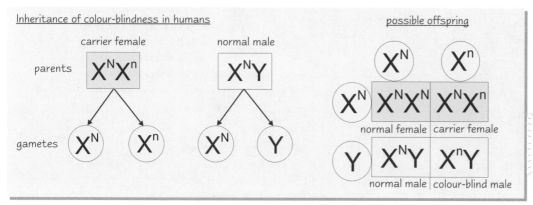

N = allele for normal eyesight
n = allele for colour-blindness

A carrier is a person carrying an allele which is not expressed in the phenotype, but which can be passed on.

The **Chi-Squared Test** Checks the Results

Genetic ratios are the result of **chance fusion** of gametes — and because it's random, the results you get are rarely exactly the same as the theory predicts. The **chi-squared** test (see page 39) can be used to find out whether the difference between **observed** and **expected** results is big enough to be **significant**, or whether it's just due to chance.

Practice Questions

Q1 What does epistasis mean?

Q2 Which chromosomes determine the gender of mammalian offspring?

Q3 Why is red-green colour-blindness much more common in males than in females?

Exam Questions

Q1 Draw a genetic diagram to show the expected results of a cross between a normal-winged grey-bodied fruit fly (genotype NnGg) and a normal-winged ebony-bodied fruit fly (Nngg). [4 marks]

Q2 The recessive allele for haemophilia is carried on the X chromosome.
Explain why you would expect haemophilia to be more common in males than females. [5 marks]

Pedigree charts aren't just for dogs... they're for royals too...

You can use pedigree charts to track the inheritance of sex-linked conditions like haemophilia. The classic example is the inheritance of haemophilia from Queen Victoria, who carried the allele and spread it through the royal families of Europe.

Mutation and Phenotype

Skip this page if you're doing AQA B. Mutation sounds quite exciting, but if you're expecting pictures of chickens with two heads or green monsters, you're going to be disappointed. Anyway, here's what happens when cell division goes wrong.

Point Mutations are Changes in the DNA Base Sequence

Before a cell divides, its DNA is replicated (copied) — look at your AS notes for more detail on DNA replication. Sometimes the base sequence of the DNA (the genetic code) gets changed. This is a **gene mutation** and it can make the DNA code for a different protein. A change of **one base** (C,G,A or T) is called a **point mutation**. The effect of a point mutation depends on exactly what happens:

No mutations

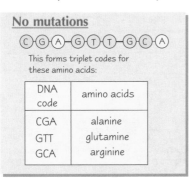

This forms triplet codes for these amino acids:

DNA code	amino acids
CGA	alanine
GTT	glutamine
GCA	arginine

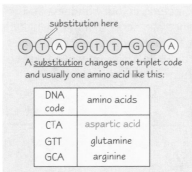

substitution here

A <u>substitution</u> changes one triplet code and usually one amino acid like this:

DNA code	amino acids
CTA	aspartic acid
GTT	glutamine
GCA	arginine

Substitution

One base is **swapped** for another in a triplet code. It means that the gene will make a **similar protein** to the normal protein, but with just **one amino acid** different. Because the structure of a protein is so important, this can have a big effect. (The **sickle-cell** allele is the result of a base substitution.)

Insertion (addition)

An **extra** nucleotide (with a base) is included in the DNA molecule. This has a much bigger effect than a substitution, because it causes **all** the following triplet codes in the gene to be altered. If there's an insertion, the gene doesn't make **any** useful protein at all, which can cause serious problems.

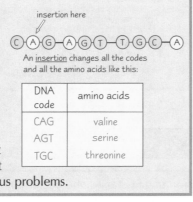

insertion here

An <u>insertion</u> changes all the codes and all the amino acids like this:

DNA code	amino acids
CAG	valine
AGT	serine
TGC	threonine

Deletion

A nucleotide (with its base) is **missed** out when the DNA is copied. Like insertion, this **shifts along** all the triplet codes after it, so it really messes things up. This is also known as a '**frame shift**' mutation.

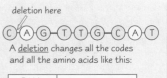

deletion here

A <u>deletion</u> changes all the codes and all the amino acids like this:

DNA code	amino acids
CAG	valine
TTG	asparagine
CAT	valine

Chromosome Mutations are caused by Errors in Cell Division

A **chromosome mutation** is where one or more **whole chromosomes** (or **parts** of chromosomes) get in the **wrong place** during cell division. If this happens during meiosis, then the chromosome abnormality will be in the **gametes**, so it gets passed to the **next generation**. There are **three main types** of chromosome mutation:

Polyploidy

All the chromosomes fail to segregate at anaphase I of meiosis. This is called **non-disjunction**. This creates a diploid gamete (with **two sets** of chromosomes), and after fertilisation, the zygote will have **three** (or even more) complete sets of chromosomes.

Polysomy

Non-disjunction of some, but not all, the chromosomes. E.g. a gamete may have just one extra chromosome, so the offspring also has one extra chromosome in each cell. About **95%** of **Down's syndrome** cases in humans are caused by a person having an extra copy (**trisomy**) of chromosome 21.

Translocation

This is where **part** of a chromosome gets broken off, and then **reattaches** in a different place. This doesn't usually have such a big effect on the phenotype as polyploidy or polysomy, but it can still have serious consequences. About **5%** of **Down's syndrome** cases are caused by this.

Mutation and Phenotype

Mutagens make Mutations more likely

Mutations are accidents, and can happen for no obvious reason.
However **mutagens** make mutations more likely to happen:

1) <u>Radiation</u>

Some types of **radiation** are mutagenic. This includes X-rays, UV rays and ionising radiation such as gamma-rays.

2) <u>Chemicals</u>

Some **chemicals** are mutagens. Most of these chemicals are also carcinogens (they cause cancer).
E.g. mustard gas, and substances in tobacco smoke. These chemicals cause **point mutations** in the DNA.

Other chemical mutagens affect the structure of **chromosomes**, making chromosome mutation more likely.
E.g. the dye **colchicine** has this effect, and is used by **plant breeders** to produce new plant species.

Mutations are often Harmful

Any random change to the DNA in a cell is probably going to be **damaging**.
Most **genetic diseases** are the result of mutations.

Changes in the cell's DNA might mean that it codes for a different protein,
or for none at all, which could stop an important process from working.

Albino wallaby Mervyn gets upset when
people call him a mutant.

> ### The Human Genome Project (HGP)
> This international project to **map the positions** of different
> genes on the chromosomes was completed in 2003.
> Information from the HGP will help scientists
> understand **mutations** and **genetic diseases**.

Mutations can lead to Evolution

Occasionally a mutation creates an **improvement**. If so, the mutant will have a selective advantage and will
probably end up having more offspring, so **natural selection** causes the mutant form to become more common.
This type of mutation is important in **evolution**.

Practice Questions

Q1　Name the three main types of gene mutation.

Q2　Explain the difference between polyploidy and polysomy.

Q3　Give examples of two chemical mutagens.

Exam Questions

Q1　Explain why the deletion of three adjacent nucleotides in a gene mutation will usually have a less severe
　　effect on the phenotype than the deletion of one nucleotide.　　　　　　　　　　　　　　　　[3 marks]

Q2　Suggest explanations for these facts:
　　a) Radiographers in hospitals stay behind a lead screen when giving X-rays to patients.　　　[2 marks]
　　b) Excessive exposure to bright sunlight can cause skin cancer.　　　　　　　　　　　　　　[2 marks]
　　c) When fruit-flies are exposed to X-rays, and then mated, some of their offspring have abnormal
　　　　white eyes or deformed wings.　　　　　　　　　　　　　　　　　　　　　　　　　[2 marks]

So you're telling me I'm a mutant...

*Loads of genetic diseases start off as just a random mutation in one person, then the mistake just keeps getting passed on
down the generations. On the other hand, a mutation could be the reason for your stunning good looks.*

Frequency of Alleles

Skip these pages if you're doing AQA B. If you're doing Edexcel or OCR you only need the first section on page 53.
Sometimes you need to look at the genetics of a whole population, rather than a cross between just two individuals.
Here goes — get ready for a bit of statistics.

AQA A ONLY

Members of a Population share a Gene Pool

A **population** is a group of organisms of the same species living in a particular area.
The **gene pool** is all the genes or alleles present in the population.

The Hardy-Weinberg Principle can Predict Allele Frequency

You can use the **Hardy-Weinberg Principle** to work out the **frequency** of certain genotypes and phenotypes
in a population. The principle can be demonstrated using the following example:

A species of plant has either **red** or **white** flowers. The
frequency of the allele for a red flower (**A**) or a white flower
(**a**) is written as a decimal between 0 and 1. So if the alleles
for red and white flowers are **equally common**, they would
each have a frequency of **0.5**.

Now you can work out the genotype frequencies:

> In the **Hardy-Weinberg equation:**
> p = the frequency of the dominant allele (A)
> q = the frequency of the recessive allele (a)
> If there are just two alleles for
> a gene, p + q should = 1

> p^2 = frequency of the genotype AA   2pq = frequency of the genotype Aa q^2 = frequency of the genotype aa

Every member of the population must have **one** of these genotypes, so: $p^2 + 2pq + q^2 = 1$.

> **Example**
> **p** (the frequency of allele A) is **0.8**, and **q** (the frequency of a) is **0.2**.
> Work out the frequencies of the different **genotypes** (and **phenotypes**) in the population:
>
> **Frequency of AA** = 0.8^2 = 0.64
> **Frequency of Aa** = 2 × 0.8 × 0.2 = 0.32 ← *These frequencies need to add up to 1.*
> **Frequency of aa** = 0.2^2 = 0.04

You can **reverse** the equation — if you know the **phenotype frequencies**
you can work out the **genotype** and **gene frequencies**:

> **Example**
> The frequency of cystic fibrosis in the UK is approximately **1 birth in 2000**.
> Cystic fibrosis is caused by a **recessive allele**, so:
> q^2 = 1/2000 = 0.0005
> ∴ q = $\sqrt{0.0005}$ = 0.0224 — this is the frequency of the cystic fibrosis allele.
> ∴ p = 1 - 0.0224 = 0.9776 — this is the frequency of the normal allele.
> The proportion of the population who are **carriers** of the cystic fibrosis allele
> is 2pq = 2 × 0.9776 × 0.0005 = 0.0438, so **4.38%** of the UK population carry the CF allele.

The Hardy-Weinberg Principle only works under Certain Conditions

The Hardy-Weinberg principle only applies in certain conditions:
- A **large population**.
- **Random mating** — all genotypes must be equally likely to mate with all others.
- No **immigration or emigration**.
- No **mutations** or **natural selection**.

This is because the Hardy-Weinberg principle relies on the **proportions** of the alleles
of a particular gene staying **constant** from one generation to the next. If any of these
conditions **aren't** met, it'll cause the **allele frequencies** in the population to start **changing**.

Frequency of Alleles

Selection affects Allele Frequencies

Selection can have **different effects** on a population:

1) **Stabilising** selection is where individuals with traits towards the **middle** of the range are more likely to survive and reproduce. It's the **commonest** type, which occurs when the environment is **not** changing. It helps to keep the population **stable**.

2) **Directional** selection is where individuals of **one extreme type** are more likely to survive and reproduce. This happens when the environment changes, and it causes corresponding **genetic change** in the population.

3) **Disruptive** selection is where **two different extreme types** are selected for, perhaps because they live in **two different habitats**. This leads to **two distinct types** developing, and eventually these may become **different species**.

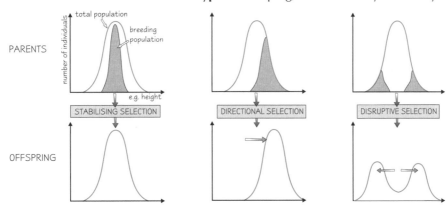

Selection depends on the Environment

Certain traits only become more **common** in a population because they give individuals a better chance of **surviving** and **reproducing** in the particular environmental **conditions** at the time.

Example

Sickle-cell anaemia is a genetic disorder common in tropical countries where malaria is endemic. People with **two copies** of the allele die quite young if they're not receiving treatment. One copy of the **sickle-cell allele** gives carriers some **protection** against malaria. In **non-malarial** parts of the world, the allele would be removed by **selection**, but in malarial areas it gives carriers an advantage so it stays in the population.

When the environment **changes**, so do the selection pressures. Over the past century, **humans** have started using **antibiotics** against bacteria, and insecticides against agricultural pests. Many populations of bacteria are becoming resistant to antibiotics, and pesticides are becoming less effective against the pests. These changes are the result of **directional selection**.

Practice Questions

Q1 What is the Hardy-Weinberg equation?

Q2 Explain three different conditions needed, if the Hardy-Weinberg Principle is to hold true.

Q3 Name the three different types of selection, and explain what effect each has on allele frequencies.

Exam Questions

Q1 In one of the human blood group systems, there are three blood groups, dependent on a single pair of codominant alleles. The genotypes and phenotypes are as follows: genotype MM - blood group M, genotype NN - blood group N, genotype MN - blood group MN. If the frequency of blood group M in a population is 0.36, use the Hardy-Weinberg equation to work out the frequencies of blood groups MN and N. [4 marks]

Q2 If malaria were eradicated from the world, explain what you'd expect to happen to the frequencies of the sickle-cell and normal alleles in the human population. What type of selection would cause the changes? [4 marks]

The best kind of selection comes in a box at Christmas...

The Hardy-Weinberg Principle looks a lot worse than it is, so don't be put off by the calculations. Like most equations in Biology, all you have to do is stick the figures that you have into the equation, then it's just simple maths.

Classification

Classification is all about grouping together organisms that have similar characteristics.
The system of classification in use today was invented by a Swedish botanist, Carolus Linnaeus, in the 1700s.

Classification is the way Living Organisms are Divided into Groups

The classification system in use today puts organisms into one of five **kingdoms**:

KINGDOM	EXAMPLES	FEATURES
Prokaryotae	bacteria	unicellular, no nucleus, less than 5 μm, naked DNA in circular strands, cell walls of peptidoglycan
Protoctista	algae, protozoa	eukaryotic cells, usually live in water, unicellular or simple multicellular
Fungi	moulds, yeasts and mushrooms	eukaryotic, heterotrophic, chitin cell wall, saprotrophic
Plantae	mosses, ferns, flowering plants	eukaryotic, multicellular, cell walls made of cellulose, photosynthetic, contain chlorophyll, autotrophic
Animalia	nematodes (roundworms), molluscs, insects, fish, reptiles, birds, mammals	eukaryotic, multicellular, no cell walls, heterotrophic

You can Classify Organisms according to how they Feed

There are **three** main ways of getting **nutrition** —

1) **Saprotrophic** organisms, e.g. **fungi**, absorb substances from **dead** or **decaying** organisms using **enzymes**.
2) **Autotrophic** organisms, e.g. **plants**, produce their **own** food using **photosynthesis**.
3) **Heterotrophic** organisms, e.g. **animals**, consume complex organic molecules, i.e they consume **plants** and **animals**.

All Organisms can be organised into Taxonomic Groups

Taxonomy is the branch of science that deals with **classification**.

A **species** is the **smallest** unit of classification (see p. 56-57 for more about species). Closely related species are grouped into **genera** (singular = genus) and closely related genera are grouped into **families**. The system continues like this in a hierarchical pattern until you get to the largest unit of classification, the **kingdom**.

> The Hierarchy of Classification
> Kingdom
> Phylum
> Class
> Order
> Family
> Genus
> Species

For Example, Humans are Homo sapiens

This is how **humans** are classified:

		FEATURES
KINGDOM	Animalia	animal
PHYLUM	Chordata	has nerve cord
CLASS	Mammalia	feeds young on milk, has hair / fur
ORDER	Primates	five digits with opposable thumb, modified claws, reduced snout, binocular vision
FAMILY	Hominidae	upright walking, short face, small incisors
GENUS	*Homo*	no prehensile tail, menstrual cycle, narrow nasal septum
SPECIES	*sapiens*	cranial capacity > 700 cm³, jaw / teeth adaptations, pelvis adapted for bipedalism, precision grip with thumb and index finger

This column shows the features that have been used to classify humans into each of these groups.

Classification

The **Binomial System** is used to **Name** organisms

The full name of a human is **Animalia Chordata Mammalia Primate Hominidae** *Homo sapiens*. The name gives you a lot of information about how humans have been classified. Using full names is a bit of a mouthful so it's common practice to just give the **genus** and **species** names — that's the **binomial** ('two names') **system**.

The binomial system has a couple of **conventions**:

1) Names are always written in *italics* (or they're <u>underlined</u> if they're **handwritten**).

2) The **genus** name is always **capitalised** and the **species** name always starts with a **lower case** letter.

e.g.

Human	*Homo sapiens*
Polar bear	*Ursus maritimus*
Sweet pea	*Lathyrus odoratus*

Cladograms show **Evolutionary Relationships**

When taxonomy was first developed organisms were classified according to characteristics that were **easy to observe**, for example, number of legs. Thanks to modern **scientific techniques** like **DNA technology**, **genetics**, **biochemical analysis** and **behavioural analysis**, many more **criteria** can now be used to classify organisms.

A **cladogram** is a diagram which emphasises **phylogeny** (the genetic relationship between organisms). Cladistics focuses on the features of organisms that are **evolutionary developments**. The **advantage** of cladograms is that you can see points where **one** species split into **two**.

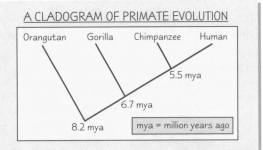
A CLADOGRAM OF PRIMATE EVOLUTION

Orangutan Gorilla Chimpanzee Human

5.5 mya
6.7 mya
8.2 mya

mya = million years ago

99.4% of a chimp's DNA is identical to human DNA. The fact that humans and chimps diverged most recently shows that humans are more closely related to chimps than they are to the other ape species in the cladogram.

Practice Questions

Q1 Name the five kingdoms of classification, giving an example organism in each.

Q2 What do the phrases saprotrophic, autotrophic and heterotrophic mean?

Q3 Explain the difference between fungi and plants in terms of how they get their nutrition.

Q4 What are the two rules for using the binomial system?

Exam Questions

Q1 Explain the difference between phylogenetic classification and traditional classification. [2 marks]

Q2 The King Penguin has the scientific name *Aptenodytes patagonicus*. Fill out the missing words a) – e) in the table. [5 marks]

Kingdom	Animalia
a)	Chordata
b)	Aves
Order	Sphenisciformes
c)	Spheniscidae
Genus	d)
Species	e)

I prefer scantilycladograms...

The good thing about this is everything is pretty straightforward — don't be put off if lots of the words are new to you (and if 'cladogram' is part of your day-to-day vocabulary then I suggest you get out more). You need to learn this thoroughly. In the exam you'll be glad that you did cos there's often some easy marks to be had about this kind of stuff.

Speciation

Speciation is how new species appear — you need to know about two kinds, allopatric and sympatric speciation.

Speciation is the Development of a New Species

A **species** is defined as a group of organisms that can **reproduce** and produce **fertile young**. Every true species we know of has been named using the **binomial system**. When new species are discovered they are also classified using it.

Sometimes two individuals from different species can breed and produce offspring. These **hybrid** offspring aren't a new species because they're **infertile**. For example, **lions** and **tigers** have bred together in zoos to produce **tigons** and **ligers** but they aren't new species because they **can't** produce offspring.

Speciation (development of a new species) happens when **populations** of the **same species** become **isolated**. Local populations of a species are called **demes** (be careful with this — demes aren't always isolated).

Geographical Isolation causes Allopatric Speciation

1) Geographical isolation happens when a **physical** barrier **divides** a population of a species.

2) **Floods**, **volcanic eruptions** and **earthquakes** can all be barriers that cause some individuals to become **isolated** from the main population.

3) **Conditions** on either side of the barrier will be slightly **different**. For example, there might be a different **climate** on either side of the barrier.

4) Environmental conditions like this put **pressure** on the organisms, forcing them to **adapt** — the **natural selection processes** differ in each isolated group.

5) **Mutations** will take place **independently** in each population and, over a **long** period of time, the gene pools will **diverge** and the **allele frequencies** will **change**.

6) Eventually, individuals from different populations will have changed so much that they won't be able to breed with one another to produce **fertile** offspring — they'll have become **two separate species**.

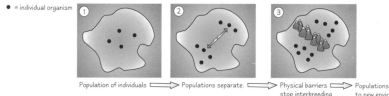

● = individual organism

Population of individuals ⇒ Populations separate. ⇒ Physical barriers stop interbreeding between populations. ⇒ Populations adapt to new environments. ⇒ Gene pools diverge, leading to development of new species.

Isolation doesn't have to be Physical

Reproductive isolation happens when something **prevents** some members of a population breeding with each other. There are **several causes** of reproductive isolation:

1) **Seasonal isolation** — where a mutation means that some individuals of the same species have different **flowering** or **mating** seasons, or become **sexually active** at different times of the year.

2) **Mechanical isolation** — where mutations cause changes in genitalia which prevent successful mating.

3) **Behavioural isolation** where a group of individuals develop **courtship rituals** which are **not attractive** to the main population of a species.

4) **Gametic isolation** — mutations mean that male and female **gametes** from different populations of the species can't create new individuals — so the individuals can mate, but fertilisation fails or the foetus is aborted.

> If two populations have become so different that they can't breed then a **new species** will have been created — this is called **sympatric speciation**.

Plant Speciation can occur through Polyploidy

Sometimes cell division doesn't happen when gametes form, and some gametes end up being **diploid** rather than **haploid**. If these gametes fuse with other gametes you end up with individuals that have one or more **extra sets of chromosomes** — that's **polyploidy**. Sometimes, the chromosome set doubles **after fertilisation** — the chromosomes replicate as they would before mitosis, but then the cell doesn't divide. This **post-fertilisation polyploidy** becomes important if two closely related species are crossed. The offspring would be **sterile**, because the chromosomes would be **non-homologous** and so couldn't pair up during meiosis. But if the diploid number **doubles**, each chromosome **will** have a homologous one to pair with and meiosis **can** happen. This is thought to have happened to produce the modern **wheat** plant.

Speciation

Darwin's Finches are a good example of Allopatric Speciation

Darwin studied **finches** that live on the Galápagos Islands, a small group of islands 1000 km west of Ecuador, to develop his theory of evolution. He based his theory on his observations:

1) On the Galápagos islands, there are **fourteen** species of **finch** belonging to **four genera**.

2) Each species of finch inhabits a different ecological niche (see p. 20) on the islands and some are only found on one island.

3) The main difference between the finches is the **shape** and **size** of their **beaks**. The birds feed on a variety of different foods from grubs to hard-shelled seeds — each finch has a beak suited to the food it eats.

<u>main food</u>	fruits	insects	insects	cacti	seeds	seeds
<u>feeding adaptation</u>	parrot-like beak	grasping beak	uses cactus spines	large crushing beak	pointed crushing beak	large crushing beak

Despite these differences, Darwin thought that all the finches had a **common ancestor**.
Since then more research has been done which has proved that **geographical isolation** did cause **speciation** on the Galápagos islands. Finches are small birds and it's unusual for them to fly over water, so once a population gets onto an island (perhaps because they were blown off course by a storm) they are effectively **isolated** from the finches on other islands. The differing environmental conditions on each island put **selection pressures** on the birds — and the birds gradually became **adapted** by natural selection to the conditions on the different islands.

Convergent Evolution is when Unrelated Species have Similar Features

Convergent evolution happens when **unrelated** species have **evolved** so that they look very **similar**. For example, **sharks** and **dolphins** look pretty similar and swim in a similar way but they're totally different species — sharks are cartilaginous **fish** and dolphins are **mammals**. They have different evolutionary roots but they have developed similar bone structures to make them well **adapted** for swimming.

Practice Questions

Q1 Define the term 'species'.

Q2 What is a hybrid? Give an example.

Q3 What is the difference between allopatric and sympatric speciation?

Q4 Name four causes of sympatric speciation.

Q5 What was Darwin researching when he proposed his theory of evolution?

Exam Question

Q1 Charles Darwin studied different species of finch on the Galápagos Islands.
a) Describe Darwin's observations. [3 marks]
b) Give an explanation of how Darwin believed the different species developed. [4 marks]

I wish there were biology field trips to the Galápagos Islands...

It's easy to learn the basics of these pages — what a species is and how new ones develop. Then it's just a matter of learning the detail and the correct words for everything. It's important that you know words like 'sympatric' and 'convergent evolution' because they might be used in the exam questions and you'll be stuck if you forget what they mean.

Natural Selection and Evolution

You can skip these two pages if you're doing AQA A.

Darwin's book on his theory of evolution 'On The Origin of Species' is probably the most important biology book that's ever been published. Apart from this one of course.

Darwin wrote his Theory of Evolution in 1859

In **1831** Darwin was invited to join the ship **HMS Beagle** on a map-making trip around the world. Darwin was a keen **naturalist** and on the voyage he collected lots of **data** and many **samples** of **plants** and **animals**. When he returned he spent twenty years studying his data and eventually he came up with the **theory of evolution** which was published in **1859** in a book called '**On The Origin of Species by means of Natural Selection**'. Darwin's theory was based on five main assumptions:

1) **More individuals** are **produced** than can **survive**.
2) There is a **struggle** for **existence**.
3) Individuals within a species show **variation**.
4) Those with **advantageous features** have a greater chance of **survival**.
5) Those individuals who **survive** produce **similar offspring**.

I'm not sure I'd like to try and sail in a beagle

Darwin used the term '**natural selection**' to describe the way that individuals with variations that help them survive in their habitat have advantages which make it more likely that they'll be able to pass on their genes. Darwin believed that natural selection **caused evolution**.

The Latest evolutionary theories still include Natural Selection

Obviously, there has been a lot more scientific research into natural selection done since Darwin published his book. The latest developments in genetics have been incorporated into Darwin's theory to update it:

1) There are changes in the **genetic composition** (gene frequencies) of a population from one generation to the next.
2) These changes are brought about by **mutation**, **genetic drift**, **gene flow** and **natural selection**:

Mutation

The **mutation** of genes can produce different **allele** and **phenotype frequencies** (see p. 50).

Genetic Drift

This is the **alteration** of **gene frequencies** through **chance**. For example, if two **heterozygous** individuals breed, their offspring might not have the exact **Mendelian** ratio — the gene frequencies in populations change over time.

Gene Flow

This happens when new genes enter or leave a population by **migration**.

Natural Selection

1) As **conditions** change, or organisms **move** into a new environment, the organisms that are better **adapted** to the new conditions because of the **alleles** they carry will **survive**.
2) **Variation** between **isolated populations** increases as **gene pools diverge**.
3) Changes occur in **allele**, **genotype** and **phenotype** frequencies.
4) Eventually a **new species** will evolve.

Limiting Factors affect Survival and Reproduction Rates

There's more about the type of things that act as limiting factors on pages 24 and 25.

All organisms tend to **overproduce** — this inevitably brings about **intraspecific competition** for resources. **Limiting factors** like **parasites** put **selection pressures** on the organisms. This is when natural selection comes into play — the individuals that are best adapted to the conditions because of the genes that they carry are more likely to survive and reproduce. This process changes the **allele frequencies** in the population. For example, there would be a **greater number** of **individuals** with **resistance** to certain **parasites** in the population.

Natural Selection and Evolution

Peppered Moths are a classic Natural Selection example

Remember there's three types of natural selection – **stabilising**, **directional** and **disruptive** (see page 53). The peppered moth is a good example of **directional** selection:

The **peppered moth** (*Biston betularia*) has always had **two phenotypes** — a **peppered** form and a **melanic** (black) form. The moth lives on the bark of trees. In areas with lots of **industrial pollution**, soot was deposited on the tree trunks darkening them. In these areas, the melanic form was much more common — that's because the peppered form was much more obvious to predators so it was far less likely to **survive** and **reproduce**. In areas where there's **little** pollution, the **peppered** form is more common.

Peppered and Melanic moths on tree bark in unpolluted area

Peppered and Melanic moths on tree bark in polluted area

There are Three main Sources of Evolutionary Data

Macroevolution is a term used to describe **long-term**, **large scale evolutionary change** — you could study the macroevolution of dinosaurs for example. There are three main sources of information about macroevolution:

1) Studies of **comparative morphology** (differences in what organisms look like) — especially comparisons from **embryology** and **anatomy** which gives evidence of **related species**.

2) **Geographical distribution** of organisms, e.g. the study of **marsupial** mammals in Australia, gives evidence of evolutionary development (i.e. it is likely that they have a common ancestor).

3) **Fossil records** give details of evolution in animals and early plants, from fossils of bones, leaves and spores. Fossil records are **incomplete** and cannot give evidence of the **earliest** life on Earth.

Practice Questions

Q1 How has modern biology updated Darwin's theory of evolution?
Q2 What does genetic drift mean?
Q3 What is macroevolution?
Q4 Describe the three main sources of information about macro-evolution.
Q5 Write a paragraph explaining natural selection in your own words.

Exam Question

Q1 In 1976-1977, a severe drought struck the Galapagos islands. No rain fell for over a year. During the drought a number of plant species died out. Some others did not produce seeds, causing a food shortage for the seed-eating ground finch *Geospiza fortis*.

The seeds of one plant species that survived the drought were stored in large, tough fruits.
Only *Geospiza fortis* individuals with a beak depth greater than 10.5 mm were able to feed from this plant.

A biologist conducted a survey of the finches and recorded the mean beak depths and lengths of birds that survived and birds that died during the drought. The results of the survey are recorded in this table:

	Beak Length	Beak Depth
Surviving birds	11.07	9.96
Dead birds	10.68	9.42

As the population of *G. fortis* recovered after the drought, the mean beak depth of the population was greater than before (an increase of 4–5%). Explain this change with reference to evolutionary change. [6 marks]

Lack of chocolate is a revision-limiting factor...

Natural selection is the most important thing on this page so make sure that you read that bit thoroughly. You need to be able to describe how natural selection happens and what impact it has on future generations of the species. Also, check that you know how genetics has been incorporated into modern thinking on evolution.

Adaptations to the Environment

These pages are for OCR, and AQA B and AQA A need to know the bits on xerophytic plants and kangaroo rats.
Skip these pages if you're doing Edexcel.

Natural selection eventually gives organisms special features that mean they are well adapted to their environments.

Penguins are adapted to **Cold Environments**

Homeothermic (warm blooded) animals can **maintain** their **body temperature** even when the external temperature is much lower.

> In the Antarctic, temperatures reach **-40°C** and wind speeds of up to 80 kph can make the wind chill temperature as cold as **-70°C**. Penguins have various adaptations to help them cope with these conditions:
>
> 1) They have a **large body size**.
> 2) Their **surface area : volume ratio** is **small**.
> 3) They have **thick, waterproof, insulating feathers**.
> 4) They have a **thick** layer of **insulating fat** under their skin.

A small penguin living in a temperate environment

In order to keep warm, small animals need to produce more heat than larger ones. Only larger penguins such as Emperor penguins are found in the coldest regions of the Antarctic. Small penguins (e.g. Humboldt penguins) are found in more temperate areas like the south of South America. Small penguins find it harder to survive in the extreme cold of the Antarctic because they have a larger surface area : volume ratio which means they lose heat faster than the larger species.

Plants can adapt to **Dry Conditions**

In **deserts** (where water is **scarce**) and in **arctic** regions (where it's **frozen**), plants need special adaptations to avoid dehydration. Natural selection has meant that plants which are successful in arid environments are **efficient** at **water uptake** and have facilities to **reduce transpiration** and **store water**.

Plants that are specially adapted to live in dry conditions are called <u>xerophytes</u>.

> Cacti are adapted to desert conditions:
>
> 1) **Transpiration** is reduced by various leaf adaptations — a **thick waxy cuticle**, few stomata, sunken stomata, stomata that **open at night** and **close by day**, and a leaf surface covered with **fine hairs**.
> 2) **Water uptake** is **increased** by having an **extensive root system** which covers a wide area.
> 3) Water is **stored** in **fleshy**, **succulent** leaves or stems. Cacti also have **spines** to **protect** the plant from predators.

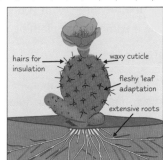

hairs for insulation
waxy cuticle
fleshy 'leaf' adaptation
extensive roots

Kangaroo Rats can tolerate **Dry Conditions**

Kangaroo rats live in the **deserts** of north America. They are so well adapted to the dry conditions that they never drink... pretty amazing huh? — the only water they ingest is a small amount of water from their food. Instead of drinking they use a combination of **physiological** and **behavioural** adaptations to **make** and **retain** water:

Other desert animals have features so that they can cope with the dry conditions. For example, some desert frogs have waterproof skin.

1) They use the water produced in respiration — metabolic water. And they metabolise **fat** because it produces **twice** as much water as **carbohydrate** does.
2) When they breathe through their **noses** the inhalation has a cooling effect on the nasal passages. The water vapour in the exhaled air **condenses** against the nasal passages, reducing water loss.
3) They don't **sweat**.
4) They spend most of their day in a deep **burrow**, feeding at **night** when the temperature is much **cooler**.
5) They produce **dry faeces** and very **little urine** — their urine is **highly concentrated** (see p. 67).

Adaptations to the Environment

In Hot Places some Plants use C_4 Photosynthesis

Plants that grow in **hot**, **tropical** climates are **adapted** to those conditions — they use a different type of **photosynthesis** to the usual one. The difference is in the **Calvin cycle**, which is described on page 18. Normal C_3 **plants** use **ribulose bisphosphate** (**RuBP**) to accept carbon dioxide in the Calvin cycle, but C_4 **plants** (e.g. **maize** or **sugar cane**) use a different carbon dioxide acceptor called **phosphoenolpyruvate** (**PEP**). This means that a **4-carbon** compound called **oxaloacetate** is formed instead of the usual **3-carbon** compound, and that's why plants that use this method are called C_4 **plants**.

Using C_4 photosynthesis has a number of **advantages** for plants growing in hot, tropical conditions:

1) C_4 plants use a different **enzyme** to catalyse the fixation of **carbon dioxide**. C_3 plants use an enzyme called **rubisco**, but in C_4 plants the enzyme is **PEP carboxylase**. This enzyme has some useful features:
   - it has an **extremely high affinity** for carbon dioxide, even at **low concentrations**.
   - it works much **faster** and more **efficiently** than rubisco.
   - it works particularly well at **high temperatures** and in **bright light**.

2) C_4 plants can keep their **stomata** closed to reduce **water loss**. In C_3 plants this would lead to a **shortage of CO_2** for photosynthesis, but PEP carboxylase has such a **high affinity** for CO_2, that this doesn't become a problem.

3) In tropical conditions, CO_2 levels can get very **low** at the hottest part of the day when all of the dense vegetation is **photosynthesising rapidly**. The high affinity PEP carboxylase has for CO_2 comes in handy again.

4) The difference in the pathway also means that C_4 plants only need **half** as much **water** for photosynthesis as C_3 plants. This is useful in hot conditions with lots of competition for water from other plants.

Artificial Selection chooses Useful Characteristics

Once people realised that species were adapted to certain conditions, they started selectively breeding organisms for their own benefit. **Artificial selection** involves **selectively breeding** plants or animals to make them more **useful** to humans. Farmers aim to increase the **volume** and **quality** of their produce, and so **increase** their **profits**.

1) Animals can be selectively bred for certain **characteristics** e.g. **appearance** and **meat** production. For example, farmers are breeding pigs to meet demand from the supermarkets for less fatty bacon. They **deliberately breed** the leanest ones together. Over **many generations**, the mean percentage of fat on the pigs decreases.

2) Plants are selectively bred. **Tomatoes** are interbred to make plants with high yields and tasty, attractive fruit.

Continuous **inbreeding** like this can cause **problems**. Related individuals are more likely to share the same kind of alleles and be homozygous for certain traits. Lots of harmful conditions are recessive, so only homozygous individuals will have them. This can lower the fitness of populations with lots of inbreeding (see p. 126). To prevent problems like this breeders are now encouraged to use **heterozygous** animals in their breeding programmes — this maintains **genetic diversity**.

Practice Questions

Q1 Why do species of small penguins not live in the coldest parts of the Antarctic?

Q2 What is the proper name for plants that are adapted to living in very dry conditions?

Q3 Give three features that make the kangaroo rat suited to desert conditions.

Q4 How does the enzyme PEP carboxylase benefit tropical plants?

Exam Questions

Q1 Guernsey cows have been selectively bred to produce especially rich milk.
a) Briefly explain the principles behind selective breeding.   [3 marks]
b) Explain how farmers can maintain genetic diversity when selectively breeding animals.   [3 marks]

Q2 Explain how polar bears are adapted to very cold conditions.   [4 marks]

I'm perfectly adapted to the conditions on my sofa...

I'm sorry about that bit of C_4 photosynthesis, it's not easy — you might need to look back at the stuff about photosynthesis in section one if you're unsure about it. At least the other stuff isn't too difficult. Adaptation is mostly common sense really — I mean it's hardly surprising that kangaroo rats come out at night when the temperatures are coolest.

Homeostasis and Temperature Control

This section is all about keeping things constant inside the body.
Everything has to be carefully balanced — otherwise your body would be totally out of control.

Homeostasis - the ability to maintain a constant internal environment within fluctuating external conditions.

Homeostasis *keeps the Internal Environment* Constant

Homeostasis keeps the **blood** and the **tissue fluid** that surrounds the cells
(the **internal environment**) within **certain limits**, so the cells can function normally.
Changes in the external environment can affect the internal one, which can **damage cells**:

> 1) **Temperature** changes affect the rates of metabolic reactions, and high temperatures can denature proteins.
>
> 2) **Solute concentrations** affect **water potentials** of solutions
>   and therefore the loss or gain of water by cells due to osmosis.
>
> 3) Changes in **pH** can affect the function of proteins by changing their shapes.

Homeostasis keeps the internal environment **constant**, avoiding **cell damage**.

A *Homeostatic System* detects a *Change* and *Responds* to it

1) A **receptor detects** a change (the **stimulus**).

2) The receptor communicates with the part of the body that brings
   about a response (the effector), via the **nervous system** or **hormones**.

3) An **effector** brings about the response. Glands and muscles are effectors.

Negative feedback keeps the internal environment **constant**:

> Changes in the environment trigger a response that
> **counteracts** the changes — e.g. a **rise** in temperature
> causes a response that **lowers body temperature**.
>
> This means that the **internal environment** tends to stay
> around a **norm**, the level at which the cells work best.
>
> This only works within **certain limits** —
> if the environment changes too much, then
> the effector may not be able to **counteract** it.

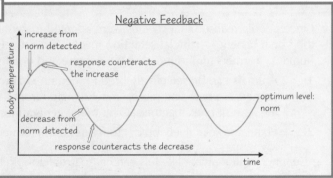

Mammals often use several **different responses** to keep a system in homeostasis. This lets them control
things better, e.g. temperature control and control of blood glucose concentration (see pages 86-87).

Animals *are* Ectotherms *or* Endotherms

All organisms release **heat energy** from their metabolism, which increases body temperature. Body heat can be
exchanged with other solid objects by **conduction** or **radiation** and lost to air by **convection** or by **evaporation**.
Animals are either **ectothermic** or **endothermic**. Endothermic animals have more control over their body temperature:

Ectotherms
1) E.g. invertebrates, fish, amphibians and reptiles.
2) Their body temperature depends on the temperature of the external environment.
3) They have a **variable** metabolic rate, and often generate very little heat.
4) The temperature of their surroundings determines their **activity levels**: they're more active and eat more at higher temperatures.
5) They can regulate their temperature only by changing their behaviour, e.g. many reptiles are **heliotherms** — they gain heat by basking in the sun.

Endotherms
1) E.g. mammals and birds.
2) Their body temperature is largely **independent** of external temperature (within certain limits).
3) They have a constantly **high** metabolic rate so they generate a lot of heat.
4) They have efficient mechanisms for **thermoregulation** (regulation of body temperature), so are less affected by the temperature of their surroundings.

AQA A ONLY

Homeostasis and Temperature Control

Mammals can Regulate their Body Temperature

The **skin** has a surface **epidermis**, and a thicker, deeper **dermis** with features for temperature control:

1) The skin has lots of **blood capillaries**. When you're too hot, the arterioles dilate (**vasodilation**), and more blood flows through the capillaries in the surface layers of the dermis to release more heat by **radiation**. When you're too cold, the arterioles constrict (**vasoconstriction**), reducing heat loss.

2) In mammals, **sweat** is secreted from **sweat glands** when the body is too hot. Sweat evaporates from the surface of the epidermis, using **body heat** and so cooling the skin.

3) Mammals have a layer of hair to provide **insulation** by trapping air, which is a poor heat-conductor. When it's cold, the **erector pili muscles** contract, which raises the hairs, trapping more air and preventing heat loss.

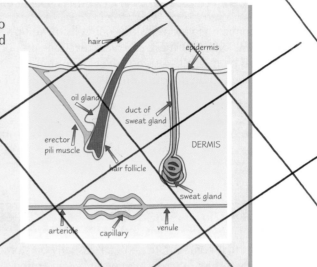

Thermoregulation is controlled by the Autonomic Nervous System

Body temperature is regulated by the unconscious actions of the **autonomic nervous system** (see pages 80-81):

1) If the external temperature **rises**, this stimulates thermoreceptors in the skin dermis, which send action potentials along **sensory neurones** to the hypothalamus. The hypothalamus sends **action potentials** (see p. 72) along **motor neurones** to effectors in the skin. The effectors cause **vasodilation** of arterioles and more sweat secretion, so **more heat** is lost from the skin. The production of certain **hormones** is **decreased**, which decreases the metabolic rate so less heat is generated.

2) If the external temperature **falls**, the hypothalamus causes dermal arterioles to constrict and erector pili muscles to contract, erecting the hairs and trapping more air. The hypothalamus stimulates the production of a hormone to **increase** the **metabolic rate**, generating heat from **increased respiration**.

Practice Questions

Q1 Define homeostasis.

Q2 Give two factors that are controlled by homeostasis in the body of a mammal.

Q3 For one of your chosen factors, explain why it is beneficial to control it.

Q4 What is the difference between the external environment and the internal environment?

Exam Questions

Q1 a) Explain what is meant by the term 'negative feedback'. [2 marks]
b) Give two examples of factors that are controlled by a negative feedback mechanism. [2 marks]

Q2 The rate of food consumption of a lizard (reptile) and a mouse (mammal) were compared.
It was found that overall the mouse consumed more food as a % of its own body weight per day than the lizard.
However, the rate of food consumption by the lizard increased as temperature increased from 20°C to 30°C.
There was no significant change in rate of food consumption in the mouse as temperature increased.
Explain the differences between the lizard and the mouse. [6 marks]

My biology teacher often gave me negative feedback...

The key to understanding homeostasis is getting your head round negative feedback. It's not complicated — if one thing goes up, the body responds to bring it back down, or vice versa. Look at pages 86-87 for more negative feedback loops.

The Liver and Excretion

These next few pages deal with all the ways that the body gets rid of the things we don't need, like nitrogenous waste, carbon dioxide and Westlife albums.

Chemical Reactions in cells produce Waste Products

All the chemical reactions inside your cells make up your **metabolism**. Many of these chemical reactions produce substances not needed by the cell — some are even poisonous. These **waste** products need to be removed from the body by **excretion**.

Don't get excretion mixed up with secretion, which is the way a cell releases useful substances, such as hormones, that work outside cells.

> **Example**
>
> **Carbon dioxide** is a toxic **waste product** of **respiration**. It is removed from the body by the **lungs** (in land animals) or **gills** (in aquatic ones). The lungs and gills act as **excretory organs**.

Excess Amino Acids Can't be Stored in the Body

Substances that contain nitrogen can't usually be stored by the body for use later on. Proteins contain **amino acids**, and nitrogen is part of the **amino group**, NH_2.

Animals often eat protein with **more amino acids** than the body can use at once. Some amino acids are used to make **useful proteins**, but the **excess** ones need to be **converted** to other things. This happens in the **liver**.

deamination in the ornithine cycle.

Structure of an amino acid

organic acid

amino group of amino acid → NH_2 — C — COOH, with H above and R below

this R just represents the rest of the molecule, which varies depending on the amino acid.

Excess Amino Acids have their Amino Groups Removed

In mammals, excess amino acids are changed by **metabolic reactions** in the **liver**:

1) Nitrogen-containing **amino groups** from the excess amino acids are removed, forming **ammonia** and **organic acids**. This is **deamination**. The organic acids are respired or converted to carbohydrate and stored as glycogen.

2) Ammonia is too poisonous for mammals to excrete it directly, so the **ammonia** reacts with **carbon dioxide** to form safer **urea**.

3) The urea is released into the blood, then **excreted** from the body by the **kidneys**.

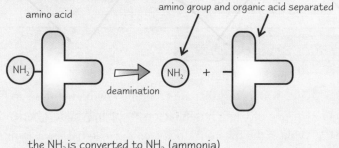

amino acid

amino group and organic acid separated

NH_2 [amino acid] → deamination → NH_2 + [organic acid]

the NH_2 is converted to NH_3 (ammonia)

then: $NH_3 + CO_2$ ⟹ urea

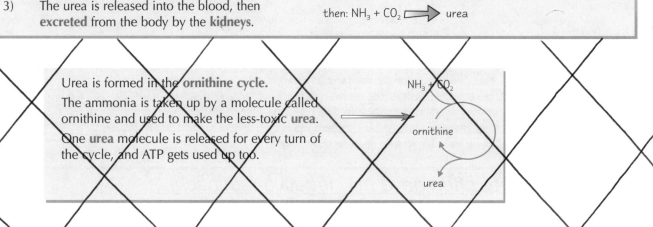

Urea is formed in the **ornithine cycle**.

The ammonia is taken up by a molecule called ornithine and used to make the less-toxic **urea**.

One **urea** molecule is released for every turn of the cycle, and ATP gets used up too.

$NH_3 + CO_2$

ornithine

urea

OCR physiology and behaviour option only

The Liver and Excretion

Not All animals excrete Urea

The livers of **mammals** produce **urea**, but other animals produce
different types of **nitrogenous** (nitrogen-containing) waste.

1) **Ammonia** is very toxic, but it's also the simplest nitrogenous waste product to produce.
   The excess amino acids get deaminated, but don't get converted to urea, so this method uses up
   less energy. This is what happens in lots of aquatic animals, like freshwater fish. Their bodies are
   surrounded by lots of water and **ammonia** is **very soluble**, so it easily **diffuses** into the surrounding
   water. In marine fish and land animals, too much water would be needed to dilute it.

2) Animals that live in dry environments or need to conserve water can't excrete nitrogenous waste
   in **solution** (with lots of water). Birds and insects excrete **uric acid** — it's less soluble than urea, so
   it can be removed as a more **solid** paste. This means that they don't lose much water.

The Liver converts one Amino Acid into another by Transamination

20 different amino acids are used in the body to make **proteins**. About half of these are **essential amino acids**.
This means that you need to get them from your food, as they can't be made in the body. The other amino
acids are **non-essential**. They're just as important for making proteins, but you don't need them from your diet
because they can be **made** in the liver by a process called **transamination**. This uses an enzyme called
transferase to transfer the **amino group** from one molecule to another to make another kind of amino acid.

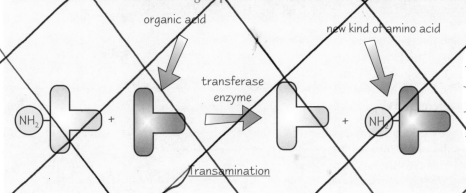

organic acid

new kind of amino acid

transferase enzyme

NH_2 +   →   + NH_2

Transamination

*Essential amino acids are the ones
that can't be made by
transamination, so they have to
come from the food we eat. That's
one reason why it's so important to
eat a balanced diet.*

Practice Questions

Q1 Define excretion and give two examples of excretory products.

Q2 Which organs of the body are responsible for excreting these products?

Q3 Describe the reactions involved in the formation of urea from excess amino acids.

Q4 What happens to amino acids that are *not* excess to the body's requirements?

Exam Questions

Q1 a) Explain why the urea concentration in the urine will rise after consumption of a meal that is
      rich in protein. [6 marks]
   b) Suggest why deamination in the liver could be associated with the accumulation of glycogen. [3 marks]

Q2 Explain why urea is not formed in the body of a freshwater fish. [4 marks]

Better out than in...

*The best thing about this page is that it tells you why snake pee is solid. If you've ever kept snakes, you'll know
I'm telling the truth. And now you know the science behind it… if only all of A2 biology was made up of
fascinating facts like that (sigh). Still, there's only a few things to learn on this page, and they're not so hard, really.*

The Kidneys and Excretion

You've learnt how urea is produced, so now complete the picture by learning exactly how the kidneys excrete all the stuff our bodies don't need any more. Then have some steak and kidney pie.

The **Kidneys** are **Organs** of **Excretion**

1) Urea produced by the liver is **excreted** from the body by the **kidneys**. Urea is dissolved in the blood plasma.

2) When the blood passes through the kidney nephrons (the tubes that run through the kidneys — see below for more on these), liquid is filtered out of the blood, carrying small solutes with it, including **urea**.

3) The useful solutes are reabsorbed, and the waste products are removed from the body in **urine**.

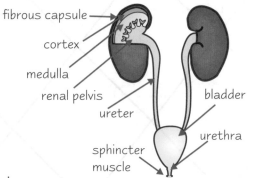

fibrous capsule
cortex
medulla
renal pelvis
ureter
bladder
urethra
sphincter muscle

You need to know about **two main stages** of kidney function — **ultrafiltration** in the renal capsules (also called the Bowman's capsules) and **selective reabsorption** in the medulla.

Don't get mixed up — urea is a specific kind of substance, but urine is a mixture that contains urea.

Ultrafiltration in the kidneys takes place in the **Renal Capsules**

Blood enters the kidney cortex through the **renal artery**, then goes through millions of knots of capillaries in the kidney cortex. Each knot, (**glomerulus**) is a bundle of capillaries looped inside a hollow ball called a **renal (or Bowman's) capsule**. An **afferent arteriole** takes blood into each glomerulus, and an **efferent arteriole** takes blood out.

Ultrafiltration takes place in all body tissues that have capillaries. Blood pressure squeezes liquid from the blood through the capillary wall. Small molecules and ions pass through, but larger ones like proteins and blood cells stay behind in the blood. In most parts of the body, this liquid gathers between the cells as tissue fluid. In the kidney, the liquid collects in microscopic **tubules**. Useful substances are **reabsorbed** back into the blood, and waste substances (like dissolved urea) are excreted in the urine.

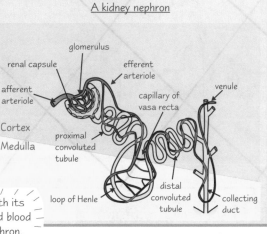

A kidney nephron

glomerulus
renal capsule
efferent arteriole
afferent arteriole
venule
capillary of vasa recta
Cortex
Medulla
proximal convoluted tubule
loop of Henle
distal convoluted tubule
collecting duct

One tubule system with its associated capsule and blood supply is called a nephron.

Ultrafiltration - the forcing, under high pressure, of fluid from the blood out of the glomerulus and into the bowman's capsule.

Ultrafiltration in the capillary and renal capsule membranes

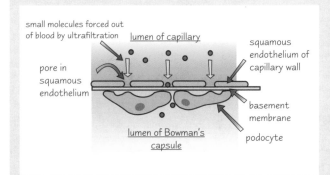

small molecules forced out of blood by ultrafiltration
lumen of capillary
squamous endothelium of capillary wall
pore in squamous endothelium
basement membrane
lumen of Bowman's capsule
podocyte

Small molecules and ions pass from blood in the glomerular capillary into the renal capsule through:

1) Pores in the **capillary wall** (the wall is made of one layer of flat cells — squamous endothelium).

2) A **basement membrane** made up of collagen fibres and glycoprotein.

3) A specialised epithelium of the renal capsule, made up of cells called **podocytes**. These support the membrane while letting the filtrate pass through.

The Kidneys and Excretion

Useful substances are Reabsorbed

Filtrate from the renal capsule enters the **proximal convoluted tubule** in the cortex of the kidney. The wall of this tubule is made of **cuboidal epithelium**, with **microvilli** facing the filtrate to increase the surface area.

Blood leaving the glomerulus along the efferent arteriole enters another capillary network, called the **vasa recta**, that's wrapped around the proximal convoluted tubule. This provides a big surface area for **reabsorption** of useful materials from the **filtrate** (in the tubules) into the **blood** (in the capillaries) by:

1) **active transport** of glucose, amino acids, vitamins and some salts

2) **osmosis** of water

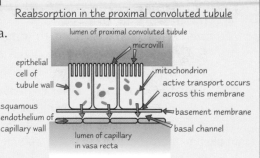

Reabsorption in the proximal convoluted tubule

Water is reabsorbed from the Whole Tubule

Water enters the blood by **osmosis** because the water potential of the blood is **lower** than that of the filtrate. Most of the water is reabsorbed from the **proximal convoluted tubule**. Reabsorption from the **distal convoluted tubule** and **collecting duct** is controlled by **hormones** (see pages 68-69 for more on this).

Urine is a Mixture of substances dissolved in Water

Urine usually contains the following things:

1) **Variable amounts** of **water** and **dissolved salts**, depending upon much you've drunk (see pages 68-69).

2) **Variable amounts** of **dissolved urea**, depending upon how much protein you've eaten (see pages 64-65).

3) Other substances such as hormones and water-soluble vitamins.

It **doesn't** usually contain:

1) **Proteins**, because they're too big to be filtered out in the renal capsule (they can't pass through the basement membrane).

2) **Glucose, amino acids or vitamins**, because they're actively reabsorbed back into the blood from the proximal convoluted tubule.

3) **Blood cells**.

Glucose if diabetic can be found in urine.

Practice Questions

Q1 Describe the process of ultrafiltration.

Q2 Name three components of the filtrate of a nephron that will be reabsorbed back into the blood from the proximal convoluted tubule.

Q3 What are the names of the processes that account for the reabsorption of substances?

Exam Questions

Q1 a) Suggest how the features of the proximal convoluted tubule of a kidney nephron maximise the rate of absorption of glucose. [5 marks]

   b) Suggest why a smaller quantity of urea passes from the tubule into the blood than glucose. [3 marks]

Q2 Occasionally some people produce urine that contains traces of protein. Suggest an explanation for this. [2 marks]

It's steak and excretion organ pie for dinner...

The kidneys are pretty complicated organs. That's why it's so serious when they go wrong — all that toxic urea would just stay in your blood and poison you. If your kidneys fail you'll end up hooked up to a machine for hours every week so it can filter your blood, unless some kind person donates a new kidney for you.

Water and Metabolic Waste Control

You should have a rough idea about how the kidneys work by now, so here's a bit more detail on how they help with homeostasis. Read. Learn. Enjoy.

The **Kidneys** regulate the body's **Water Content (Osmosregulation)**

Mammals excrete urea in solution, so **water** is lost too. The kidneys regulate the levels of water in the body.

1) If the body is **dehydrated** (e.g. if the body has lost a lot of water by sweating), then more water is **reabsorbed** by osmosis from the tubules of the nephron, so less water is lost in the urine.

2) If the body has a **high water content** (e.g. from drinking a lot), then **less** water is reabsorbed from the tubules, so more water is lost in the urine.

This regulation takes place in the middle and last parts of the nephron — the **loop of Henle**, the **distal convoluted tubule** and the **collecting duct**. The **volume** of water reabsorbed is controlled by hormones.

The **Loop of Henle** has a **Countercurrent Multiplier Mechanism**

This sounds really scary, but stick with it, it's not too complicated really:

Just between the proximal and distal convoluted tubules is a part of the nephron called the **loop of Henle**. It's made up of two 'limbs'.

1) Water leaves the **descending limb** of the loop by osmosis, and Na^+ and Cl^- ions diffuse into the loop. This makes the **ion concentration** of the tubule **higher** towards the **base** of the loop.

2) The Na^+ and Cl^- ions are then actively pumped out of the top of the **ascending limb** of the loop into the medulla. The high concentration of Na^+ and Cl^- ions in the medulla causes water to leave the collecting duct and descending limb by **osmosis**.

3) The water is then **reabsorbed** into the blood through the **capillary network**.

This mechanism is called the **countercurrent multiplier**.

The countercurrent multiplier mechanism lets land-living mammals produce urine with a **solute concentration** higher than that of the blood, so they can avoid losing too much water. The volume of water reabsorbed can be **regulated** depending on the needs of the body.

Another hard day studying reading, writing and the countercurrent multiplier mechanism.

Diagram labels:
- Bowman's capsule
- distal tubule
- CORTEX
- ⇐ = movement of water by osmosis
- ⇐ = movement of Na^+ and Cl^- by active transport
- active transport of Na^+ and Cl^- into tissue of medulla
- collecting duct
- MEDULLA
- loop of Henle
- water moves by osmosis from collecting duct into tissue of medulla — because of the high Na^+ and Cl^- concentrations created by the loop of Henle
- descending limb
- ascending limb
- to renal pelvis then bladder

✱ Descending limb - impermeable to ions
✱ Ascending limb - impermeable to water

Water Reabsorption is controlled by Hormones

A hormone released from the **posterior pituitary gland**, called **antidiuretic hormone (ADH)**, makes the walls of the distal convoluted tubules and collecting ducts **more permeable** to **water**. More water is then **reabsorbed** from these tubules into the medulla (and then into the blood) by **osmosis**. This means that **less water** is lost in the urine.

It's called <u>anti</u>diuretic hormone because diuresis is when lots of dilute urine is produced.

Water and Metabolic Waste Control

Blood ADH levels are High when you're Dehydrated

Dehydration is what happens when you **lose water**, e.g. by sweating during exercise:

1) The water content of the blood drops, so its **water potential drops**.
2) This is detected by osmoreceptors in the hypothalamus.
3) This stimulates the **pituitary gland** to release **more ADH** into the blood.
4) The ADH **increases** the **permeability** of the walls of the **collecting ducts** in the kidneys, so **more water** is **reabsorbed** back into the blood by **osmosis**.
5) **Less water** is lost in the urine.

Blood ADH levels are Low when your body is Hydrated

If you drink lots of water, more water is absorbed from the gut into the **blood**, and the **excess** is lost in the **urine**:

1) The water content of the blood rises, so its **water potential rises**.
2) This is detected by the osmoreceptors in the hypothalamus.
3) This stimulates the **pituitary gland** to release **less ADH** into the blood.
4) Less ADH means that the collecting ducts are less permeable, so **less water** is **reabsorbed** into the blood by **osmosis**.
5) **More water** is lost in the urine.

* Juxtamedullary nephron - long loop
* Cortical nephron - short loop

Desert Animals need to Conserve Water

Animals in a **hot**, **dry** environment have to control their water loss to survive, so they often have **special adaptations** (see the section on kangaroo rats on page 60 for more on this).

Desert **mammals** have very long loops of Henle, so there's a greater **surface area** to accumulate **more sodium chloride** ions in the medulla. This means their medullas have especially **low water potentials** so **more** water can be reabsorbed from the collecting ducts, making their urine more concentrated.

Practice Questions

Q1 What are the main ways in which water can be lost from the body of a terrestrial animal?
Q2 Describe how the loop of Henle increases the salt concentration of the medulla of the kidney.
Q3 Explain the importance of the medulla in allowing reabsorption of water from urine.
Q4 Which gland releases ADH?
Q5 What is the effect of ADH on kidney function?

Exam Questions

Q1 Levels of ADH in the blood rise during strenuous exercise.
Explain the cause of the increase and the effects it has on kidney function. [10 marks]

Q2 Suggest why mammals adapted to life in dry deserts have longer loops of Henle. [5 marks]

If you don't understand what ADH does, urine trouble...

Seriously, though, there are two main things to learn from these pages — the countercurrent multiplier mechanism and the role of ADH in controlling the water content of urine. You'll need to be able to identify the different parts of the kidney nephron too. Keep writing it down until you've got it sorted in your head, and you'll be just fine.

The Nervous System and Receptors

The nervous system is really important to animals. Plants don't have one at all, and that's why they tend to sit around doing nothing all day. Our nervous system lets us respond to our environment and tells us what's going on.

Organisms have **Receptors** that are sensitive to **Stimuli**

A **stimulus** is any change in the environment that brings about a **response** in an organism — for example, a vibration that's detected by receptors in an organism's ears as a sound, or light that's detected by receptors in the organism's eyes as an image. A **receptor** is the part of the body of an organism that **detects** the stimulus.

You can classify receptors in animals depending on the **type of stimulus** they detect:

1) **Thermoreceptors** are sensitive to **temperature** — they're stimulated by **heat energy**.

2) **Photoreceptors** are sensitive to **light** — they're stimulated by **electromagnetic energy**, e.g. the cells that contain the pigments in the retina of the eye.

3) **Mechanoreceptors** are sensitive to **sound**, **touch**, **pressure** or even **gravity** — they're stimulated by **kinetic energy**.

4) **Chemoreceptors** are sensitive to **chemicals** — they're stimulated by **chemical energy**. They're involved in the senses of smell and taste.

Some kinds of chemicals in plants, such as **phytochromes** (see page 89) are receptors — they detect **light**.

Receptor cells have **Excitable Membranes**

1) Receptor cells are **excitable**. This means that in their resting (unstimulated) state, their cell membranes have a **potential difference** across them — i.e. the receptor cells have a difference in **charge** across their cell membranes. There's a **negative** charge on the **inside** of the membrane, and a **positive** charge on the **outside**. This is generated by a combination of protein **ion pumps** and **channels** (see page 72).

2) When the receptor cell is stimulated, changes inside the cell affect the charge across the cell membrane. **Ions** (charged atoms) move into or out of the cell and alter the **charge** on each face of the membrane. Charges are **reversed**, creating a **generator potential**. The larger the stimulus, the larger the generator potential.

3) When the receptor cell is stimulated (**excited**) like this, it can transmit a signal to an **effector**, as long as the **generator potential** is big enough. Stimulated receptors that set up **nerve impulses** in nerve cells are called **transducers**.

4) The **minimum** size of stimulus needed to transmit a signal is called the **threshold stimulus**. Some kinds of receptor cells need a bigger stimulus than others to get a response (they have a **higher threshold**), so they're **less sensitive**.

AQA A ONLY

A **Pacinian Corpuscle** is a **Skin Receptor**

The **skin** has lots of different types of **receptors**. **Pacinian corpuscles** detect **pressure** applied to the skin. When a corpuscle is deformed by something pushing on it, '**stretch-mediated**' sodium channels in the cell membrane open. **Sodium ions** move into the cell by **facilitated diffusion**, creating a **generator potential**. Each Pacinian corpuscle contains a **sensory nerve ending** from a **sensory neurone**. When the **threshold stimulus** is reached, an **action potential** (nerve impulse) is set up in the membrane of the sensory neurone. (See pages 72-73 for more on this).

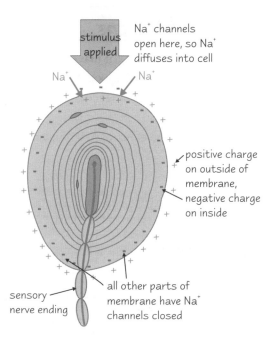

stimulus applied

Na$^+$ channels open here, so Na$^+$ diffuses into cell

Na$^+$ Na$^+$

positive charge on outside of membrane, negative charge on inside

sensory nerve ending

all other parts of membrane have Na$^+$ channels closed

Pacinian corpuscle

Tip no. 31: Pictures of pigs are great for disguising boring pages.

The Nervous System and Receptors

The Receptors Communicate with Effectors

An **effector** is a part of the body that brings about a **response** to the signal from a receptor.
In animals, effectors are usually **muscles** (the response is **contraction**), or **glands** (the response is **secretion**).
There are two main ways that a receptor **communicates** with an effector:

1) The receptor produces a **chemical** (hormone) which binds to the effector. The chemical moves from receptor to effector by **diffusion** for short distances, or **mass flow** (transport in bulk) for long distances.

2) The receptor triggers a **nerve** (electrical) **impulse** in nerve cells, which stimulates the effector to respond.

It takes longer for a **chemical signal** to reach the effector than it does for an **electrical signal**. Unlike plants, animals have **nervous systems** (electrical) as well as hormone systems, so they can respond more **quickly** to changes than plants can.

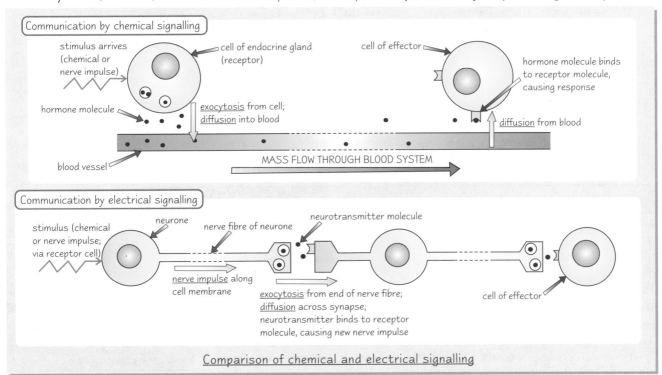

Comparison of chemical and electrical signalling

Practice Questions

Q1 Why is it important that organisms can respond to stimuli in their environment?

Q2 What do receptors and effectors do in the nervous system?

Q3 What kinds of energy stimulate mechanoreceptors and photoreceptors?

Q4 Resting receptor cell membranes carry a potential difference across them. What does this mean?

Q5 How does the potential difference change when a stimulus is applied?

Q6 What is the threshold level of a stimulus?

Exam Questions

Q1 a) Outline the two main ways in which the receptors of the nervous system communicate with the effectors. [3 marks]

b) Explain why the absence of one of these systems means that plants can respond only slowly to changes in their environment. [2 marks]

Q2 Chemoreceptors detect chemical stimuli. Suggest why exposure of a chemoreceptor cell to various different kinds of chemicals results in the generation of a nervous impulse for some chemicals, but not for others. [4 marks]

This page isn't as stimulating as I expected...

You should have some idea now about what receptors do and why they're important. It's basically so that organisms can respond to their environment. Being able to see, hear, feel, smell and taste allows them to find food and shelter, escape predators, avoid injury and illness — survive, really. The nitty gritty of how organisms respond is covered on the next page.

The Nervous System — Neurones

It's no good being aware of your environment if you can't do anything about it. That's where neurones come in.

Nerve Cells have Polarised Membranes so they can carry Electrical Signals

1) The **nervous system** is made up of nerve cells called **neurones**. Each neurone consists of a **cell body** and extending **nerve fibres**, which are very thin cylinders of cytoplasm bound by a cell membrane. Neurones carry waves of electrical activity called **action potentials** (nerve impulses). They can carry these impulses because their cell membranes are **polarised** (see below) — there are different **charges** on the inside and outside of the membrane.

2) The nerve fibres let the neurones carry action potentials over **long distances**. There are tiny gaps, called **synapses**, between the different nerve fibres. Action potentials can't cross, so a chemical called a **neurotransmitter** is secreted at the tip of each nerve fibre to cross the gap. This stimulates a **new** action potential in the next nerve fibre on the other side of the synapse (see pages 74-75).

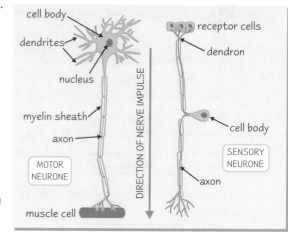

3) Between the receptors and the effectors, the **central nervous system** (i.e. brain and spinal cord) **coordinates** the action potentials passing through the nervous system. **Sensory neurones** carry action potentials from receptors to the central nervous system. **Relay neurones** carry action potentials through the central nervous system, and **motor (effector) neurones** carry them from the central nervous system to effectors.

Neurone cell membranes are Polarised when they're Resting

Resting neurones have a **potential difference** (a difference in **charge**) of about -65 millivolts (mV) across their cell membranes. This is because the **outer** surface of the membrane is **positively** charged and the **inner** surface **negatively** charged — the -65 mV is the **overall difference** in charge between them. This is the **resting potential** of the membrane, which is said to be **polarised**.

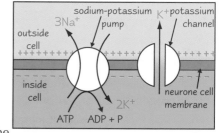

Sodium-potassium pump and potassium channel

The resting potential is generated by a **sodium-potassium pump** and a **potassium channel** in the membrane. The sodium-potassium pump moves **three sodium ions** out of the cell by **active transport** for every **two potassium ions** it brings in. The potassium channel then allows **facilitated diffusion** of potassium ions back out of the cell. The outer surface of the membrane becomes more positive than the inner surface because overall, more positive ions move **out** of the cell than move **in**.

movement of potassium/sodium ions by active transport   *movement of potassium ions by diffusion*

Neurone cells become Depolarised when they're Stimulated

The **sodium-potassium pumps** work pretty much all the time, but **channel proteins** (like the potassium channels) can be opened or closed. **Depolarisation** of neurone cell membranes involves another type of channel protein, **sodium channels**. If a neurone cell membrane is stimulated, sodium channels **open** and **sodium ions** diffuse in. This **increases** the positive charge **inside** the cell, so the charge across the membrane is **reversed**. The membrane now carries a potential difference of about **+40 mV**. This is the **action potential** and the membrane is **depolarised**.

When sodium ions diffuse into the cell, this stimulates nearby bits of membrane and **more** sodium channels open. Once they've opened, the channels automatically **recover** and close again.

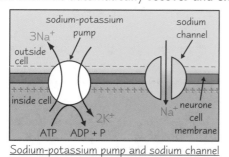

Sodium-potassium pump and sodium channel

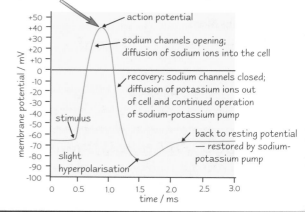

The Nervous System — Neurones

Remember these important features of Action Potentials

1) **Nerve axons** in vertebrates are usually covered in a layer of **myelin sheath**, which is produced by **Schwann cells**. Myelin is an **electrical insulator**. Between the sheaths there are tiny patches of bare membrane called **nodes of Ranvier**, where sodium channels are **concentrated**. Action potentials **jump** from one node to another, which lets them move **faster** (this is called **saltatory conduction**).

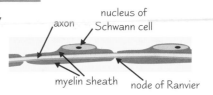

2) Action potentials also go faster along axons with **bigger diameters**, because there's less **electrical resistance**.

3) They go faster as **temperature** increases too, up to around **40°C**. After that, the proteins begin to **denature**.

4) Action potentials have an **all-or-nothing** nature. This means that the values of the resting and action potentials for a neurone are **constant**, and you can't get anything in between. (A **bigger stimulus** just increases the **frequency** of the action potentials. The **strength** of the action potentials stays the same.)

5) A **threshold stimulus** must be applied to get an action potential (see page 70).

6) Straight after an action potential has been generated, the membrane enters a short **refractory period** when it can't be stimulated, because the sodium channels are **recovering** and can't be opened. This makes the action potentials pass along as **separate signals**.

7) Action potentials are **unidirectional** — they can only pass in one direction.

Practice Questions

Q1 What do sensory, relay and motor neurones do in the nervous system?
Q2 What happens to sodium ions when a neurone membrane is stimulated?
Q3 Give two factors that increase the speed of conduction of action potentials.
Q4 What is meant by the 'all-or-nothing' nature of action potentials?

Exam Questions

Q1 The graph shows an action potential 'spike' across an axon membrane following the application of a stimulus.

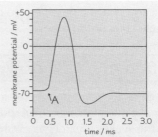

a) What label should be added at point A? [1 mark]
b) Explain what causes the change in potential difference from -65 to +40 mV. [3 marks]
c) Another stimulus was applied at 1.5 ms, but failed to produce an action potential. Suggest why. [2 marks]

Q2 Multiple sclerosis is a disease of the nervous system characterised by damaged myelin sheath. Suggest and explain how this will affect the transmission of action potentials. [5 marks]

I'm feeling a bit depolarised after all that...

The nervous system can seem like a really hard subject at first, but once you've gone over it a couple of times it starts to make sense. Nerves work because there's a charge across their membranes, and it's a change in this charge that sends the message along the nerve. The charge is set up using ions, which can then be pumped in and out to change the charge.

Synapses and the Reflex Arc

This page is all about synapses, which are the little gaps between the end of one neurone and the start of the next one. Seems like quite an insignificant little thing to fill a whole two pages with, but never mind.

There are Gaps between Neurones

A **synapse** is a gap between the end of one **neurone** and the start of the next. An action potential arrives at the end of the axon of the **presynaptic neurone** (the neurone before the synapse), where there's a swelling called a **bouton** or **synaptic knob**. This has **vesicles** containing a chemical **neurotransmitter**, and the impulse passes across the synapse as follows:

1) The action potential opens **calcium channels** in the membrane, allowing calcium ions to diffuse **into** the bouton. Afterwards these are pumped back out using ATP.

2) The increased concentration of calcium ions in the bouton causes the **vesicles** containing the **neurotransmitter** to move up to and to fuse with the **presynaptic membrane**. This also requires ATP (it's an active process).

3) The vesicles **release** their neurotransmitter into the **synaptic cleft** (this is called **exocytosis** and it's an active process too).

4) The neurotransmitter **diffuses** across the synaptic cleft and binds to **receptors** on the **postsynaptic membrane** of the other neurone.

5) This stimulates an **action potential** in the postsynaptic membrane by opening the **sodium channels** (see page 72).

6) An **enzyme** is sometimes used to **hydrolyse** the neurotransmitter, so the response doesn't keep on happening. The neurotransmitter may also be taken back up into the presynaptic bouton, ready to be used again.

Because the receptors are only on the **postsynaptic** membranes, a signal can only pass across a synapse in **one direction** (it's **unidirectional**). The postsynaptic cell behaves as a **transducer**, just like **receptor cells** (see page 70 –71), because the **chemical** stimulus (neurotransmitter) is converted into an **electrical** one (action potential).

There are lots of **mitochondria** in the bouton of a neurone. These provide the **ATP** to make more neurotransmitter, power exocytosis and pump calcium ions out of the bouton.

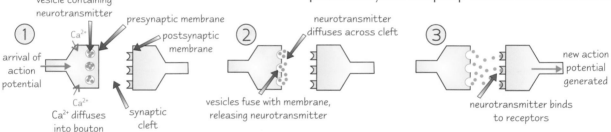

There's a Synapse between a Motor Neurone and Muscle Fibres

Motor neurones carry action potentials to **muscle fibres**. There's a synapse between the presynaptic membrane of the motor axon and the postsynaptic membrane of the muscle fibre (the **sarcolemma** — see pages 76-77). This region is called the **neuromuscular junction**. The synapse functions in the same way as a synapse between two neurones, and an action potential is generated on the sarcolemma in the same way too. What happens afterwards to bring about **muscle contraction** is described on pages 76-77.

There are Different Kinds of Neurotransmitters

There are many kinds of neurotransmitters. A lot of neurotransmitters are **excitatory** (as described above), but some are **inhibitory**. These make the membrane **hyperpolarised**, creating a resting potential value that's even more **negative** than the usual -65 mV. This makes it harder to **excite**.

Some kinds of neurones in particular parts of the body, only have receptors for certain **types** of neurotransmitter, e.g:

1) The neurotransmitter **acetylcholine** binds to **cholinergic receptors**. It's an excitatory neurotransmitter at many neurones and neuromuscular junctions.

2) **Noradrenaline** binds to **adrenergic receptors**, and it's generally **excitatory**.

Some **drugs** and **poisons** affect the action of neurotransmitters:

- **Agonists** are chemicals that **mimic** the effects of neurotransmitters. An example is **nicotine**, which binds to **cholinergic** receptors and stimulates **neuromuscular junctions**.

- **Antagonists block** the effects of neurotransmitters. An example is **curare**, a chemical that binds to **cholinergic receptors**, but blocks – rather than mimics – the acetylcholine molecules. This **paralyses** the muscle.

Synapses and the Reflex Arc

Synapses are sites of Interaction between Different Signals

The main function of a synapse is to bring about the passage of signals between neurones.
However, another important point about synapses is that signals can **interact** there:

1) **Inhibition.** A neurotransmitter can **inhibit** a postsynaptic membrane
and make it harder to excite by **hyperpolarising** it (see page 74).

2) **Summation.** This means that the **overall effect** of lots of different neurotransmitters from lots
of different neurones on one postsynaptic membrane is the **sum** of all their **individual effects**
(bearing in mind that some might be **excitatory** and some might be **inhibitory**).

Reflex responses involve Reflex Arc Pathways

A **reflex** is an **involuntary stereotypical** response of part of an organism to an applied stimulus. This means that there's
usually no **conscious control** over it, and it always produces the **same** kind of effect. It works because there are special
patterns of neurones that make up **reflex arcs**. The simplest is a **monosynaptic reflex**, where the sensory neurone
connects **directly** to the motor neurone so there's only **one synapse** within the central nervous system in the arc.
An example is the **knee jerk reflex**. These reflex arcs often play a role in controlling **muscle tone** and **maintaining
posture**. Action potentials don't pass to the **brain**, and so no conscious thought is needed for them to happen.

A **polysynaptic reflex** has at least **two
synapses** within the central nervous system,
due to the presence of a **relay neurone**.
Action potentials **can** pass to the brain, so
some conscious thought might be involved.
An example of such a reflex is the quick
response if you touch something hot.

Ventral roots and dorsal roots are just the parts
of spinal nerves that join with the spinal cord.

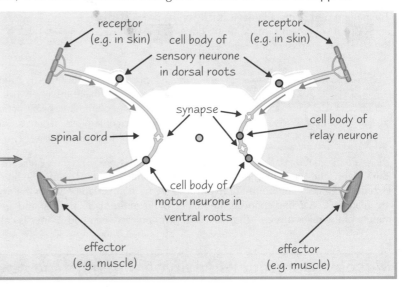

Transverse section through spinal cord
showing monosynaptic reflex arc (left)
and polysynaptic reflex arc (right)

Practice Questions

Q1 Why are there a lot of mitochondria in the bouton (synaptic knob) of the neurone?

Q2 Give two examples of a neurotransmitter.

Q3 What effect do antagonists have on neurotransmitters? Give an example of an antagonist.

Q4 What is summation at a synapse?

Q5 What is a reflex response?

Exam Questions

Q1 Describe the sequence of events leading from the arrival of an action potential at a
bouton to the generation of a new action potential on a post-synaptic membrane. [8 marks]

Q2 Explain how the structure of the synapse ensures that signals can only pass through it in one direction. [4 marks]

Bouton — like button, but in a weird French accent...

Not the most exciting page in the book, but pleasantly dull I'd say. There's nothing too hard there. Some chemicals with
annoyingly long names cross a gap so action potentials can move from neurone to neurone through the body. Reflex arcs
usually don't have many synapses, so the response can happen quickly. Some, like the knee jerk reflex, only have one.

The Nervous System — Effectors

If the receptors in your eyes detect a hungry-looking lion coming towards you, your nervous system sends a message to your legs telling them to run away. This is where muscles come in handy. **If you're doing AQA A you can skip these two pages.**

Muscle *is a tissue made up of cells that are* Contractile

Contraction of a muscle gives a 'shortening force' that causes **movement** or sets up **tension**. The process needs lots of **energy** from ATP. There are three main types of muscle:

1) **Striated muscle** (also called **skeletal** or **voluntary muscle**) is attached to bone via **tendons** and is controlled by the motor neurones of the **voluntary nervous system**.

2) **Smooth muscle** (also called **unstriated** or **involuntary muscle**) is found in walls of tubular organs and is controlled by the **autonomic nervous system**.

Voluntary means it's consciously controlled, and autonomic means it's not.

These two types of muscle only contract when stimulated by an **action potential** from the nervous system. They're **neurogenic**.

3) **Cardiac muscle** is found only in the **heart**. It's **myogenic**, which means it contracts **spontaneously**, without input from the nervous system. But the **autonomic nervous system** does control the **rate** of contraction.

Striated Muscle *is made up of* Muscle Fibres

Each **muscle fibre** contains lots of nuclei and is bound by a cell membrane called a **sarcolemma**. Contraction depends on protein **myofilaments**, which are arranged in bundles called **myofibrils**. The pattern of **thin** myofilaments (made of the protein **actin**) and **thick** myofilaments (made of the protein **myosin**) gives the **striations** (stripes) in the muscle.

Striated muscle
sarcolemma — nucleus
thick myosin myofilament
myofibrils
thin actin myofilament

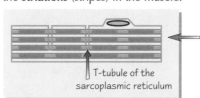
T-tubule of the sarcoplasmic reticulum

The cytoplasm of muscle fibres is called **sarcoplasm** and it's penetrated by **transverse (T) tubules** which make up a network called the **sarcoplasmic reticulum**. They let the sarcolemma transmit action potentials in towards the myofilaments. There are lots of **mitochondria** to provide ATP for contraction.

There are alternating dark and light bands in a muscle fibre. The dark ones are called **A bands** (think d<u>a</u>rk), and the light ones are called **I bands** (think l<u>i</u>ght). The middle bit of the A band, the **H zone**, is lighter than the rest of it, because there's no **overlap** between the myosin and actin myofilaments. A **Z line** connects the middle of the **actin** myofilaments, and an **M line** connects the middle of the **myosin** ones. The section of myofibril between two Z lines is called the **sarcomere**.

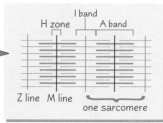
I band
H zone A band
Z line M line
one sarcomere

Muscles Contract *when* Myofilaments *slide over one another*

Muscle contraction is explained by **Huxley's sliding filament hypothesis**. It depends on bits of myosin called '**myosin heads**' binding to sites on actin filaments:

1) When muscle is stimulated it sets up an **action potential** in the **sarcolemma**, which spreads down the membranes of the **T-tubules**.

2) The **sarcoplasmic reticulum** membranes now become much more **permeable** and **calcium ions** diffuse out rapidly.

3) The Ca^{2+} quickly reaches the **actin filaments** and binds to a protein called **troponin**. This causes another protein called **tropomyosin** to change position and **unblock** the **binding sites** on the actin filaments.

4) ATP (from the myosin head) is hydrolysed and the energy released causes the myosin heads to alter their **angle** and attach to the binding sites forming **actomyosin cross bridges** between the two filaments.

5) The myosin head then **changes angle**, pulling the actin over the myosin towards the **centre** of the sarcomere.

6) ATP provides energy for the cross bridges to detach and then reattach, this time **further along** the actin filament. In effect the myosin heads '**walk**' along the actin filaments until they reach the end.

The process keeps repeating so that the whole muscle **contracts**. This changes the width of the bands in the muscle (see next page).

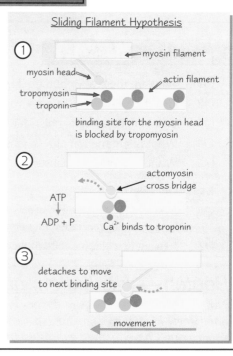
Sliding Filament Hypothesis
① myosin filament
myosin head
actin filament
tropomyosin
troponin
binding site for the myosin head is blocked by tropomyosin
② actomyosin cross bridge
ATP
ADP + P
Ca^{2+} binds to troponin
③ detaches to move to next binding site
movement

The Nervous System — Effectors

Muscles *Relax* when *Excitation Stops*

When **excitation** of the sarcomere stops (i.e. the membrane has recovered its **resting potential**), calcium ions are **actively pumped** back out of the cytoplasm into the **T tubules**. This means that the troponin is released, and the tropomyosin moves back to **block** the myosin binding site again. The cross bridges are broken and the myosin **detaches** from the actin filament. The sarcomere returns to its resting position and the width of the bands in the muscle change back to their relaxed appearance.

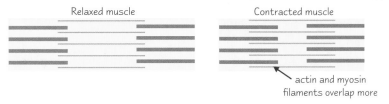

Relaxed muscle   Contracted muscle

actin and myosin
filaments overlap more

Muscles *have many Features like those of Neurones*

1)   When a muscle fibre is stimulated by an action potential, it only contracts if the stimulus is above a certain **threshold level**.

2)   Muscle fibre membranes have an **all-or-nothing** nature, and so increasing the strength of a stimulus above the threshold level doesn't give a bigger force of contraction in the muscle fibre. But an increase in stimulus strength does increase the **number of muscle fibres** that contract.

3)   Bigger forces of contraction happen in muscles with bigger **cross-sectional areas**.

4)   After the response, the muscle goes through an **absolute refractory period** (when no contraction can happen), followed by a **relative refractory period** (when it takes a **bigger stimulus** to get a response). The refractory period is the time it takes for the muscle to go back to its **resting potential** and be capable of further stimulation.

5)   As muscles can only produce a **shortening force**, at least **two** sets of muscles have to be used to move a bone into position and back again. Pairs of muscles acting in this way are called **antagonistic pairs**.

Muscles contain *Fast* and *Slow* Fibres

Fast (twitch) fibres release energy from **glycogen** and rely less on glucose delivered by the **blood**. They contain **less myoglobin** (a red pigment similar to haemoglobin) and so are white, have **fewer mitochondria** and **fatigue** more quickly. Contraction is **faster**. **Slow (tonic) fibres** use **glucose** from the **blood**, have **more myoglobin** (so are red) and have **more mitochondria**. Their **slower** contractions are concerned with things like **maintaining posture**, involving muscles that are contracted for **longer periods**.

EDEXCEL and AQA A ONLY

Practice Questions

Q1   Name the three main types of muscle, and say which part of the nervous system controls each one.

Q2   Which type(s) of myofilament (thick or thin) are present in the following bands on a muscle fibre:
a) A band   b) I band   c) H zone?

Q3   What is the refractory period of a muscle?

Exam Questions

Q1   Describe the sequence of events that lead to the contraction of muscle following the arrival of an action potential along the T tubule.   [10 marks]

Q2   Predators have many muscle fibres with particularly well developed sarcoplasmic reticulum. Suggest the advantage of this.   [5 marks]

Muscles — pretty, but very boring...

Similar to models in that respect. Not that I know many models, so I'm probably being very unfair. Probably many models have degrees in psychology or philosophy, and enjoy sky-diving and oil-painting in their spare time. Male models are often quite muscular, which leads us neatly back to the original subject. You need to learn everything on this page about muscles.

The Mammalian Eye

For some reason the eye seems more interesting than most of the other subjects in this section. Maybe it's because we're very aware of our own eyes — more so than our muscles or our nerves, anyway.

The **Eye** is an organ with **Photoreceptors** for detecting light

1) The mammalian eye is a fluid-filled ball bound by a tough external **sclera**, which forms the transparent **cornea** in front.

2) A thinner transparent **conjunctiva** covers the cornea.

3) The inner lining of the back of the eyeball is the photoreceptive **retina**.

4) Between the sclera and the retina is the **choroid**, which is a layer rich in blood vessels to supply the retina and covered with pigment cells to prevent internal reflection of light.

5) The shape of the eyeball is maintained by the hydrostatic pressure of the **aqueous humour** behind the cornea (a clear salt solution) and the jelly-like **vitreous humour** behind the lens.

6) Light rays pass through the **pupil** (hole in the front) and are focused by the **lens** onto the **fovea** of the retina.

7) Action potentials are then carried from the retina to the brain by the **optic nerve**, a bundle of sensory neurones.

8) The **blind spot** is where the optic nerve leaves the eye. There are no photoreceptors there so it's **not** light sensitive.

> The **iris** is a muscular diaphragm surrounding the **pupil**. It controls the amount of light entering the eye. In bright light the **circular muscle** of the iris **contracts** and the **radial muscle relaxes**, making the pupil **smaller**. Less light enters the eye, preventing **damage** to the **retina**. The opposite happens in dim light — the iris makes the pupil **dilate** to allow more light in.

Light Rays are **Refracted** to focus on the **Retina**

Light rays are **refracted** (bent) as they pass through the **cornea** and the **lens**. Most refraction happens at the cornea, and then the lens **fine-tunes** the direction of the light to focus it onto the **retina**. It does this by changing its **shape** so that the light is refracted more or less. Surrounding the lens is a radial array of **suspensory ligaments** that are connected by a ring of **ciliary muscles**:

1) When the eye focuses on a **distant** object the ciliary muscles **relax**, which pulls the **suspensory ligaments** taut. This pulls the lens **flat**.

2) When the eye focuses on a **near** object the ciliary muscles **contract**, so the **suspensory ligaments** are slack. This gives the lens a more **rounded** shape.

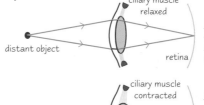

Light from a Distant Object
circular ciliary muscles relaxed;
suspensory ligaments taut;
lens pulls in; light focused on retina.

Light from a Near Object
circular ciliary muscles contracted;
suspensory ligaments slack;
elastic lens more convex;
light focused on retina.

The **Retina** is made up of **Three Main Layers**

1) **Photoreceptors** (rods and cones) which have **outer segments** and **inner segments**:
   - The outer segments have lots of **flattened vesicles** containing **pigments** that absorb light energy. When light is detected, there's a **chemical change** inside the cell which creates a **generator potential** on the cell membrane.
   - The **inner segment** of each receptor has **mitochondria** and a **nucleus**, and connects with a **synapse**.

   Rods and cones differ in shape (see diagram). There are **three types** of cones, each with a different type of **pigment**.

2) There's a layer of **bipolar neurones** (with some **cross connections**). The cross connections between the bipolar neurones allow the signals sent from the cells of the retina to be **coordinated**.

3) A layer of **sensory neurones** have axons leading to the **optic nerve**.

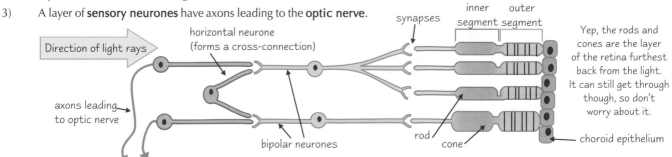

Structure of the retina

Yep, the rods and cones are the layer of the retina furthest back from the light. It can still get through though, so don't worry about it.

The Mammalian Eye

Outer Segments contain Visual Pigments that are Proteins

Rods use a visual pigment called **rhodopsin**. **Cones** use three forms of another pigment called **iodopsin**.
This is how **rhodopsin** is used to send impulses to the brain:

1) Light energy is absorbed by a part of the rhodopsin pigment called **retinal**. This changes shape from one form to another, and detaches from the protein part called **opsin**. This process is known as **bleaching**.

2) **Bleaching** causes the cell membrane of the **outer segment** to become **hyperpolarised**. This means that an excess of **sodium channels** close, so the **resting potential** value gets more negative (from about -65 mV to about -120 mV).

3) Less **inhibitory neurotransmitter** is released across the synapse, so there's less **inhibition** of the **bipolar neurone**. This means the cell membrane of the bipolar neurone becomes **depolarised**.

4) An **action potential** is formed in the bipolar neurone membrane, and is transmitted to the brain through the **optic nerve**.

Retinal and **opsin** then join back together using an enzyme-catalysed reaction. This **regenerates** the pigment so that it's ready to be stimulated again.
The same sort of thing happens with the **iodopsin** in **cones**, but it breaks apart **less easily** and joins back together **more slowly**. This means that cones are best suited to **higher light intensities**, and we depend more on rods in **dim light**.

Rods are more Sensitive, but Cones let you see more Detail

1) There are about **twenty times** more rods in the human eye than there are cones. The cones are mostly found packed together in the **fovea**, which is where most of the light that enters the eye tends to focus. They give better **visual acuity** (clarity) than rods do, and let us see images more **accurately** and in more **detail**. This is because each cone synapses with its **own individual bipolar synapse**, so it can send more detailed information to the brain. Cones also give **colour vision** (see below).

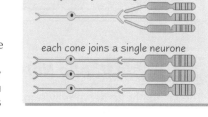
many rods join a single neurone

each cone joins a single neurone

2) The rods are found outside the fovea, in the more peripheral parts of the retina. They're a lot more **sensitive** than cones, because lots of rods converge onto the **same bipolar neurone**. This means that even **small** responses from rod cells can be detected in the brain, because responses from lots of rod cells are **combined**. But having lots of rods converging onto the same bipolar neurone also means that they can't provide as much **clarity** or **detail** as the cones can. This is why an object you can see 'out of the corner of your eye' isn't very clear. You have to move your eyes so that the light from the object focuses on your **fovea** to see it clearly.

3) Colour vision is explained by the **trichromatic theory**. It states that all the colours we see come from mixing just the **three primary colours** of light — **red**, **blue** and **green**, in different proportions. Each of the three types of **cone** found in the retina is sensitive to **one** of these wavelengths of light. Any particular colour is then experienced because the wavelength stimulates one, two or all three of these types to a different degree.

Lenses Change as people get Older

OCR ONLY

1) As you get older, your eye's **lens** becomes **less elastic**, so it can't easily spring back to a rounded shape. This means light can't be **focused** very well for near viewing.

2) The lens may become **less transparent** as proteins in it **coagulate**, giving a **cataract**. A lens affected by cataracts can be **removed surgically** and replaced by an artificial one.

Practice Questions

Q1 Give the function of the following parts of the eye: a) choroid   b) ciliary muscle   c) iris.

Q2 What shape is the lens when the eye focuses on a distant object?

Q3 What happens to rhodopsin when it absorbs light energy?

Exam Question

Q1 Explain, with reference to rods and cones, how the human eye has both high sensitivity and high acuity.   [8 marks]

Be thankful for the eye — without it, you wouldn't see this lovely page...

OK, so maybe that's not the best reason to be thankful for the eye. In fact, it might be enough to make you wish it had never evolved. But too late, it has, so there's no excuse not to learn about it. Those rod and cone diagrams are just to illustrate a point, but the diagram of the eye and those two about the lens have to be learnt. The best way is to practise drawing them yourself.

The Autonomic Nervous System

Skip these pages if you're doing Edexcel. These pages deal with the autonomic nervous system, which controls all the stuff in your body that happens without you thinking about it. If only revision happened that way...

The **Autonomic Nervous System** controls **Unconscious Activities**

The nervous system is divided into **two parts**:

1) Conscious activity is controlled by the **voluntary (somatic) nervous system** (see pages 70-77).
2) **Unconscious activities**, like the actions of the heart and the digestive system, are controlled by the **autonomic nervous system**, which sends impulses to the involuntary (smooth) muscle and glands. Unconscious, involuntary reactions to stimuli are 'reflex' reactions. They're stereotypic — there's always the same reaction to the stimulus.

The **Autonomic Nervous System** is divided into **Two Parts**

The autonomic nervous system is made up of the **sympathetic** and the **parasympathetic** nervous systems, which have **opposite effects** on organ activity.

These two systems can have **different effects** on the same kinds of muscles because the motor neurones involved secrete different kinds of neurotransmitter at the synapses (see pages 70-77 for more on these). The sympathetic nervous system uses **noradrenaline**, and the parasympathetic system uses **acetylcholine**.

The **Sympathetic Nervous System** causes the '**Fight or Flight**' response

The **sympathetic nervous system** is the part of the **autonomic nervous system** that increases the overall physical activity of the body, a response called '**fight or flight**'. In a 'fight or flight' response blood is diverted from the gut to the **lungs**, **heart** and **voluntary muscle**:

1) It **increases** the heart beat rate, to increase **oxygen supply** to the muscles.

2) It **increases** the ventilation rate, so that **more oxygen** can be absorbed in the lungs.

3) It **decreases** peristalsis in the gut so there's more blood available for the heart and lungs.

Other parts of the body also respond to enable **increased sensory awareness**. For example, the radial muscles of the iris contract, causing the pupils of the eye to dilate (see p. 78). The sympathetic nervous system triggers the same kinds of responses as **adrenaline**, the 'fight or flight' **hormone**.

The **Parasympathetic Nervous System** prepares the body for **Rest**

The **parasympathetic nervous system** is the part of the autonomic nervous system that decreases overall physical activity, so it's associated with rest:

1) It **decreases** heart beat rate.

2) It **decreases** ventilation rate.

3) It **increases** peristalsis in the gut, so that food can be digested.

The Autonomic Nervous System

The **Sympathetic** and **Parasympathetic** systems have **Opposite Effects**

The two systems have a range of effects on the body. One example of the two systems working together with opposite effects is that **pupil diameter** can be changed to control the amount of light reaching the retina of the eye:

1) If there's **too much** light, the **parasympathetic** nervous system causes the circular muscles to contract, to **reduce** pupil diameter.

2) If light levels are **low**, the **sympathetic** nervous system contracts the radial muscles, **increasing** the diameter of the pupil.

Comparison of the sympathetic and parasympathetic nervous systems

Feature	Sympathetic	Parasympathetic
transmitter substance at synapses	noradrenaline	acetylcholine
heart rate	speeds up	slows down
iris	dilates	constricts
movements of digestive tract	slows down	speeds up
sweating	stimulated	not stimulated

Bladder Control can be Learned

Some of the functions controlled by the autonomic nervous system can be **learned** over time, e.g. children learn bladder control:

1) The **sympathetic nervous system** causes a **sphincter** (a circular muscle) to contract at the bladder opening, letting the bladder **fill up**.

2) As the bladder becomes full, **stretch receptors** in the walls produce **nerve impulses** which go to the central nervous system (p.82-83).

3) The **parasympathetic nervous system** sends impulses to the spincter, **relaxing** it.

4) The bladder **empties**.

Harry wished that his parasympathetic nervous system would stop making such a mess on the floor.

Young children learn how to **recognise** the messages from the stretch receptors, and then control the sphincter **consciously** — this is potty training.

Practice Questions

Q1 Distinguish between the autonomic and voluntary (somatic) nervous system.
Q2 What is meant by the 'fight or flight' response?
Q3 Give three effects of stimulation of the body by the parasympathetic nervous system.

Exam Question

Q1 The size of the pupil of the eye decreases in bright light. This is an example of a reflex response and is brought about by the parasympathetic nervous system, which stimulates contraction of the circular muscles of the iris and effects relaxation of the radial muscles.

a) Explain why this kind of behaviour is described as a reflex. [2 marks]

b) Acetylcholine is the neurotransmitter released at the end of parasympathetic nerve fibres. Suggest how a single type of neurotransmitter can affect the different muscles in different ways. [2 marks]

No one will be sympathetic if you don't learn this...

*They're weird names, really. How can a nervous system be sympathetic? It doesn't offer you a tissue when you're upset. Basically, you just have to learn that the sympathetic system gets things ready for fight or flight, and the parasympathetic system calms things down. Try remembering it like this — **s**ympathetic for **s**tress, and **p**arasympathetic for **p**eace.*

The Central Nervous System

Skip these pages if you're doing AQA A. And now ladies and gentlemen, the most important bit of all — the central nervous system, the controller of almost everything you do. It's like a giant computer, only much more complicated...

The **Central Nervous System** has **Two Parts**

The central nervous system is made up of your **spinal cord** and your **brain**:

1) The **spinal cord** runs through the protective, bony vertebral column.
   The sensory and motor **spinal nerves** are connected to it.
   Quick **reflex** reactions like sneezing can be processed by the spinal cord, without using the brain.

2) The **brain** is a massive bunch of **relay neurones** (see p. 72) connected to the spinal cord. The brain is inside in the protective bony **cranium** (skull), so the nerves connected to the brain are called **cranial nerves**.

Nervous Tissue contains **White** and **Grey** Matter

White matter is mainly made up of the **nerve fibres** of the neurones — the **myelin sheaths** (see page 73) glisten, so it looks white.

Grey matter is a concentration of **cell bodies**, which appears darker because of the absence of the sheaths. In the spinal cord, the grey matter is in the centre, with white matter around the outside, but in the brain, the grey matter forms a thin surface coating called the **cortex**.

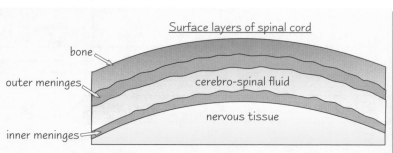

Surface layers of spinal cord

bone
outer meninges
inner meninges
cerebro-spinal fluid
nervous tissue

Running right through the spinal cord is a fluid-filled tube called a **sinus**, which carries **cerebro-spinal fluid**, produced by layers of connective tissue called **meninges**. The fluid absorbs mechanical shocks and supplies food and oxygen.

The **Lower Parts** of the brain deal with **Unconscious Activities**

The nervous pathways going through the **spinal cord** deal with simple **reflex responses** (see page 75). You don't consciously control them. The frontal parts of the brain often deal with **conscious** decisions that influence behaviour.

1) **Hypothalamus**

   This is just beneath the middle part of the brain (above the medulla). It controls important **conscious** activities, but it also **regulates** blood temperature and water potential. In some cases (e.g. water potential regulation), it communicates with the **endocrine system** (pages 84-85). The hypothalamus is the main part of the body where nervous and endocrine systems meet, because it is connected to the **pituitary gland** (see below). **Neurosecretory cells** in the hypothalamus can carry action potentials, and produce chemicals like ADH and oxytocin. Blood vessels carry these chemicals to the **pituitary gland** where they are stored, and later released in response to action potentials from the hypothalamus.

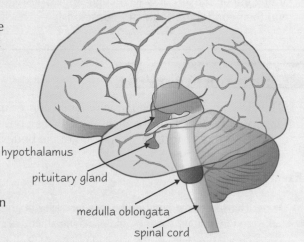

hypothalamus
pituitary gland
medulla oblongata
spinal cord

2) **Pituitary gland**

   This controls the **endocrine system**, and releases a range of **hormones** (see pages 84-85 for more on this). It responds to **neurosecretions** from the hypothalamus.

3) **Medulla oblongata**

   This is at the base of the brain. It's the coordination centre of the **autonomic nervous system** (see page 80). It controls the unconscious activities of vital organs like the heart and lungs — it controls breathing movements, heart rate and also the action of smooth muscle in the gut.

The Central Nervous System

Upper Parts of the brain allow Complex Conscious Behaviour

The upper parts of the brain are the **cerebellum** and the **cerebrum**. They each have a highly folded **cortex** (the outer bit). They coordinate sensory input with motor output. Action potentials arriving along sensory nerve fibres pass along lots of possible pathways via the relay neurones, so each **input** has lots of **different potential responses**. This is the basis of **memory** and learning from experience.

The **cerebellum** is important for balance and controlled muscle movements. It helps make movements coordinated. Over a period of time, **learning experiences** affecting the cerebrum influence the **neurone pathways** in the cerebellum. This means that the cerebellum becomes involved in controlling more **skilled coordinated sets of movements** that become routine (such as walking, maintaining posture or playing a piano).

The Cerebrum is divided into Two Cerebral Hemispheres

The **cerebrum** is the largest part of the brain. It deals with the **voluntary activities** of the body. It's associated with advanced mental activity, like **emotion**, **memory** and **language**. It's divided into **two parts** (hemispheres). The right hemisphere deals with actions for the left side of the body, and vice versa.

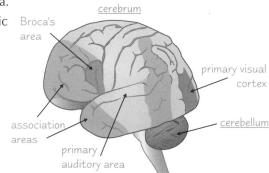

1) Action potentials from particular parts of the body arrive in specific areas of the cerebral cortex (the outer bit) called **primary sensory areas**, e.g. impulses from the optical nerve of the eye go to the visual cortex, where they're interpreted by the brain. The more sensory cells in a part of the body, the bigger the sensory area in the brain.

2) Sensory areas pass action potentials to **association areas** where information is integrated.

3) **Motor areas** send impulses to effectors (muscles), causing movement.

AQA B ONLY →

Example: Understanding language
Receptors in the ear send information to a **sensory area** of the cerebrum — the **auditory area**. This sends information to the **auditory association area** (**Wernicke's area**), where words are identified, using the memory. The information then goes to the **auditory motor area** (**Broca's area**) which coordinates muscle movement to produce speech for a response.

Alzheimer's Disease affects the cerebral cortex
About 5% of people over the age of 65 develop **Alzheimer's disease**, which is a form of **dementia** involving **loss of memory** and **personality change**. Sufferers develop deposits (**plaques**) of **beta amyloid protein** in the cerebral cortex, and also **tangles** of protein fibres within the neurones. The cerebrum becomes less able to produce neurotransmitters. Scientists still don't know what causes it, and there's no treatment at the moment.

Practice Questions

Q1   Describe the structure of the central nervous system.
Q2   How do the kinds of behaviour controlled by the brain differ from those controlled by the spinal cord?
Q3   Name two factors in the body that are controlled by the hypothalamus, and two that are controlled by the medulla oblongata.

Exam Questions

Q1 a) Describe the functions of the sensory areas and the association areas of the brain's cerebrum.   [8 marks]
   b) Describe how the cerebrum is affected by Alzheimer's disease and describe the effects on its function.   [4 marks]
Q2   Explain why the pituitary gland is connected to the hypothalamus of the brain.   [4 marks]

I'm not sure that I have a brain...

It goes without saying that the brain is a pretty complicated organ. That doesn't mean that it has to be too stressful to learn about it for your exam. All you have to do is know what each of the main areas do, and understand that the brain processes information from receptors and sends impulses to effectors. Don't worry about exactly what's going on there.

Hormonal Control

Skip these pages if you're doing AQA A or OCR. All this physiology stuff talks a lot about hormones, but they haven't really had a page to themselves yet. Well here you go, here's two whole pages devoted to the weird chemicals...

The *Endocrine System* secretes *Hormones*

The endocrine system is made up of endocrine glands that secrete chemicals called hormones. A **gland** is any structure that is specialised for the **secretion** of one or more types of substance. Endocrine glands secrete **hormones directly** into the **blood** without using ducts.

1) Many types of hormones are **proteins** or smaller **peptides**, e.g. **insulin** and **adrenaline**.

2) Other hormones are fatty **steroids**, e.g. **oestrogen**, **progesterone** and **testosterone**.

Hormones are *Chemical Communicators*

Chemical communication happens when one cell produces and releases a substance (the **chemical communicator**), which then binds to a **receptor** on another cell (the **target**). The chemical communicator molecule has a shape that fits into the shape of the receptor molecule on the target cell membrane. This brings about a response in the target cell. There are **two main kinds** of chemical communicator:

1) Those that work only on cells that are very close to the cell that released the chemical, and often move by **diffusion**. E.g. **neurotransmitters**, which act on postsynaptic neurones (see page 74).

2) **Hormones** — these work on targets that are much further away because they enter the blood and are carried by **mass flow** due to the pumping action of the heart.

Hormones are secreted into the **Blood**

1) Hormones are secreted when the **endocrine gland** is **stimulated**. Some glands are stimulated by a change in concentration of a specific solute (sometimes another type of hormone). Others are stimulated by **action potentials** arriving from the nervous system.

2) The hormone is **secreted** and **diffuses** into **blood capillaries**.

3) The hormone is circulated around the body by **mass flow** through the **bloodstream**.

4) The hormone diffuses out of the blood capillaries at different parts of the body. However, it will only bind to **cell-surface receptors** with complementary-shaped **binding sites**. This means that only **target cells** with the correct receptors will respond. Very **small** concentrations of hormones are needed to give a response in target cells. This response usually activates **enzymes** inside the cells.

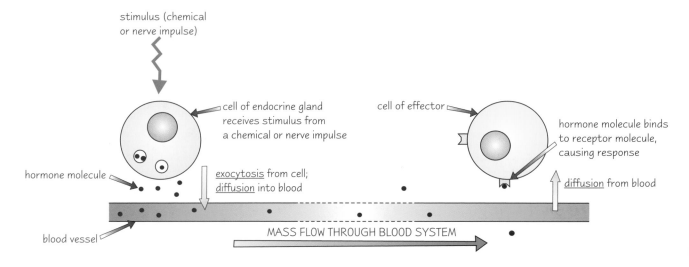

Hormonal Control

Hormonal Control is often by Negative Feedback

Hormones that **regulate** factors in the body, such as **glucose concentration,** often work by **negative feedback** (see pages 62-63). This happens when a change in a factor is detected and the response **counteracts** the change.

Negative feedback control happens when hormones are working in **pairs**, e.g. insulin and glucagon (pages 86-87). Here, an **increase** in blood glucose concentration stimulates the secretion of **insulin** to bring the glucose concentration back down. Then when blood glucose concentration **drops**, this stimulates secretion of **glucagon** to bring it back up. These pairs of hormones are said to work **antagonistically**.

Different Hormones affect Different Parts of the Body

Here's a handy chart that tells you everything you need to know about specific hormones for your exam:

	HORMONE	SOURCE	TARGET	EFFECT ON TARGET
Polypeptide hormones and their derivatives	gastrin	gastric mucosa	gastric glands	secretion of gastric juice
	secretin	duodenal mucosa	pancreas	secretion of hydrogen carbonate
	cholecystokinin-pancreozymin	duodenal mucosa	pancreas	secretion of pancreatic enzymes
	FSH	anterior pituitary gland	follicle cells / testes	development of follicles / stimulates testosterone production
	LH	anterior pituitary gland	follicle cells / testes	controls growth of ovaries and testes / ovulation stimulates testosterone production
	oxytocin	produced in hypothalamus, stored in posterior pituitary gland	smooth muscle	oviduct contractions, uterine contractions at birth, secretion of milk
	prolactin	anterior pituitary gland	mammary glands	initiates and maintains milk production
	vasopressin (ADH)	produced in hypothalamus, stored in posterior pituitary gland	renal collecting duct and distal tubules	increases permeability to water: less water lost in urine
	adrenaline	adrenal gland	widespread	"fight or flight" response: mimics sympathetic nervous system.
	noradrenaline	adrenal gland	widespread	similar to adrenaline
	insulin	pancreas (Islets of Langerhans, β-cells)	widespread, especially liver	glucose converted to glycogen
	glucagon	pancreas (Islets of Langerhans, β-cells)	liver	glycogen converted to glucose
Steroid hormones	oestrogen	follicle cells	endometrium, pituitary gland, widespread	thickening of endometrium, inhibits secretion of FSH, development of female sexual characteristics
	progesterone	corpus luteum	endometrium, pituitary gland	maintains endometrium, inhibits secretion of FSH
	HCG	placenta	corpus luteum	maintains corpus luteum
	testosterone	Leydig cells	widespread	production of spermatozoa, development of male sexual characteristics

Practice Questions

Q1 Explain what is meant by the term endocrine gland.

Q2 Explain how a hormone reaches its target cell.

Q3 Give two examples of hormones that are proteins and two that are steroids.

Q4 For each of the hormones listed in Q3, state one function in the body.

Exam Questions

Q1 Explain why hormonal control in the body brings about slower responses than nervous control. [2 marks]

Q2 Explain what happens when an endocrine gland is stimulated to produce a hormone. [5 marks]

Bet you never knew you had so many hormones...

Or maybe you did know that. In fact maybe you know everything in this section already, and you're just reading it so that you can chuckle secretly to yourself as you watch lesser mortals trying desperately to learn information that you've known instinctively since you were born. If so, you should probably be writing this book.

Blood Glucose Control

Ever felt a 'sugar rush' after eating eighteen packets of Refreshers? Ok, maybe no one else eats that many at once.
These two pages are about how your body deals with all the glucose you get from your food.

Many Factors can change the Blood Glucose Concentration

Glucose enters the blood from the **small intestine** and dissolves in the blood plasma. Its concentration usually stays around **100mg per 100cm³ of blood**, but some factors can change this level:

1) Blood glucose concentration **increases** after consuming food, especially if it's **high** in carbohydrate.

2) Blood glucose concentration **falls** after exercise, because more glucose is used in respiration to release energy.

Big changes in the concentration of blood glucose can **damage cells** by changing the water potential.

The Pancreas secretes Hormones to control Blood Glucose Levels

Two hormones — **insulin** and **glucagon**, regulate blood glucose concentration. They're secreted by clusters of cells in the **pancreas** called the **Islets of Langerhans**. These cells detect changes in blood glucose concentration.

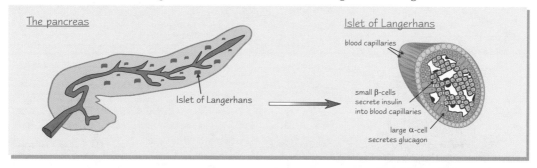

Controlling Blood Glucose Levels is an example of Negative Feedback

Insulin and glucagon work together to regulate the blood glucose concentration by **negative feedback** (pages 62-63):

When there's a rise in blood glucose concentration...

1) The glucose molecules bind to receptors in the cell membranes of small **beta (β) cells** in the Islets of Langerhans.

2) These cells secrete **insulin** into the blood.

3) Insulin molecules bind to receptors in the cell membranes of **hepatocytes** (liver cells) and other cells, e.g. muscle cells.

4) This **increases** the **permeability** of the hepatocyte cell membranes to glucose, so more glucose is absorbed.

5) Inside the hepatocytes, the insulin activates an **enzyme** that catalyses the condensation of glucose molecules into **glycogen**, which is stored in the cytoplasm of the hepatocytes, and in the muscles. This process is called **glycogenesis**.

6) Insulin also increases the rate of respiration of glucose in other cells. The blood glucose concentration **decreases**.

When there's a fall in blood glucose concentration...

1) The larger **alpha (α) cells** of the Islets of Langerhans secrete the hormone **glucagon** into the blood.

2) Glucagon binds to receptors on the **hepatocytes**.

3) This activates an **enzyme** inside the hepatocytes that catalyses the **hydrolysis** of stored glycogen into **glucose**. This process is called **glycogenolysis**.

4) The blood glucose concentration **increases**.

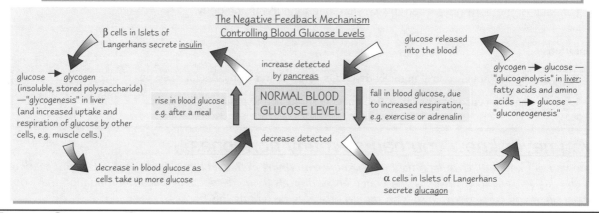

Blood Glucose Control

Other Hormones can affect Blood Glucose Level too

Other hormones can affect glucose levels in similar ways, although they bind to different receptors. For example, **adrenaline** is secreted during exercise (during a 'fight or flight response') and activates enzymes that **hydrolyse** stored glycogen into glucose, ready for increased muscle activity. It also increases the rate of respiration taking place inside the cells.

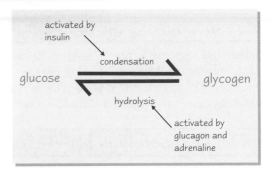

Diabetes causes problems with Glucose Regulation

Diabetes mellitus is a condition where blood glucose concentration can't be regulated properly. There are **two** main types of diabetes:

1) In **Type 1** diabetes, the pancreas doesn't produce enough insulin. After eating, the blood glucose level rises and stays high (**hyperglycaemia**). The kidneys can't reabsorb this glucose, so it's **excreted** in the urine. The symptoms include tiredness, thirst and weight loss. It can be treated by regular **injections** of insulin, and **controlling carbohydrate intake** from the diet. The amount of insulin injected has to be carefully controlled, because too much can produce a dangerous drop in blood glucose levels (**hypoglycaemia**).

2) **Type 2** diabetes is usually acquired later in life. This type of diabetes is linked with obesity. Blood insulin levels are normal or high, but the **receptor proteins** on the hepatocyte cell membranes don't work properly. The symptoms are the same as for Type 1 diabetes. It can usually be controlled by reducing **carbohydrate** intake.

Insulin used to treat diabetes is produced by **genetic engineering** (see p. 104). Genetically engineered insulin is considered better than the old method of using insulin from pigs because:

* There's less chance of the body producing anti-insulin **antibodies** against the 'foreign' peptide.
* There's less risk of contracting a **virus** than from the animal insulin.
* The 'human' insulin works more **quickly** in humans than pig insulin.

Practice Questions

Q1 Give one factor that will increase blood glucose concentration, and one that will reduce it.

Q2 What are the roles of the alpha and beta cells of the Islets of Langerhans?

Q3 State two effects of insulin on liver cells.

Q4 Name two hormones that increase the blood glucose concentration.

Q5 Describe two ways in which diabetes mellitus can be controlled.

Exam Questions

Q1 a) Explain the role of the endocrine system in returning the blood glucose level to normal after consuming a meal that has a high glucose content. [10 marks]

   b) Some sufferers of diabetes can produce insulin, but still cannot regulate their blood glucose levels effectively. Suggest a reason for this. [2 marks]

Q2 Explain why glucagon levels in the blood increase during exercise. [5 marks]

My glucose levels are low — pass the chocolate...

Learn this carefully, or you'll end up getting your hormones confused. And make sure you're clear on the different types of diabetes too — people with Type 1 can't produce enough insulin, people with Type 2 produce it fine, but can't respond to it. Now eat a huge meal, wait for your blood glucose levels to rise, then draw that diagram from page 86 until you know it.

Communication Systems in Plants

Skip these pages if you're doing AQA A or B. Plants don't have brains, but they do communicate. They move too, but it's so slow you can't see it. Just as well, because plants that ran around the place would be pretty scary things.

Plants respond more Slowly than Animals

Coordination and **response** in plants happens by **chemical communication** between cells. It takes time for the chemicals to move to their targets, so plant response times are **slower** than those of animals. Many chemical communicators in plants are called **plant growth regulators**, because they control certain aspects of the plant growth, making cells grow bigger or stimulating cell division.

Auxins and Gibberellins stimulate Cell Growth

Plants can move because different parts of them grow at different rates, due to chemicals called **auxins** and **gibberellins**.

Auxins are continuously made at the **shoot tips** and in **young leaves**. They diffuse from cell to cell and loosen the cellulose fibres in the plant cell walls. This allows more water to be taken up by the cells by **osmosis**. The cell then makes new material, and **elongates**.

The shoot tip is the dominant part of the plant growing upwards because the auxins inhibit growth of lateral (side) branches. This is called **apical dominance**.

A **drop** in auxin levels is also what causes leaves to fall (**abscission**).

Gibberellins are also produced in parts of the plant associated with growth, like **young leaves** and the **embryos** in seeds. They stimulate the synthesis of **amylase**, which **hydrolyses** stored starch to **maltose** for respiration. Many dormant seeds secrete gibberellins when they are soaked in water, which encourages them to germinate. Gibberellins also allow stem elongation, which enables plants to grow tall.

Auxins and gibberellins are synergistic — they work together to produce the same effect. Antagonistic substances oppose each other's actions.

Auxins are used **commercially** by farmers and plant breeders for the following uses:

1) **Selective weedkillers.** Auxins can kill fast-growing weeds. This is because they make the weed use up its stored chemical energy to produce a long stem, instead of leaves for photosynthesis.

2) **Rooting hormones.** Specific kinds of auxins promote the rooting of cuttings.

3) Applying low concentration of auxin in the early stages of **fruit production** prevents the fruit from falling. Applying a high concentration at a later stage promotes fruit drop.

Tropisms are a Plant's Response to a Stimulus

A **tropism** is the movement of a part of the plant in response to an **external stimulus**. The movement almost always involves growth, and so it's influenced by plant growth regulators.

A **positive tropism** is movement **towards** the stimulus. A **negative tropism** is **away** from the stimulus:

1) **Phototropism** is movement in response to **light**. Shoots are positively phototropic, but roots are negatively phototropic. Experiments with **coleoptiles** (the unfolded leaves of monocotyledonous shoots) show that there's a higher concentration of auxins in the shaded part of the stem, so there's more growth on the dark side. This means that the stem starts to curve towards the light.

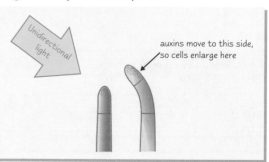

Unidirectional light

auxins move to this side, so cells enlarge here

2) **Geotropism** is movement due to **gravity**. Shoots are **negatively geotropic** (they grow upwards) but roots are **positively geotropic** and grow downwards.

Communication Systems in Plants

Abscisic Acid and Ethene are Antagonistic to Auxins and Gibberellins

1) **Abscisic acid** is produced in most parts of the plant and prepares the plant for dormancy (the period when they don't grow) by **inhibiting growth**. It stimulates the **closure** of the **stomata** in leaves when water is in short supply, and **inhibits germination** in seeds.

2) **Ethene** is also produced in most plant parts — it breaks bud dormancy and stimulates the **ripening** of fruit.

EDEXCEL ONLY   EDEXCEL ONLY

> **Cytokinins stimulate cell division**
> Chemicals called **cytokinins** stimulate **mitosis** by increasing DNA and RNA synthesis. They are produced in **seeds** to encourage embryo growth, but are also made in the roots and carried to other parts of the plant in the transpiration stream.

Phytochromes control Flowering Times

When **light** intensity or wavelength determines the timing of a biological process, this is called **photoperiodism**. The flowering time of many plants is determined by the length of the day (the **photoperiod**) and controlled by chemicals called **phytochromes**. There's more on phytochromes on page 158.

Different species react in different ways to photoperiods:
1) **Long-day plants** flower in the summer (long days, short nights).
2) **Short-day plants** flower in the winter (short days, long nights).
3) **Day-neutral plants** flower independently of day length.

Phytochrome exists in two forms:
1) **Phytochrome red (PR)** absorbs **red light** (660nm wavelength); and changes into the **PFR form**.
2) **Phytochrome far red (PFR)** absorbs **far red light** (730nm wavelength) and changes slowly back into **PR** in the dark.

Example — long-day plants

Sunlight contains more **red** than **far red** wavelengths, so at the end of the longer **summer** days a lot of PR has been changed into PFR. Because the nights are **short**, there's not enough time for PFR to be changed back into PR, so there's more **PFR**, which **stimulates flowering in long-day plants**.
In winter (when days are **short**), the nights are long enough for a lot of PFR to be converted back to PR. So there's a **lack** of PFR (which inhibits flowering in short-day plants).

PR — this form builds up in winter → sunlight or red light → PFR — this form builds up in summer
← far red light or darkness

In short-day plants, PFR may inhibit flowering, so they flower in winter.

Practice Questions

Q1 Distinguish between the effects of auxin and gibberellin on plant cell growth.
Q2 Auxins control apical dominance. Explain what this means.
Q3 Explain how the distribution of auxin in a growing shoot enables it to grow towards a light source.
Q4 Give one commercial application of auxins.

Exam Questions

Q1 Explain how the production of gibberellins in seeds stimulates them to germinate. [5 marks]

Q2 With reference to phytochromes, suggest how increased day length stimulates the flowering of some species of plants in the summer. [5 marks]

It's all gibberellin-ish to me...

Phytochromes are probably the most complicated things on these pages. The most confusing thing about it is that not all plants respond to phytochromes in the same way. If you concentrate on how long-day plants react to phytochromes, and get that part clear in your head first, then you should be well on your way to understanding the pesky little chemicals.

Water and Mineral Uptake in Plants

These pages are only for AQA A and Edexcel. Everyone else can skip them.

Water Enters a Plant through its Root Hair Cells

Water has to get from the **soil**, across the **root** and into the **xylem**, which takes it up the plant. The bit of the root that absorbs water is covered in **root hairs**. This increases its surface area and speeds up water uptake. Once it's absorbed, the water has to get through two root tissues, the **cortex** and the **endodermis**, to reach the xylem.

> Water always moves from areas of **higher water potential** to areas of **lower water potential** — it goes down a **water potential gradient**. The **soil** around roots has a **high water potential** (i.e. there's lots of water there) and **leaves** have a **low water potential** (because water constantly **evaporates** from them). This creates a water potential gradient that keeps water moving through the plant in the right direction, **from roots to leaves**. Plants also actively transport ions such as nitrates into their root hair cells — this lowers the **water potential** so even more water moves in by **osmosis**.

There are Two Routes Water can take through the Root

Water can travel through the roots into the xylem by two different paths:

1) The **apoplast pathway** — goes through the **non-living** parts of the root — the **cell walls**. The walls are very absorbent and water can simply diffuse through them, as well as passing through the spaces between them.

2) The **symplast pathway** — goes through the **living** cytoplasm of the cells. The **cytoplasm** of neighbouring cells connects through **plasmodesmata** (they're little strands that pass through small gaps in the cell walls).

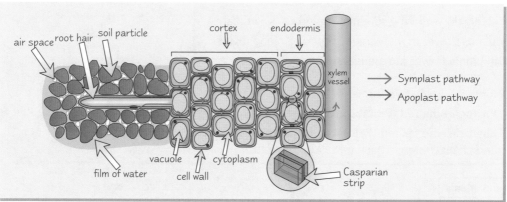

Both pathways are used, but the main one is the **apoplast pathway** because it provides the **least resistance**. When the water gets to the **endodermis** cells, though, the apoplast pathway is blocked by a **waxy strip** in the cell walls, called the **Casparian strip**, which the water can't penetrate. Now the water has to take another pathway. This is useful, because it means the water has to go through a **cell membrane**. Cell membranes are able to control whether or not substances in the water get through. Once past the endodermis, the water moves into the **xylem**.

Xylem Transports Water and Provides Support

There are **three** ways that water moves up through the xylem **against** the **force of gravity**:

1) The **cohesion-tension theory** explains how water moves up plants from roots to leaves, against the force of gravity.
   - Water evaporates from the leaves at the 'top' of the xylem (through transpiration).
   - This creates a tension, which pulls more water into the leaf.
   - Water molecules **stick together** ('cohesion'), due to hydrogen bonds, so when some are pulled into the leaf others follow.
   - This means the whole **column** of water in the xylem, from the leaves down to the roots, moves upwards by mass flow.

2) **Root pressure** also helps move the water upwards. Water is transported into the xylem from the roots, which creates a pressure and tends to shove water already in the xylem further upwards. This pressure is weak, and couldn't move water to the top of bigger plants by itself. It helps though, especially in young, small plants where the leaves are still developing.

Xylem also provides the plant with support. In **stems**, which need to resist bending, the xylem is **near the outside** to provide a sort of 'scaffolding'.

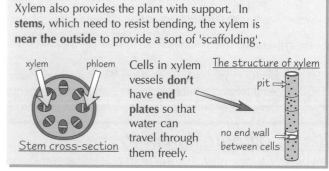

Cells in xylem vessels **don't** have **end plates** so that water can travel through them freely.

3) **Capillarity** is when water moves up a thin tube. In plants it happens because of the molecular attraction (adhesion) between water and the inside of the xylem tube.

Scientists reckon that water movement actually takes place through a **combination** of these **three** methods — none of them is **powerful** enough to move it on their own.

Water and Mineral Uptake in Plants

You can measure Water Uptake using a Potometer

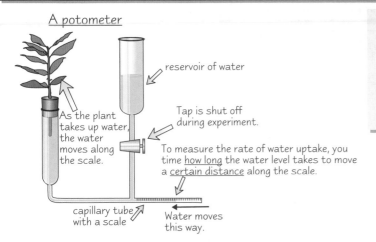

A potometer

reservoir of water

Tap is shut off during experiment.

As the plant takes up water, the water moves along the scale.

To measure the rate of water uptake, you time how long the water level takes to move a certain distance along the scale.

capillary tube with a scale

Water moves this way.

A **potometer** is a piece of apparatus used to **measure water uptake**. You can use a potometer to do a **quantitative study** comparing rates of water uptake in different environmental conditions. (Remember that the rate of water uptake is closely linked to the amount of water that is lost via **transpiration**.)

Xerophytic plants are plants which are especially suited to growing in dry conditions. If you're doing AQA A then you need to know about them for the exam — you can find the stuff you need on p. 60.

Plants Need to Absorb Mineral Ions as well as Water

Plants absorb **dissolved mineral ions** from the soil. This can happen by **diffusion**, but plants often **only** need **certain ions** and not others — diffusion **isn't** a selective process. So, ions needed by a plant are also absorbed by **active transport**. This means a plant can absorb more of a certain ion, even if it already has a **high concentration** of them **inside** its cells — so it doesn't have to rely on there being a **diffusion gradient** into the plant. It also means that the plant can **pump out** any ions it **doesn't need**.

Mineral ions travel up the plant with the water, in the **xylem**. Scientists know this because of experiments using **radioactive tracers**. This uses radioactive forms of ions, so that the radioactivity can be detected and the scientist will know where the ion has moved to.

Different Mineral Ions have different Functions in plants

Mineral Ion	Function
nitrate	Needed to make amino acids for protein synthesis, and for organic bases for nucleic acids.
phosphate	Needed in photosynthesis and respiration reactions — for example there are phosphate groups in ATP and NADP. Also needed for phospholipids and in the backbones of DNA and RNA.
magnesium	Needed to make chlorophyll.

Practice Questions

Q1 What is the difference between the apoplast and symplast pathways?
Q2 What are the two roles of xylem in plants?
Q3 Explain the cohesion-tension theory of water movement in plants.
Q4 Explain how you use a potometer.

Exam Question

Q1 Explain why phosphate, nitrate and magnesium are important plant nutrients and describe the method via which they are taken up by plants. [4 marks]

Don't worry if it takes you a long time to learn these pages

There's absolutely masses to learn here, and lots of it isn't easy. So I'm afraid you'll need to spend a bit of time on these pages before you'll know it all well enough for the exam. Try learning it in small chunks and test yourself as you go along. Don't forget to look at the diagrams too — they contain lots of information that you need to know.

EDEXCEL ONLY

EDEXCEL ONLY

Gas Exchange and Water Loss

These pages are only for AQA A.
Many organisms have developed adaptations that enable them to exchange gases efficiently. Insects and plants always have to make a compromise between the need for respiratory gases and problems with water loss.

Gas exchange surfaces have **Four** Major **Adaptations**

Most gas exchange surfaces have four things in common:

1) They have a **large surface area** to **volume** ratio.
2) They are **thin** — often they're just one layer of epithelial cells.
3) There are **short diffusion pathways** between the gases and the internal tissues.
4) **Steep concentration gradients** between the tissues where gases are absorbed are maintained.

Organisms that have a blood transport system have a good <u>capillary supply</u> to gas exchange surfaces.

Fick's Law *is used to Calculate* **Diffusion Rates**

The rate at which a substance diffuses can be worked out using **Fick's law**:

$$\text{rate of diffusion} \; \alpha \; \frac{\text{surface area} \times \text{difference in concentration}}{\text{thickness of membrane}}$$

α means "is proportional to"

The body surface of a **Protoctist** is **Adapted** to its **Environment**

Protoctists are **small**, **soft bodied**, **unicellular** organisms which evolved around 1.5 billion years ago. There are about 60,000 protoctistan species all of which are **aquatic**. Algae are **immobile**, **autotrophic** protoctists and **protozoa** are **heterotrophic** protoctists.

Protoctists are well adapted to aquatic environments which only contain around **1% oxygen** —

1) They have all the usual features for efficient gas exchange — a **large**, **thin surface**, and an ability to maintain **high concentration gradients**.

2) The **short diffusion pathway** in **unicellular** organisms means that oxygen can take part in **biochemical reactions** as soon as it has **diffused** into the cell — there is **no need** for a circulatory system.

Fish *are* **Adapted** *to live in an* **Aquatic Environment**

The gills of a fish are a bit like the lungs of a mammal, except lungs are inward penetrating sacs and gills are **outward projecting filaments**. Gill filaments, called **lamellae**, increase the surface area for diffusion in just the same way that **alveoli** do in the lungs. The walls of gill lamellae are also made from very thin **squamous epithelium** to minimise the diffusion distance. A **blood system** carries gases between the **gaseous exchange surface** and the respiring **cells** — there are many **capillaries**, which carry blood **close** to the **surface** of the **gill lamellae**.

Water constantly flows over the gills and oxygen diffuses into the blood — that's because oxygen is more concentrated in the water than in the blood inside the capillaries. Some of the fastest moving fish have a **counter current system** where the blood and the water flow in opposite directions. The advantage of a counter current system is that it maintains a **high concentration gradient** of oxygen between the water and the blood.

The counter current method allows 90% of the available oxygen in the water to diffuse into the blood.

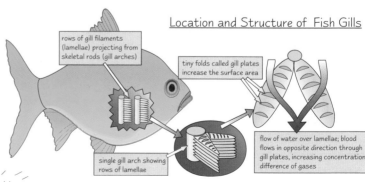

Location and Structure of Fish Gills

rows of gill filaments (lamellae) projecting from skeletal rods (gill arches)

tiny folds called gill plates increase the surface area

single gill arch showing rows of lamellae

flow of water over lamellae; blood flows in opposite direction through gill plates, increasing concentration difference of gases

Gas Exchange and Water Loss

Insects use Tracheae to Exchange Gases

Insects deal with gaseous exchange by having microscopic air-filled pipes called **tracheae** which penetrate the whole of the body from pores on the surface called **spiracles**. The tracheae branch off into smaller **tracheoles** which have **thin**, **permeable walls** and go to individual cells. This means that oxygen diffuses directly into the respiring cells — there's no need for a circulatory system. Insects use **rhythmic abdominal movements** to move air in and out of the spiracles.

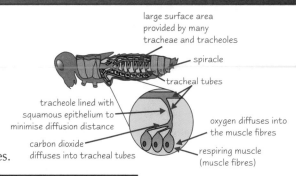

Plants exchanges gases at the surface of the Mesophyll Cells

Plants need CO_2 for photosynthesis and they produce oxygen as a waste gas. The main gas exchange surface is the **surface of the mesophyll cells** in the leaf. This is well adapted for its function — there's a **large surface area**.

The mesophyll cells are inside the leaf. Gases pass back and forth from the outside through special pores in the **epidermis** called **stomata** (singular = stoma). The stomata can open to allow exchange of gases, and close if the plant is losing too much water.

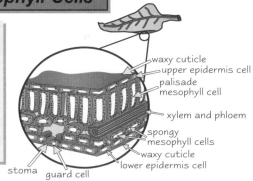

Insects and Plants can Control Water Loss

The problem with openings like stomata and spiracles that are designed to allow gases in and out is that they can lead to **water loss**. Plants and insects have **adaptations** which prevent dehydration:

1) **Insects** have muscles that they can use to **close** their **spiracles** if they are losing too much water. They also have tiny hairs around their spiracles which reduce evaporation.

2) In plants, the stomata are usually kept **open** to allow **gaseous exchange**. **Proton pumps** in the guard cells pump H^+ ions **out** of them. This opens **potassium channels**, allowing K^+ ions to enter the guard cells. This **lowers** their **water potential** and so water enters the guard cells by **osmosis**, which **opens** the stomata.

3) If the plant starts to get **dehydrated**, high light levels and temperatures cause **abscisic acid** to be released. This **stops** the proton pump working and so no water enters the guard cells by **osmosis**, the guard cells become **flaccid** and the **pore closes**.

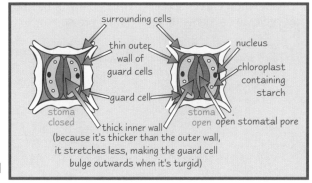

Practice Questions

Q1   What are the four essential adaptations for efficient gas exchange systems?
Q2   What is the equation for Fick's Law?
Q3   How are protoctists adapted for efficient gas exchange?
Q4   Describe how insects and plants stop excess water loss through gas exchange surfaces.

Exam Question

Q1   Using Fick's Law predict whether the values of these three variables will be high or low when the rate of diffusion through the membrane is at its maximum:
a) surface area   b) difference in concentration   c) thickness in membrane.   [3 marks]

Learn the principles — they apply to fish, plants and insects

Once you've learnt those principles at the top of page 90, you'll have a really good basis for understanding the rest of the stuff. Don't you think that it's pretty amazing that organisms as different as plants, fish and insects have evolved to use the same mechanisms to get the gases they need and get rid of the ones that they don't? No, I didn't think you would.

Transport of Respiratory Gases

These pages are only for AQA A.

Hmm, so how exactly does blood transport oxygen and carbon dioxide?

\\ \ \ \ \ | | | | / / /
'Affinity' for oxygen
means <u>willingness to</u>
<u>combine</u> with oxygen.
/ / / | | | \ \ \ \

Oxygen *is Carried Around the Body as* Oxyhaemoglobin

Oxygen is carried around the body by **haemoglobin** (Hb), in red blood cells. When oxygen joins to it, it becomes **oxyhaemoglobin**. This is a **reversible reaction** — when oxygen leaves oxyhaemoglobin (dissociates from it), it turns back to haemoglobin.

1) **Haemoglobin** is a large, **globular protein** molecule made up of four polypeptide chains.

2) Each chain has a **haem group** which contains **iron** and gives haemoglobin its **red** colour.

3) Haemoglobin has a **high affinity for oxygen** — each molecule carries **four oxygen molecules**.

$$Hb \quad + \quad 4O_2 \rightleftharpoons \quad HbO_8$$
Haemoglobin + oxygen $\rightleftharpoons$ oxyhaemoglobin

Partial Pressure *Measures* Concentration *of* Gases

The **partial pressure of oxygen (pO_2)** is a measure of **oxygen concentration**.
The **greater** the concentration of dissolved oxygen in cells, the **higher** the partial pressure.
Similarly, the **partial pressure of carbon dioxide (pCO_2)** is a measure of the concentration of carbon dioxide in a cell.

> Where there's a **high pO_2**, oxygen **loads onto** haemoglobin to form oxyhaemoglobin.
> Where there has been a **decrease in pO_2**, oxyhaemoglobin **unloads** its oxygen.

1) Oxygen enters blood capillaries at the **alveoli** in the **lungs**. Alveoli cells have a **high pO_2** so oxygen **loads onto** haemoglobin to form oxyhaemoglobin.

2) When our **cells respire**, they use up oxygen. This **lowers pO_2**, so red blood cells deliver oxyhaemoglobin to respiring tissues, where it unloads its oxygen.

3) The haemoglobin then returns to the lungs to pick up more oxygen.

Dissociation Curves *show how* Affinity for Oxygen *Varies*

Dissociation curves show how the ability of haemoglobin to combine with oxygen varies, depending on partial pressure of oxygen (pO_2).

\ \ \ \ \ | | | | / / / /
100% saturation means every
haemoglobin molecule is
carrying the maximum
of 4 molecules of oxygen.
/ / / | | | \ \ \ \ \

\ \ \ \ \ | | | | / / / /
0% saturation means none
of the haemoglobin molecules
are carrying any oxygen.
/ / / | | | \ \ \ \ \

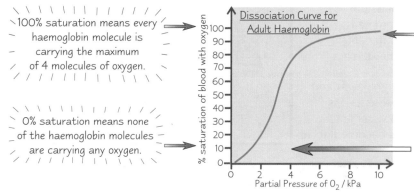

Where **pO_2 is high** (e.g. in the lungs), haemoglobin has a **high affinity** for oxygen (i.e. it will **readily combine** with oxygen), so it has a **high saturation** of oxygen.

Where **pO_2 is low** (e.g. in respiring tissues), haemoglobin has a **low affinity** for oxygen, which means it **releases oxygen** rather than combines with it. That's why it has a **low saturation** of oxygen.

The graph is '**S-shaped**' because when haemoglobin (Hb) combines with the **first O_2 molecule**, it **alters the shape** of the Hb molecule in a way that makes it **easier** for other molecules to join too. But as the haemoglobin starts to become fully saturated, it becomes harder for more oxygen to join. As a result, the curve has a **steep** bit in the middle where it's really easy for oxygen molecules to join, and **shallow** bits at each end where it's harder for oxygen molecules to join.

Carbon Dioxide *is transported to the* Lungs *as Sodium Hydrogencarbonate*

1) Carbon dioxide is a waste product of respiration. CO_2 diffuses into red blood cells where the enzyme **carbonic anhydrase** converts it into **carbonic acid** by combining it with water.

2) Carbonic acid then **dissociates** into **hydrogencarbonate ions** (HCO_3^-) and **hydrogen** ions (H^+). The hydrogen ions cause oxyhaemoglobin to **dissociate** and the oxygen **diffuses** into the **cells** for **respiration**.

3) The HCO_3^- ions are pumped through the red blood cell membrane into the plasma where they combine with sodium to form **sodium hydrogencarbonate**. To make sure that the red blood cells remain electrically neutral, chloride ions pass into the red blood cells — this is known as the **chloride shift**.

4) The sodium hydrogencarbonate ions are carried in the **plasma** to the lungs where they combine with a **H^+ ion** and form **H_2O** and **CO_2**. The CO_2 is released during expiration.

Transport of Respiratory Gases

Carbon Dioxide Levels Affect Oxygen Unloading

To complicate matters, haemoglobin gives up its oxygen **more readily** at **higher partial pressures of carbon dioxide** (pCO_2). It's a cunning way of getting more oxygen to cells during activity. When cells respire more they produce more carbon dioxide, which raises pCO_2, increasing the rate of oxygen unloading.

The reason for this is linked to how CO_2 affects blood pH:

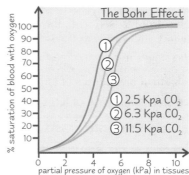

1) Remember from the previous page... CO_2 from respiration enters red blood cells, is converted to carbonic acid and then dissociates to give **hydrogen ions** and **hydrogencarbonate ions**.

2) Hydrogen ions combine with the haemoglobin, displacing oxygen and forming **haemoglobinic acid**.

3) This means that more oxygen is offloaded and the haemoglobin molecules act as **buffers, mopping up** H⁺ and preventing changes in pH.

When carbon dioxide levels increase, the dissociation curve 'shifts' to the right, showing that more oxygen is released from the blood (because the lower the saturation of O_2 in blood, the more O_2 is being released). This is called the Bohr effect.

Haemoglobin is Not the Same in All Animals

The **chemical composition** of haemoglobin and its **oxygen carrying capacity** is different in different species.

Organisms that live in environments where little oxygen is available have dissociation curves to the **left** of human ones.

Organisms that are active and have large available oxygen supplies have curves which are to the **right** of the human one.

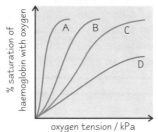

A = animal living in depleted oxygen environment e.g. a lugworm.

B = an animal living at high altitude where the partial pressure of oxygen is lower e.g. a llama living in the Andes.

C = a human dissociation curve.

D = an active animal with a high respiratory rate where there is plenty of available oxygen e.g. a hawk.

Practice Questions

Q1   Describe how haemoglobin carries oxygen.

Q2   Does haemoglobin have a high or low affinity for oxygen? What does 'affinity' mean?

Q3   What is the Bohr effect?

Q4   Name an organism that has an oxygen dissocation curve that is to the right of the human one.

Exam Question

Q1   Look at the graph below. It shows an oxygen dissociation curve for a human.
   a) Draw a line on the graph to represent the dissociation curve of a pigeon. Label the line with a Y.   [1 mark]
   b) Draw a line on the graph to represent the dissociation curve of a yak, an animal that lives at high altitude. Label the line with an X.   [1 mark]

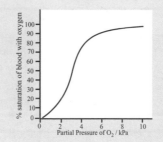

I have an affinity for fruit pastilles...

Dissociation graphs are definitely confusing — it's hard to imagine what's going on. The thing to remember is that when tissues contain lots of oxygen (i.e. the partial pressure is high), haemoglobin readily combines with oxygen, so the blood gets saturated with oxygen. When the partial pressure is low, oxygen moves from the blood into the tissues.

The Human Gut

These pages are for AQA A, and for the OCR optional module, Mammalian Physiology and Behaviour.

Well, food goes in one end, and it comes out of the other. You'd think that's pretty simple but I'm afraid those examiners have set about making it as complicated as they can...

Each Part of the Digestive System has a Specific Function

Animals are **heterotrophic** (see p54), they **ingest** (eat) their food, **digest** it (break it down into smaller molecules), **absorb** it, and **egest** any wastes as faeces.

There are two stages in human digestion:

1) **Mechanical breakdown** of large pieces of food into small pieces.
2) **Chemical breakdown** of large molecules into small molecules.

Each part of the **digestive system** has a **role** in **breaking down** food and **absorbing** its **nutrients**:

The Mouth

In the **mouth**, **mastication** (chewing) of food by teeth **mechanically** breaks up food so there's a **larger surface area** for enzymes to work on. Mixing food with saliva (water, **amylase** and mucus) partly digests food so it can be swallowed easily. The amylase **hydrolyses** starch into maltose.

The Stomach

The stomach is a small sac beneath the diaphragm. It can hold up to 4 litres of food and liquid — it has many folds which expand when there is lots of food. The entrance and exit of the stomach are controlled by **sphincter muscles**. **Gastric juice** produced by the stomach walls is responsible for chemical digestion. **Gastric juice** consists of hydrochloric acid (HCl), **pepsin** (an enzyme) and mucus. Pepsin is an **endopeptidase** — it hydrolyses peptide bonds in the **middle** of polypeptide molecules (proteins), breaking them down into smaller polypeptide chains (which gives a larger surface area for **exopeptidases** to work on later on in digestion). It only works in **acidic conditions**, which are provided by the HCl. **Peristalsis** (rhythmic muscular movements) of the stomach turns food into an acidic fluid, called **chyme**.

The Small Intestine

The small intestine has two main parts — the **duodenum** and the **ileum**. Chyme is moved along the small intestine via peristalsis.

In the **duodenum**, alkaline bile and pancreatic juice **neutralise** the **acidity** of the chyme. These chemicals also break the chyme down into smaller molecules.

In the **ileum** the small, soluble molecules of digested food (glucose, amino acids, fatty acids and glycerol) are **absorbed** through the **microvilli** lining the gut wall. Absorption is through diffusion, facilitated diffusion and active transport.

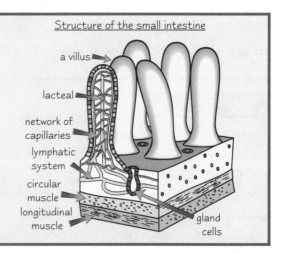

Structure of the small intestine

a villus
lacteal
network of capillaries
lymphatic system
circular muscle
longitudinal muscle
gland cells

The Pancreas

The pancreas is an organ which acts as a gland releasing **pancreatic juice** into the duodenum via the **pancreatic duct**. These chemicals are produced by the epithelial cells which line the ducts in the pancreas. Pancreatic juice contains **amylase**, **trypsin**, the exopeptidase **chymotrypsin**, and **lipase** (the functions of these chemicals are listed on the page opposite). Pancreatic juices also contain **sodium bicarbonate** which neutralises the acidity of the hydrochloric acid from the stomach.

The Colon

The colon (large intestine) absorbs **water**, **salts** and **minerals** — like the other parts of the digestive system the colon has a **folded wall** to provide a **large surface area** for absorption. In the colon bacteria decompose some of the leftover nutrients. Faeces are stored in the rectum and then pass through **sphincter** muscles at the **anus** during **defaecation**.

The Human Gut

Enzymes and *Digestive Juices* have different functions

You need to know all the information in this table for the exam:

	site of production	site of action	function
amylase	salivary gland	mouth	starch to maltose
pepsin	stomach	stomach	protein to polypeptides
amylase	pancreas	duodenum	starch to maltose
bile salts	liver (stored in the gall bladder)	duodenum	emulsify fats and neutralise the acidity of chyme
lipase	pancreas	duodenum	fats to fatty acids and glycerol
maltase	duodenal glands	duodenum	maltose to glucose
trypsin	pancreas	duodenum	protein to polypeptides
chymotrypsin	pancreas	duodenum	protein to polypeptides
exopeptidase	pancreas	duodenum	polypeptides to amino acids

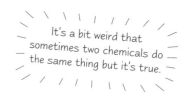

It's a bit weird that sometimes two chemicals do the same thing but it's true.

Not all organisms can *Digest Cellulose*

Cellulose is a major component of vegetation (it forms the cell walls). Because plant material generally has a low nutritional value it is important that herbivorous animals can access the **nutrients** and **energy** in the **cellulose**.

Ruminants are herbivores that have **four stomach sections** and specialised teeth to grind up plant materials — deer, giraffes, antelopes, cattle, sheep and goats are all ruminants.

1) Ruminants use **incisors** (large teeth at the front of the mouth) to crop vegetation.

2) The first section of the stomach, the reticulum, forms balls of **cud** from the swallowed material.

3) The animal **regurgitates** this cud and chews it again to break it down mechanically. Ruminants have a special horny pad inside their mouth which they use with their premolars and molars to mash the cud.

4) When the ruminant swallows again, the food bypasses the reticulum. It gradually makes its way through the next three sections of the stomach — the **rumen**, the **omasum** and the **abomasum**. The rumen is the largest part of the stomach and it contains millions of **mutualistic microorganisms** that **ferment** the vegetation in **anaerobic conditions** — it's these bacteria that break down the cellulose.

Carnivores like **tigers** have specialised **teeth** — their canines can rip flesh and their carnassial teeth can break bones and tendons. Most **carnivores** can't digest cellulose because they don't have the enzyme that's needed to break it down — **cellulase**. Many carnivores show other adaptations for eating meat — for example tigers have **sharp retractile claws** that they use to catch their prey.

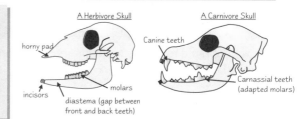

Practice Questions

Q1 What do these words mean: a) heterotrophic   b) ingest   c) mastication?

Q2 What is the difference between mechanical and chemical digestion?

Q3 Explain the structure and function of the stomach.

Q4 Name three chemicals in pancreatic juice.

Q5 Write a paragraph describing how tigers are adapted for eating meat.

Exam Questions

Q1 Compare the digestive adaptations of a ruminant with a carnivore. [4 marks]

Q2 Name 6 digestive enzymes and their functions. [6 marks]

There's urea in cow food...

Honestly, they put it in because it's a source of nitrogen for the bacteria in the rumen. Well anyway, I don't actually think these pages are too bad. There is quite a lot of stuff, but none of it is too hard. And it's all broken down into little sections so you can just work through it by setting yourself little targets like 'I'll learn about the colon in the next ten minutes'. Ooh, what fun...

Control of Digestive Secretions

These pages are for AQA A and OCR.

Once you eat some food, digestion isn't just a passive process — the unconsciously controlled parts of your body are busy making sure it's all working efficiently all the time.

Two Systems are involved in Controlling Digestion

Digestion is controlled by both the **nervous** and **hormonal** systems. They can work **together** to bring about certain effects, for example, nervous impulses can cause hormones to be secreted.

1) The **hormones** that control digestion are **gastrin, secretin** and **cholecystokinin-pancreozymin** — they're released by cells in the **mucosa** of the **stomach** and **small intestine**. These hormones are only released when food is present in the digestive system.

2) **Two types of nerves** help to control the digestive system — **extrinsic and intrinsic nerves.**

Extrinsic and Intrinsic Nerves do different jobs in the Digestive System

Extrinsic Nerves

Extrinsic nerves come to the digestive organs from the subconscious part of the brain and spinal cord. The chemical transmitters **acetylcholine** and **adrenaline** are released across the synapses. Acetylcholine **increases** the muscle action in the stomach and intestines and **increases** the production of **digestive juices** and enzymes. **Adrenaline inhibits** the muscles of the digestive system.

Intrinsic nerves

Intrinsic nerves are triggered to act when the walls of the digestive organs are **stretched**. Intrinsic nerves trigger the release of **hormones** and **digestive juices**.

Remember that the three sites where digestive chemicals are secreted are the mouth, the stomach and the small intestine.

Food can bring about Simple or Conditioned Reflexes

Reflexes are **rapid** responses to the presence of food that are controlled by the **autonomic nervous system** (ANS see p. 80-81).

Conditioned reflexes are ones that are **learnt** through **association**. For example, you learn to associate certain **stimuli** like the **sight** and **smell** of food with eating. These **stimuli** trigger **receptors** in the eyes and nose which send **nerve impulses** to the brain. The **sympathetic** nervous system (see p. 80-81) then sends **impulses** via the **vagus nerve** to the salivary glands, stomach and gall bladder and **saliva, gastric juice** and **bile** are released in **anticipation** of food.

Ivan Pavlov first researched conditioned reflexes by experimenting on dogs (see p 140).

The thought or taste of food stimulates nerve impulses that control the release of gastric juices in the stomach. About 200cm^3 of gastric juices are released. When the food arrives in the stomach over 600cm^3 of gastric juices are released. There are about 40 million cells in the stomach lining which release a total of 2 – 3 dm^3 of gastric juice per day.

Simple reflexes are the ones that can't be controlled and aren't learnt — they happen **subconsciously**. Simple reflexes are controlled by the **parasympathetic** nervous system.

1) The presence of **food** in the **mouth** sends nervous impulses (via the central nervous system) to the **stomach** bringing about the release of gastric juice.

2) The presence of food in the **stomach** stretches the walls and stimulates the **release of HCl.**

3) The **autonomic nervous system** brings about a simple reflex in the stomach which causes **peristalsis** (waves of muscular contraction) in the **small intestine**. This moves the food along making room for **chyme** to be **released** from the stomach.

Control of Digestive Secretions

Three Hormones are released in the Digestive System

1) **Gastrin** release is stimulated by food **distending** (expanding) the stomach wall. Gastrin brings about the secretion of **hydrochloric acid**.

2) **Cholecystokinin pancreozymin** (CCK-PZ) is released by the presence of **fatty food** in the **small intestine** — it stimulates bile to be released from the gall bladder. It also increases the amount of digestive **enzymes** that the **pancreas** releases.

There's more about bile on page 135.

3) **Secretin** is released when **chyme** reaches the duodenum. Secretin stimulates the pancreas to release **hydrogencarbonate ions** which neutralise stomach acid. Secretin also stimulates bile production in the liver.

There are differences between Nervous Control and Hormonal Control

In the digestive system, the **nervous impulses** and the **hormones** work together to stimulate both fast and slow reactions to the food received in the body:

These differences between hormonal and nervous control don't just apply to the digestive system — they're important throughout the whole of biology.

Hormonal Control

The hormonal system works by releasing **chemicals**, which are carried in the blood, from **glands**. Hormones take some **time** to work, but their effects last a relatively **long time**. Hormones act in fairly general ways, e.g a single hormone might bring about several different responses.

Nervous Control

The nervous system uses **chemical transmitters** that create **electrical impulses** and transmit information. These impulses are **quick-acting** and have a **short-lived** effect. One of the main jobs of the nervous system in digestion is to stimulate the release of **hormones**. It also controls **muscular activities** like peristalsis. Nervous impulses are very **specific** — e.g. some act on **single** muscle fibres.

Practice Questions

Q1 Where are gastrin and secretin released?

Q2 What is the difference between the extrinsic and intrinsic nerves that affect the digestive system?

Q3 Describe how a conditioned reflex brings about the release of saliva.

Q4 Which nervous system are simple reflexes controlled by?

Q5 What role do the following hormones play in the digestive system? a) gastin b) CCK-PZ c) secretin

Exam Questions

Q1 Describe the ways that the nervous system acts upon the digestive system. [5 marks]

Q2 State whether each of the following are brought about nervously or hormonally. (Some answers may be a combination of the above).
a) secretion of gastric juices
b) production of saliva
c) stimulation of release of bile [3 marks]

Cholecystokinin pancreozymin — just rolls off the tongue, doesn't it...

Other than that these pages aren't too bad really. Make sure that you've got the hang of the differences between nervous and hormonal control of the digestive system. And remember that there are two kinds of nerves — intrinsic and extrinsic, and two kinds of reflexes — simple and conditioned.

Metamorphosis and Insect Diet

These pages are only for AQA A.

Metamorphosis in insects happens when they take on different forms. You need to learn the example of lepidopterous insects (butterflies and moths) for the exam.

A **Lepidopterous Insect** has more than one **Body Form**

Lepidopterous insects have life cycles that go through **four stages**:

1) An adult lays an **egg**.
2) The egg hatches into a **caterpillar** or **larva**.
3) The caterpillar forms the **chrysalis** or **pupa** (pupation).
4) The caterpillar hatches into a **butterfly** or a **moth** (imago).

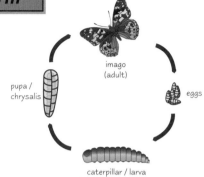

Eggs are laid on **Plants**

1) The eggs are laid by the **female** on a **plant** that will provide **food** for the caterpillar.
2) The parent butterfly recognises the plant according to its **shape**, **colour** and **chemical composition**.
3) Once the eggs are laid they are **abandoned** by the parent.
4) A caterpillar (larva) emerges after about 5 days.

Caterpillars **Eat** a **Lot**

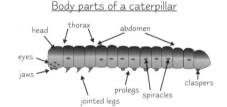

Body parts of a caterpillar

1) A newly emerged caterpillar has a **segmented body** consisting of a **head** with **biting jaws**, a **thorax** with **three** pairs of **jointed legs** with hooks, and an **abdomen** with **five** pairs of stumpy **prolegs** (one pair is modified to give a pair of **claspers**). Caterpillars have simple eyes (ocelli) which only detect changes in light.

2) Caterpillars have **powerful jaws** (mandibles) that have sharp cutting surfaces. The jaws also have two **maxillae** which guide food into the mouth. The maxillae also contain chemical detectors which enable the caterpillar to identify appropriate food.

3) Most caterpillars live for up to a month and they **eat** constantly. As caterpillars grow they **shed** their skins (**ecdysis**) — three or four moultings take place before a chrysalis or pupa is formed. Caterpillars use **silk** thread made in their **silk glands** to attach themselves onto a plant.

The **Pupa Stage** usually lasts for about **Three Weeks**

Inside the chrysalis all the tissues are broken down into a **liquid**. The only parts that aren't dissolved are the **central nervous** system and some special groups of cells called **imaginal discs**. The **adult** (butterfly / moth) **organs** are formed from the imaginal discs. Some species spend **winter** as pupae to avoid harsh weather conditions.

Eventually, the skin of the chrysalis splits and a butterfly / moth **emerges**. Butterflies / moths have **compound eyes** which are more complex than the ocelli that caterpillars have. They also have a long straw-like tube called a **proboscis** that they use for feeding on **nectar**. Butterflies and moths always have **six legs**. Once they have emerged the adults **don't grow**.

Metamorphosis and Insect Diet

Different stages have different Nutritional Requirements

The **different stages** in the **life cycle** of a lepidopterous insect have different **nutritional requirements**. For example, caterpillars need lots of protein because they are growing, whilst butterflies need lots of carbohydrate to provide energy for flying. The insects also have different **adaptations** to allow them to feed during different life stages —

	egg
source of food	yolk contains protein, phospholipids and fats
protein / energy requirements	rapid development requires high levels of energy and protein
mouthparts / digestive structures	mandibles grow as it develops
gut enzymes	peptidases, lipase
	caterpillar
source of food	leaves
protein / energy requirements	protein needed for growth, carbohydrate needed for energy
mouthparts / digestive structures	mandibles and maxillae
gut enzymes	peptidases, cellulase, amylase
	pupa
source of food	none
protein / energy requirements	uses stored energy for metamorphosis
mouthparts / digestive structures	none
gut enzymes	none
	butterfly / moth
source of food	nectar
protein / energy requirements	high energy requirement for flying
mouthparts / digestive structures	proboscis
gut enzymes	sucrase

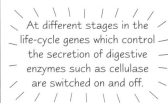

At different stages in the life-cycle genes which control the secretion of digestive enzymes such as cellulase are switched on and off.

Practice Questions

Q1 What are the four stages in the life of a lepidopterous insect?

Q2 Describe how caterpillars eat.

Q3 Describe the physical structure of a caterpillar.

Q4 What happens during pupation?

Q5 How does a lepidopterous insect feed when it is a butterfly?

Exam Questions

Q1 Name the different digestive enzymes needed at each of the four life stages of a swallowtail butterfly. [6 marks]

Q2 What are: a) mandibles   b) imaginal discs   c) maxillae? [3 marks]

I'm glad that my insides never liquefy

Well, as improbable as all this sounds, it is true. If you think back to section two, having different life-stages feeding on different foods is advantageous because it reduces intra-specific competition. Remember that this stuff applies to all lepidopterous insects, so don't be put off if they ask you about one particular species of butterfly — just apply what you've learnt here and you'll be fine.

Genetic Engineering

Skip these pages if you're doing AQA.
On these pages, you will learn how to splice rabbit DNA with that of an onion to make a new and exciting pet.

Genetic Engineering has Loads of Important Uses

Genetic engineering (or **genetic manipulation**) is when DNA is removed from one organism and joined to the DNA of another. DNA treated in this way is called **recombinant DNA**.

There are already many uses of genetic engineering, and there are potentially loads more. **Herbicide-resistant** crop plants are now created in this way. **Bacteria** are engineered to produce useful **proteins** and such as **human insulin**. It is even possible to over-ride defective alleles in **embryos** by genetic engineering. (This only works for **recessive** conditions — the normal functioning allele is added to cells, with the faulty allele still in place.)

Useful Genes can be Isolated with Restriction Endonuclease Enzymes

The first stage in genetic engineering is removing the useful DNA from the **donor** cell:

1)   Any proteins attached to the DNA are digested by **peptidase** enzymes.

2)   The useful gene is cut out of the DNA using **restriction endonuclease** enzymes. These enzymes are normally found in bacteria. Their usual job is to destroy the DNA or RNA of invading viruses by chopping it into small pieces.

3)   Restriction endonucleases leave the fragment of DNA with a small tail of unpaired nucleotide bases at each end. These tails are called **sticky ends**.

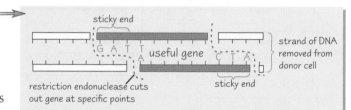

Vectors Carry the Useful Gene

1)   The useful gene is now joined to a **vector** (the DNA that carries the gene into the bacterium). The main vectors used are **plasmids** — these are small, **circular molecules** of DNA found in **bacteria**. They're useful because they **replicate** easily and can be put into other bacteria without harming them. DNA from a **bacteriophage virus** (see page 177) can also be used as a vector.

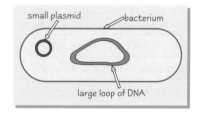

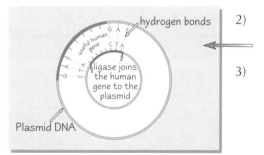

2)   The same **restriction endonuclease** that was used to remove the useful gene from the **donor** is used to cut open the plasmid. The **sticky ends** of the open plasmid are **complementary** to those of the useful gene.

3)   The useful gene is **joined** to the **plasmid vector DNA** using those clever **sticky ends**. This is known as **splicing**. The complementary bases pair by **hydrogen bonding** and the DNA is then joined by the enzyme **DNA ligase**. This process of joining is called **ligation** and the new combination of bases in the DNA (plasmid DNA + useful gene) is called **recombinant DNA**.

Recombinant DNA is put into the Host Bacterium

The plasmid containing the useful gene now has to be put into a **bacterium** so it can be **replicated** and produce its useful **protein**. A harmless strain of the bacterium *Escherichia coli* is often used.

1)   The bacteria and the recombinant plasmids are added to **ice cold calcium chloride solution**.

2)   The mixture is heated rapidly to 42 °C for 50 seconds. This **heat shock** allows the plasmids to get through the bacterial cell wall and membrane and into the cytoplasm.

3)   The mixture is quickly **cooled** on ice.

4)   Bacteria that have successfully taken up the plasmids are called **transformed bacteria**.

Genetic Engineering

Transformed Bacteria can be Located Using Marker Genes

Only a few of the bacteria will have taken up the vectors, so the next step is to find out which ones have.
Antibiotic marker genes can show which bacteria have been **transformed**.

1) When the useful gene is joined to the plasmid, a gene for **antibiotic resistance** is inserted too.

2) After transformation, the bacteria are **cultured** on an agar plate called the **master plate**.
This increases the number of bacteria.

3) When the bacteria have multiplied, a **sterile velvet pad** is pressed lightly onto the surface of
the master plate. This picks up a few from each colony of bacteria.

4) The pad is pressed onto an agar plate (the **replica plate**) containing an **antibiotic**.
Some bacteria from each colony are transferred to the replica plate.

5) Only **transformed** bacteria can survive, as they are **resistant** to the antibiotic.
They also contain the useful gene.

Bacteria can be Cultured on an Industrial Scale

1) The useful gene usually needs a '**kick-start**' before it will work, so a **promoter gene** is often
included when the donor and antibiotic resistance genes are **spliced** into the plasmid.
The promoter gene **switches on** the useful gene so it will begin producing the **protein**.

2) The bacteria are cultured in optimum conditions on an **industrial scale** (see next page).
As they reproduce **asexually** by binary fission, they make **identical copies** of the useful gene.

3) The gene causes the bacteria to produce the useful **protein**, which the bacteria can't use,
so it builds up and can be **extracted** and **purified**.

Reverse Transcriptase makes DNA from Messenger RNA

Finding a **single** copy of a useful gene is much more difficult than finding **lots** of mRNA molecules.
This is why the enzyme **reverse transcriptase** is very useful — it makes DNA from mRNA (the opposite
of normal **transcription** — which you covered at AS), so it can be used to make certain useful genes.

1) Genetic engineers can **isolate** the right mRNA, then use this enzyme to make the **complementary DNA** (cDNA).

2) The DNA made by the reverse transcriptase is **single-stranded**, so the enzyme **DNA polymerase** is used to
make the full double-stranded version.

Practice Questions

Q1 What is a sticky end?

Q2 Give two examples of vectors used in genetic engineering.

Q3 How does a replica plate differ from a master plate?

Exam Questions

Q1 *Aequorea victoria* is a species of jellyfish that fluoresces (glows). The gene that codes for the protein responsible
for this fluorescence has been discovered. It is possible to insert this gene into the bacterium *Escherichia coli*.

　a) Briefly describe the stages involved in the transformation of the bacterium. [9 marks]

　b) Suggest a use for this gene and explain your answer. [2 marks]

Q2 Explain why reverse transcriptase is so useful in genetic engineering. [6 marks]

OK, so I lied about the onion bunny (or bunnion)...

*It was a cheap trick to get your attention. But seriously, genetic engineering is really important — it means that people with
genetic diseases now have a much better chance of successful treatment. There are some dodgy ethical issues associated
with it, like designer babies and stray genes 'escaping' — but genetic engineering is still a pretty exciting breakthrough.*

Commercial Use of Genetic Engineering

Skip these pages if you're doing AQA. *Genetic engineering benefits all kinds of people, from diabetics to vegetarians. But some people worry that it could lead to a genetic underclass, new diseases, superweeds, the end of the world...*

Microorganisms can Produce Proteins on an Industrial Scale

Microorganisms include things like **bacteria**, **yeasts** (fungi) and **protoctists**. They're great for mass-production of genetically-engineered products. Here are seven reasons why:

1) Microorganisms have a very **rapid growth rate**.
   In optimum conditions they can double their biomass in 20 minutes.

2) There's a huge variety of **substrates** they can use as **food**.
   Some even feed on oil and plastic — so cheap waste can be used up too.

3) Their **genome** (the total collection of genes in an organism) is easily **manipulated**.

4) They can be easily **screened** (tested) to see if they have desirable characteristics (e.g. fast growth rate).

5) They can be grown in **fermenters** in any **climate**, unlike multicellular organisms which often need conditions that are difficult to provide.

6) **Fermenters** take up relatively little **space**.

7) The useful products can be **isolated** and **purified** quite easily.

Insulin and Human Growth Hormone are Produced for Use in Medicine

This flow chart shows how insulin is produced on an industrial scale.

Human insulin <u>extracted</u> from Islets of Langerhans in pancreas. → Gene <u>synthesised</u> from <u>nucleotides</u> using known sequence of amino acids in insulin molecule.

<u>Recombinant</u> plasmid introduced into <u>bacterium</u>. ← Gene <u>spliced</u> into bacterial <u>plasmid</u>.

Transformed bacteria <u>cultured</u> in fermentation tank (<u>batch fermenter</u>). → More processing to <u>extract</u>, <u>purify</u> and <u>package</u> hormone.

Batch fermenter

acid/alkali in to adjust pH — CO₂ out

MOTOR

sampling tube — sterile air in
drainage tube — nutrients in

temperature monitor
pH monitor

leak-proof seal to prevent contamination and/or release of bacteria

hot water out

stirrer to mix nutrient medium

water jacket to remove heat generated by metabolism of bacteria

cold water in

Human growth hormone is also produced in this way. Insulin used to be extracted from the pancreas of dead pigs and human growth hormone from the pituitary of human corpses — not nice.

Genetic engineering, batch fermentation and downstream processing ensure that **pure**, **human** proteins are produced in **large** quantities, relatively **inexpensively**.

Chymosin is a Genetically Engineered Alternative to Calf Rennin

Rennin is an enzyme used to **coagulate** the protein in milk to make **cheese**. It's found in the **abomasum** (fourth stomach) of calves.

The gene for rennin production is inserted into **yeast cells**. Cultures of yeast in fermenters can then produce large amounts of the enzyme, which is now called **chymosin** (but it's the same as rennin).

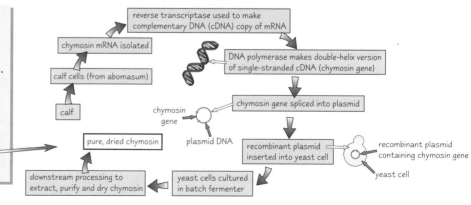

calf → calf cells (from abomasum) → chymosin mRNA isolated → reverse transcriptase used to make complementary DNA (cDNA) copy of mRNA → DNA polymerase makes double-helix version of single-stranded cDNA (chymosin gene) → chymosin gene spliced into plasmid

chymosin gene + plasmid DNA → recombinant plasmid inserted into yeast cell ← recombinant plasmid containing chymosin gene ← yeast cell

yeast cells cultured in batch fermenter → downstream processing to extract, purify and dry chymosin → pure, dried chymosin

Commercial Use of Genetic Engineering

Crop Plants can be Improved by Genetic Engineering

Genes from unrelated species can be inserted into **crop plants** to give them desirable characteristics.

<u>Transfer of Resistance to Glyphosate Herbicide to Maize</u>

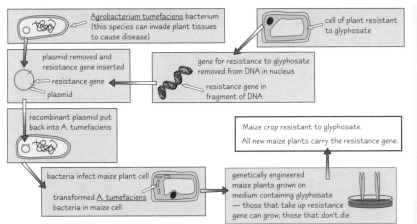

Agrobacterium tumefaciens bacterium (this species can invade plant tissues to cause disease)

plasmid removed and resistance gene inserted

resistance gene

plasmid

recombinant plasmid put back into A. tumefaciens

bacteria infect maize plant cell

transformed *A. tumefaciens* bacteria in maize cell

cell of plant resistant to glyphosate

gene for resistance to glyphosate removed from DNA in nucleus

resistance gene in fragment of DNA

Maize crop resistant to glyphosate. All new maize plants carry the resistance gene.

genetically engineered maize plants grown on medium containing glyphosate — those that take up resistance gene can grow, those that don't die

- A gene giving resistance to the herbicide **glyphosate** can be inserted into crop plants such as **maize**. When the crop is sprayed with herbicide, the weeds die but the maize is unaffected.

- Scientists have inserted genes into broccoli to increase the amount of **glucosinates** — anti-cancer chemicals.

- **Cotton plants** have been genetically modified with a gene from the bacterium *Bacillus thuringiensis* that produces an **insecticide**. Any insects that try to eat the plants are killed by the insecticide.

Many People are Worried about Genetic Engineering

There are lots of social and economic **benefits** of genetic engineering, but there are **hazards** too. Some people think all kinds of **genetic manipulation** are **unethical**.

Benefits	Hazards
Medically important proteins no longer need to be extracted from organs, but can now be produced by transformed bacteria (e.g. erythropoietin is a protein essential for regulation of red blood cell production).	No control of recombinant DNA once released into the environment (e.g. genes 'escaping' in pollen grains).
Hormones etc. can be tailor-made to exact requirements (e.g. diabetics now use human, not pig, insulin).	Viruses and bacteria containing recombinant DNA could mutate and become pathogenic, causing new diseases.
Crops can be improved and made herbicide-resistant (e.g. glyphosate-resistant maize).	Genetically modified organisms sometimes exhibit unpredicted characteristics.
Can be used to produce vegetarian products. E.g. the enzyme used in cheese-making can be mass-produced by genetically modified yeast, rather than using animal enzymes.	Weeds and pest insects could gain rogue genes from engineered crops and become resistant to chemicals designed to kill them.
Genetic diseases can be treated by adding the functioning allele. (Cystic fibrosis is already treated this way.)	Risk of creating 'GM monsters' (new animals produced by combining genes from different species).

Bioethics concerns the **life sciences** and their impact on **society**. Genetic engineering raises some important **bioethical** issues:

- Some people think it's wrong to genetically engineer animals purely for **human benefit**, especially if the animal **suffers**.
- Those who can afford it might decide which characteristics they wish their children to have (**designer babies**), creating a 'genetic underclass'.
- The **evolutionary consequences** of genetic engineering are unknown.
- There are **religious concerns** about 'playing God'.
- **Xenotransplantation** (transplanting animal organs engineered to resist rejection) can save lives, but there are **religious** implications. For example, many Muslims would find it unthinkable to receive a transplant from a pig.

Practice Questions

Q1   List four reasons why microorganisms are useful in mass production of genetically engineered products.

Q2   Where is rennin normally found? What is it used for commercially?

Q3   Give two benefits and two hazards of genetic engineering.

Exam Questions

Q1   Describe how a named GM human hormone is produced.   [5 marks]

Q2   Describe three benefits and three drawbacks of creating genetically modified herbicide-resistant crops.   [6 marks]

<u>*Designer babies — mine's Gucci, daahling...*</u>

Here are some websites you might find interesting: <u>www.bioethics-today.org</u> *(Bioethics Today – a bioethics resource for the UK) and* <u>www.newint.org/issue215/monsters.htm</u> *('Monsters of the Brave New World' – an article on GM monsters).*

The PCR and Genetic Fingerprinting

Skip these pages if you're doing AQA.

These are good pages for those of you considering a career in crime. It's a bit more tricky these days, because your own body is working against you — the tiniest hair or flake of skin it drops is enough to place you at the scene of the crime.

The **Polymerase Chain Reaction** (PCR) Creates Millions of **Copies** of DNA

Some samples of DNA are too small to analyse. The **Polymerase Chain Reaction** makes millions of copies of the smallest sample of DNA in a few hours. This **amplifies** DNA, so analysis can be done on it. PCR has **several stages**:

1) The DNA sample is **heated** to **95°C**. This breaks the hydrogen bonds between the bases on each strand. But it **doesn't** break the bonds between the ribose of one nucleotide and the phosphate on the next — so the DNA molecule is broken into **separate strands** but doesn't completely fall apart.

2) **Primers** (short pieces of DNA) are attached to tiny bits of both strands of the DNA — these will tell the **enzyme** where to **start copying** later in the process. They also stop the two DNA strands from joining together again.

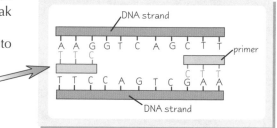

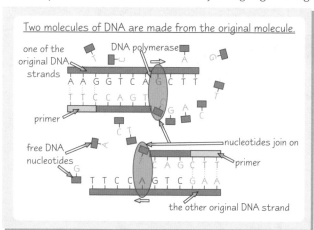

Two molecules of DNA are made from the original molecule.

3) The DNA and primer mixture is **cooled** to **40°C** so that the primers can **fully bind on** to the DNA.

4) Free **DNA nucleotides** and the enzyme **DNA polymerase** are added to the reaction mixture. The mixture is heated to **70°C**. Each of the original DNA strands is used as a **template**. Free DNA nucleotides pair with their complementary bases on the template strands. The DNA polymerase attaches the new nucleotides together into a strand, starting at the primers.

5) The cycle starts again, using **both** molecules of DNA. Each cycle **doubles** the amount of DNA.

You can **Identify People** from their **DNA** by **Cutting** it into **Fragments**

It's possible to **identify a person** from a sample of their DNA, if the sample is big enough. This is done by using **enzymes** to cut the DNA up into **fragments**, then looking at the **pattern** of fragments, which is **different** for everyone. This is called a person's **genetic fingerprint**.

1) To **cut up** the DNA into DNA fragments you add specific **restriction endonuclease** enzymes to the DNA sample — each one **cuts** the DNA every time a **specific base sequence** occurs. The **location** of these base sequences on the DNA **varies** between everyone, so the number and length of DNA fragments will be different for everyone.

2) Next you use the process of **electrophoresis** to separate out the DNA fragments by size:

How Electrophoresis Works:

1) The DNA fragments are put into **wells** in a slab of **gel**. The gel is covered in a **buffer solution** that **conducts electricity**.

2) An **electrical current** is passed through the gel. DNA fragments are **negatively charged**, so they move towards the positive electrode. **Small** fragments move **faster** than large ones, so they **travel furthest** through the gel.

3) By the time the current is switched off, all the fragments of DNA are **well separated**.

Electrophoresis

DNA moves towards the anode (as the phosphate groups give it an overall negative charge). The smallest fragments move furthest.

−ve cathode
wells
DNA fragment (invisible)
gel, with buffer solution on top
+ve anode

In electrophoresis the DNA fragments **aren't visible** to the eye — you have to do something else to them before you can **see their pattern**. Coincidentally, that's what the next page is all about...

The PCR and Genetic Fingerprinting

Gene Probes *Make the* Invisible *'Genetic Fingerprint'* Visible

DNA fragments separated by electrophoresis are invisible.
A radioactive DNA probe (also called **gene probe**) is used to show them up:

1) A **nylon membrane** is placed over the electrophoresis gel, and the DNA fragments **bind** to it.

2) The DNA fragments on the nylon membrane are **heated** to separate them into **single strands**.

3) **Radioactive gene probes** are then put onto the nylon membrane. (It's the **phosphorus** in the gene probes' sugar-phosphate backbones that's radioactive.) The probes are warmed and **incubated** for a while so that they'll attach to any bits of **complementary DNA** in the DNA fragments.

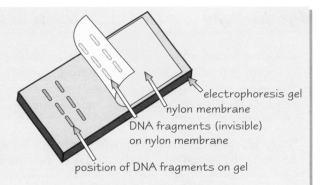

electrophoresis gel
nylon membrane
DNA fragments (invisible) on nylon membrane
position of DNA fragments on gel

4) The nylon membrane is then put on top of unexposed **photographic film**. The film goes **dark** where the radioactive gene probes are **present**, which **reveals the position** of the **DNA fragments**.

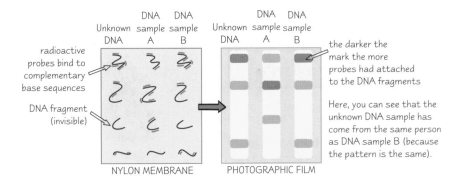

radioactive probes bind to complementary base sequences

DNA fragment (invisible)

NYLON MEMBRANE

PHOTOGRAPHIC FILM

the darker the mark the more probes had attached to the DNA fragments

Here, you can see that the unknown DNA sample has come from the same person as DNA sample B (because the pattern is the same).

Genetic fingerprinting is incredibly useful. **Forensic investigations** use it to confirm the identity of suspects from blood, hair, skin, sweat or semen samples left at a crime scene, or to establish the identity of victims. **Medical investigations** use the same technique for **tissue typing**, **paternity tests** and **infection diagnosis**.

Practice Questions

Q1 Give the full name of the technique used to increase the amount of DNA from a very small sample.

Q2 Name the enzyme used in PCR.

Q3 What types of enzyme are added to DNA before gel electrophoresis?

Q4 Which part of a gene probe is radioactive?

Q5 State three uses of genetic fingerprinting in medical investigations.

Exam Question

Q1 Police have found incriminating DNA samples at the scene of a murder.
They have a suspect in mind, and want to ascertain if the suspect is guilty.

a) Name the technique that the police could use to confirm the guilt of the suspect. [1 mark]

b) Explain how this technique would be carried out. [5 marks]

Hands up, punk — I've got a gene probe and I'm not afraid to use it...

These days, anyone who gets arrested by the police, for whatever reason, has a sample of their DNA taken and kept on file. It won't be long until most people's are recorded. Wherever you go, you're leaving little traces of your DNA behind. So one day, we'll just hoover up all the DNA at the scene of a crime and know immediately who was there. Scary thought.

Productivity

*This section covers the **Environmental Biology** option (for **OCR**) and the **Applied Ecology** option (for **AQA B**).*
This double page is for AQA B, and page 109 is also relevant to OCR.

Remember food chains? Bunny eats grass, fox eats bunny, blah blah blah. Plants are the producers in food chains. Their productivity is a measure of how much food they produce.

Productivity is a Measure of how much *Biomass* Producers make

Gross productivity (also called **gross primary production**) is the **total** amount of **energy** in the **biomass** made by the producers in a community. Some of this energy goes towards plant growth, but most gets broken down again in **respiration** to keep the plants alive. So we also have...

> Gross productivity is often described as the amount of energy in the plant biomass **in a given area and in a given time**. It's given units like kJ m⁻² year⁻¹.

Net productivity (**net primary production**). This is the amount of biomass that is left **after** respiration, which can be passed on to the next stage of the food chain. The equation for net productivity is:

> **Net productivity = gross productivity – respiratory loss**

Leaf Area Index gives an Idea of *Plant Efficiency*

The **leaf area index** measures the **surface area** of the **leaves**, compared with the area of **ground** the plant covers:

> **Leaf area index = leaf area exposed to light ÷ area of soil surface**

In measuring the area of a leaf, we only count **one side**. Even so, the leaf area index is usually **more than 1**. This is because leaves **overlap**.

Plants are most efficient when the leaf area index is about **4**. This gives enough overlap to absorb nearly **all** the light, but without **wasting** energy on leaves that are not needed.

tree

area of ground covered

Net Productivity is Less than *Gross Productivity*

Net productivity is the food that the farmer can actually **harvest**, so he or she wants this to be as high as possible.

Productivity depends on the **efficiency** of the plant — how good it is at changing **sunlight** energy into **chemical** energy.

For most crops, gross productivity is only about **1%** of the Sun's energy. So **99%** of the Sun's energy is **wasted**. Net productivity is even **less** — probably only about **0.1%**. Energy gets lost in all of the following ways:

1) Some of the Sun's rays may **miss the plant** and hit bare ground, or they may pass **straight through** the leaf.
2) Some light is **reflected** (green light especially, that's why plants look green), but also **UV rays** and **infra-red**.
3) Some light may be absorbed by the **wrong bits** of the plant — e.g. by the cell walls instead of the chloroplasts.
4) The reactions of **photosynthesis** are not completely efficient, and energy is wasted as heat during photosynthesis.

All of these things reduce **gross productivity**. And **net** productivity is even lower because some food is broken down again by **respiration**, so the energy gets released as **heat**.

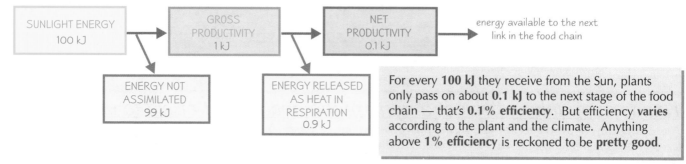

| SUNLIGHT ENERGY 100 kJ | → | GROSS PRODUCTIVITY 1 kJ | → | NET PRODUCTIVITY 0.1 kJ | → energy available to the next link in the food chain |

ENERGY NOT ASSIMILATED 99 kJ

ENERGY RELEASED AS HEAT IN RESPIRATION 0.9 kJ

For every **100 kJ** they receive from the Sun, plants only pass on about **0.1 kJ** to the next stage of the food chain — that's **0.1% efficiency**. But efficiency **varies** according to the plant and the climate. Anything above **1% efficiency** is reckoned to be **pretty good**.

Productivity

Fertilisers Increase Productivity, but Too Much is Money Down the Drain

Fertilisers make plants grow faster, because they give the plants essential minerals and nutrients. The most important mineral ions are **nitrate**, **phosphate** and **potassium**, but there are lots of **trace elements** needed in small amounts.

There are **two main types** of fertiliser. Each has its own advantages:

1) **Natural** fertilisers are **organic** matter (that's "muck" to you and me). They include manure and sewage sludge.
2) **Artificial** fertilisers are **inorganic**. They contain pure chemicals (e.g. ammonium nitrate) as powders or pellets.

Natural fertilisers supply a **wide range** of nutrients and release them slowly for a **long-lasting effect**. They're **less harmful** to the environment and are suitable for "**organic**" farming. They're also **cheaper** than artificial fertilisers and can improve the **soil structure**. However, they're expensive to **transport and apply**, and might not have the **ideal balance** of nutrients. They also contain **micro-organisms** that use up some of the nutrients.

Artificial fertilisers are **fast-acting** and easy to transport and apply. They can be used to target **particular** mineral ion needs, and the amount of each mineral supplied can be **accurately controlled**. However, they're more **expensive**, can upset the **balance** of the soil, and are more easily washed out of the soil leading to **eutrophication**.

> Farmers have to add the **right amount** of fertiliser to their crops, otherwise they're wasting money. Not enough, and the **yield** is reduced. But too much and the extra is just **wasted**. A **law of diminishing returns** comes into play — each extra amount of fertiliser makes less and less difference.

Crop Rotation Gets the Most from the Soil

Growing the **same crop** year after year in the same field can cause problems:

1) **Pests** and **diseases** start to build up. Their **eggs** or **spores** stay in the soil, ready to infect next year's crop.
2) If the crop needs a lot of one particular mineral ion, this will get **depleted**.

Farmers can get round this in one of two ways. They can either use **pesticides** and **fertilisers**. Or — the traditional method — they can use **crop rotation**. By **changing** the crop each year, crop rotation stops the pests and diseases of one crop building up in the soil. Also, it helps to stop mineral nutrients running out (because each crop has slightly **different** needs). Most crop rotations include a **legume plant** (like peas or beans), because these plants contain **nitrogen-fixing bacteria** in special **root nodules** in their root system, which means they increase the **nitrogen content** of the soil.

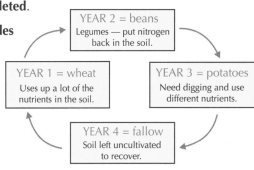

YEAR 2 = beans
Legumes — put nitrogen back in the soil.

YEAR 1 = wheat
Uses up a lot of the nutrients in the soil.

YEAR 3 = potatoes
Need digging and use different nutrients.

YEAR 4 = fallow
Soil left uncultivated to recover.

Nitrogen-fixing bacteria (e.g. **Rhizobium**) use the enzyme **nitrogenase** to convert **nitrogen gas** and **carbohydrates** into **amino acids**. They get their sugars from the bean plant. In exchange, they give the plant **amino acids** for growth.

Practice Questions

Q1 What is the difference between gross productivity and net productivity?
Q2 Define the term leaf area index.
Q3 Give four reasons why only about 1% of the Sun's energy ends up as chemical energy in plants.
Q4 What is meant by the law of diminishing returns in fertiliser use?

Exam Questions

Q1 Gross productivity in a hot desert is about 750 kJ m^{-2} y^{-1}. In a tropical rainforest, it is about 80 000 kJ m^{-2} y^{-1}. Suggest three reasons for this difference. [3 marks]

Q2 Explain why net productivity is less if: a) the leaf area index is less than about 3. [2 marks]
b) the leaf area index is more than about 6. [2 marks]

My brother's biology teacher thought his productivity was pretty gross...

She said it was disgusting how little work he did. Don't you be the same — instead get this page learned, as efficiently as possible. You'll probably need to stop reflecting rays and absorbing light with your cell walls. Those photosynthesis reactions are wasting heat, and as for your leaf area index... no, it's no good — I'm going to have to cover you with manure, I'm afraid.

Impact of Agriculture on the Environment

Agriculture produces the food we eat, but can also sometimes pollute local rivers with organic waste and pesticides.
Page 110 is for OCR only, and page 111 is for both OCR and AQA B.

BOD Measures Organic Pollution

Farms often pollute freshwater habitats with organic matter like **slurry** (animal waste), or **silage** (stored animal feed).

The **organic pollution** in a freshwater habitat can be measured by finding the **Biological Oxygen Demand** (**BOD**) of a sample of the water. The BOD is the amount of **oxygen** removed from a sample of water in a given time. Lots of **organic matter** supports a lot of **bacteria**, and they use up a lot of oxygen breaking down the organic matter.

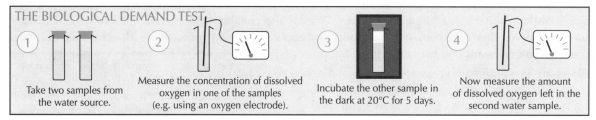

THE BIOLOGICAL DEMAND TEST

1) Take two samples from the water source.

2) Measure the concentration of dissolved oxygen in one of the samples (e.g. using an oxygen electrode).

3) Incubate the other sample in the dark at 20°C for 5 days.

4) Now measure the amount of dissolved oxygen left in the second water sample.

The **BOD** is the **difference** between the amount of oxygen in the first sample and the amount in the second sample. A **high BOD** means that the water contains a **lot** of organic pollution.

Pesticides Raise Productivity

Pesticides are chemicals that kill harmful organisms (see p.32). There are **three** main types:

1) **Herbicides** (weedkillers) kill **plants** that would otherwise compete with the crop for light, water and nutrients.
2) **Insecticides** kill **insects** that would eat the crop. Insect pests can reduce the yield in **two** main ways. They might eat the part of the plant that the farmer wants to harvest. Or, by eating leaves, they reduce photosynthesis, which reduces food production.
3) **Fungicides** kill **fungal** pests. Many fungi are **parasites**, living and feeding on plants. They cause diseases like **potato blight**.

Pesticides Can Cause Problems Too...

Pesticides can increase the **yield** of crops, but there are also **problems** with using them:

1) Pesticides get sprayed onto our **food**. Often they're not just toxic to pests, but can harm people as well. If the dose is big enough they could **poison** us.
2) They damage the **environment**. Pesticides often kill **harmless** or even **beneficial** species as well as pests.
3) Pests can become **resistant** to pesticides over time. Repeated use of the same pesticide means that **natural selection** acts to increase resistance to it in the population. The pesticide becomes less and less effective. And, if the pest's natural **predators** have been killed by the pesticide, this may cause a worse outbreak than before.

DDT (**dichloro-diphenyl-trichloroethane**) is a very effective **insecticide**. It was used a lot in the UK until the 1970s, and is still used in some parts of the world. It's now **banned** in the UK because of the dangers associated with it.

- DDT is **persistent** — it's only broken down very slowly in the environment. And it shows **bioaccumulation** — when organisms absorb DDT they **store** it in their bodies, so the concentration **increases** during their lifetime.
- Because DDT bioaccumulates, it becomes more **concentrated** in the food chain. This means that the **predators** of insect pests are likely to be more affected than the pests themselves.
- Organisms at the **top** of the food chain can be **poisoned**. The use of DDT in Britain before the 1970s was blamed for reducing **peregrine falcon** numbers.
- **Humans** can also be at risk. If you ate crops that had been sprayed with DDT, or meat from an affected food chain, you'd absorb the DDT and begin to accumulate it. DDT has been linked to **liver damage**, **reduced fertility** and **cancer**.

Impact of Agriculture on the Environment

Pollution Reduces Species Diversity

When water is badly polluted, most animals die because of the **low oxygen concentration**. Only a few **pollution-tolerant** species can survive the conditions, and these will thrive due to the lack of competition. This means polluted water has **fewer** species present — a lower **species diversity**.

Species diversity can be expressed as a **diversity index**, which uses a special formula to compare the number of different **species** in an ecosystem with the total number of **individuals** . There are **different versions** of this index — see page 26 for one common example.

Indicator Species can be Used to Check Water Quality

Some organisms can only survive in **clean water**, so if you found specimens of them in a sample of water it'd be **unlikely** that there was a problem with pollution there. Other species are adapted to cope with the extreme conditions found in **polluted water**, so if you saw a lot of them in a sample you'd know there was a problem. Organisms like this, that tend to be found in certain conditions, are known as **indicator species**. Examples include:

1) The **rat-tailed maggot**. This is found in water polluted with **organic matter**. It is adapted to survive in **low oxygen concentrations**.

2) *E. coli*. This is a bacterium that is usually found in the human **large intestine**. Its presence in fresh water suggests pollution by **human sewage**.

3) **Freshwater shrimps, stonefly larvae, mayfly larvae** and **caddis fly larvae** are all indicators of **clean** water.

The presence of mermaids usually indicates clean water

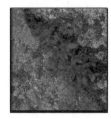

Indicator species can also be used to monitor **air quality**. **Lichens** are an important example. The better the quality of the air, the greater the **diversity** of lichens that can survive. The diagram shows different species of lichen growing on a rock in an unpolluted area.

Practice Questions

Q1 Give three ways in which the use of pesticides can improve agricultural production.
Q2 Describe three problems associated with the use of pesticides.
Q3 Name three health problems that can be caused in humans by DDT.
Q4 Why does polluted water usually have a lower diversity index?
Q5 Name three indicator species usually only seen in clean water.

Exam Questions

Q1 a) Define the term BOD.   [1 mark]
    b) Outline how you would find the BOD of a sample of river water.   [3 marks]
    c) There are some situations in which the BOD test will give a misleading measurement.
       Suggest and explain what these situations are.   [6 marks]

Those rat-tailed maggots sound lovable...

... rats, maggots and organic waste all combined into one cuddly little friend. It shouldn't take too long to learn this page. It's basically just what BOD means, how pesticides raise productivity but cause problems, pollution and indicator species. It's not too hard, and it's actually kind of interesting. I just love learning about organic pollution. Mmmm, slurry.

Harvesting Ecosystems

Fishing isn't like farming. Farms are artificial, designed just for food production. But fishing is taking food from a natural ecosystem. If the ecosystem isn't carefully conserved, the food could run out.
Page 112 is for both OCR and AQA B, and page 113 is for AQA B only.

Fisheries *Aim for* Maximum Sustainable Yield

Fish are able to reproduce very **rapidly** if the conditions are right. For example, a single pair of **cod** can produce **four million** fertilised eggs every year. But the young fish need time to develop, and their **mortality rate** is high even from natural causes. If fishermen take too many fish, it could cause the population to **collapse**.

> **Overfishing** is where the fish can't reproduce fast enough to replace what the fishermen remove. For one or two years, the fishermen may get an excellent yield from the sea, but this yield will then drop dramatically — it's **not sustainable**.
>
> **Underfishing** is where fishermen take fewer fish than they could. This is OK for the fish, of course — fewer of them die — but it means that a potential food resource is going to **waste**.

The **maximum sustainable yield** is the largest amount of fish that can be caught without causing the population to fall. It's obtained when the fish population is conserved, and the **population growth curve** is kept at its **steepest** part. In this way, the population is made to keep producing food efficiently. With **underfishing**, there are lots of large, older fish which aren't growing much, and which eat food that could be used for growth by younger fish. With **overfishing**, few individuals reach maturity, so the **breeding stock** decreases and fewer young fish are produced.

A population growth curve looks like this:

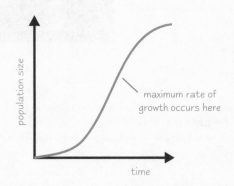

The maximum sustainable yield is obtained by keeping the population at the level where maximum growth occurs:

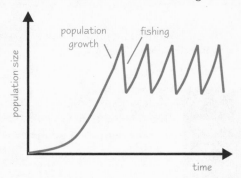

Modern Fishermen are Too Good *at their Job*

Modern fishing methods are extremely **effective**. So effective that if all the fishermen took as many fish as they could, the fish population would soon be **exterminated**. So, in popular fishing areas like the North Sea, **regulations** have been introduced to prevent **over-fishing**:

1) **Quotas** — each boat is limited to a **maximum weight** of fish that it's permitted to catch.

2) **Exclusion zones** — parts of the sea are made **off limits** for fishing boats, so that there are areas where the fish can breed safely.

3) **Close seasons** — fishing may be forbidden at certain **times of year**, usually the **spawning season**. This gives the population a chance to **recover** through reproduction.

4) **Net size restrictions** — fishing with excessively **large** nets may be forbidden. Also, fishing boats may be required to use nets with a **larger mesh size** (e.g. 90 mm minimum for North Sea cod fisheries). This prevents them catching the **smaller** fish, so that young fish have a chance to grow and **breed** before they're caught.

All these methods help to conserve fish stocks, but there are sometimes problems **enforcing** the rules. For example, if fishermen have **quotas** they might still catch more fish than they're allowed to, then just keep the best (biggest) fish, and throw the smaller ones (already dead) back into the water. Fishing restrictions are also **politically unpopular**, and depend on **international co-operation**.

Harvesting Ecosystems

Fish Farming Can Increase the Harvest

In a **fish farm**, fish are raised in a controlled situation. This helps to **maximise the yield**.
Fish farms may be **open** systems or **closed** systems:

- **Closed systems** are large tanks in which the water is continuously recirculated. This gives a lot of **control** over the conditions, but it's **expensive**. It might not be suitable for **marine** fish, especially **carnivores** that swim over a **wide area**.
- An **open system** is a section of the open sea (or river or lake) where **nets** restrict the movement of fish, but allow water to flow freely in and out. This is **cheaper**, but gives **less control** over the conditions.

The **yield** is increased under the **controlled conditions** in a fish farm because:

1) The fish are **protected** from **predators** like seals and carnivorous birds.
2) The fish are given a controlled **diet**, often in pellet form. Food is supplied in the right amount and at the right time to **maximise growth**.
3) **Competitors** are excluded, so the fish get all the food for themselves.
4) Special tanks can be used for raising the **young** in **ideal conditions**.
5) **Diseases** are controlled by the use of **pesticides** or **antibiotics**.
6) If the fish farm uses a **closed system**, the **temperature** and **composition** of the water can also be controlled.

Fish Farms Can Affect the Marine Ecosystem

Fish farming has some **disadvantages**. Some consumers think that the **quality** of the fish isn't as good, and fish farming is also relatively **expensive**. Fish farms can also damage the **environment**:

1) They produce a lot of **food** and **faeces**, attracting **scavengers** and **bacteria**. These then **deoxygenate** the water.
2) Fish farms can create high concentrations of **pests** and **diseases**. These pest species may then go out into the **open sea** and affect the **wild population**.

Practice Questions

Q1 Define the terms overfishing and underfishing.
Q2 Explain the meaning of the term maximum sustainable yield in relation to a fishery.
Q3 Give four ways that overfishing can be prevented.
Q4 Give three advantages and three disadvantages of producing salmon in a fish farm, rather than catching them wild.

Exam Questions

Q1 In the North Sea, fishermen are making use of a natural ecosystem. What are the advantages and disadvantages of this, compared with getting food from a farm on land? [4 marks]

Q2 In fish farms, bacterial diseases are often controlled using antibiotics. Suggest what damaging effects these antibiotics might have on the environment. [2 marks]

Fishermen are just too efFISHent... geddit? geddit?

Sorry. It's a touchy subject this — lots of fishermen are pretty cross about all these restrictions, and they say the scientists are talking rubbish and there's plenty of cod left. But how weird would it be if one day, the humble cod was to be found only in the zoo. And you'll take your grandkids on a day out, and you'll say "Look kids — I used to eat them with chips."

Pollution Caused by Industry

*'Industry' is basically manufacturing products or providing services in order to make a profit. Sometimes it can cause pollution too, and this has to be avoided wherever possible. **These pages are for OCR and AQA B.***

PCBs and Heavy Metals can Cause Long-Term Problems

You don't have to learn about PCBs if you're doing AQA B.

PCBs are **polychlorinated biphenyls**. They're synthetic substances first produced in the **1930s**, mainly for electrical insulation but also as heat transfer materials and for use in plastics. For many years they were widely used, and it wasn't until much later that their **harmful properties** were discovered. PCBs are now known to be **carcinogenic**, and they can also cause **liver** and **kidney** damage. As a result, PCBs haven't been used since the **1970s** — but, because they show **bioaccumulation** (see page 32), they're **still present** in our environment even now.

Heavy metals are those at the end of the periodic table (with high atomic masses). They're toxic, but due to a lack of alternatives many are still used in industry, and they sometimes get released into the environment in **industrial effluent**. The main heavy metal pollutants are **mercury**, **lead**, **cadmium** and **arsenic**.

1) **Mercury** is used in making **batteries**, and also in the **chemical industry**.
2) The use of **lead** as a **fuel additive** is now being phased out, but it still has widespread uses in the **construction industry** and in **electrical equipment**.
3) **Cadmium** and **arsenic** are also used in the manufacture of **electrical equipment**.

They're toxic because they act as **enzyme inhibitors**. In smaller doses they can cause symptoms like **nausea** and **vomiting**, but in larger amounts they can be **fatal**.

As with PCBs, the heavy metals show **bioaccumulation**, magnifying their effects. **Mercury** pollution in the **sea** is especially dangerous — micro-organisms convert it into **methyl mercury**, which is even more toxic, and which accumulates in fish. Methyl mercury damages the **nervous system**.

AQA B ONLY

Toxicity Tests Measure How Dangerous a Substance Is

The **toxicity** of a substance can be measured in the laboratory, by finding either the **lethal dose** or the **lethal concentration**. This is done by giving groups of **test animals** (or plants) different amounts of the substance.

- The **lethal dose** (**LD50**) is the **amount** that **kills 50%** of the group tested (when injected or swallowed).
- The **lethal concentration** (**LC50**) is the **concentration** that **kills 50%** of the group (when breathed in).

This can help scientists to decide whether a new substance is safe to use. But it's obviously a test that should be used sparingly. As well as harming the test organisms, the results can be **misleading**. The lethal amount will depend on **how long** the test is carried on for, and on other conditions such as **temperature** or the presence of **other pollutants**.

Burning Fossil Fuels Causes Acid Rain

Aah, your old friend acid rain from GCSEs. Clearly it's a terrible problem, but isn't it lovely and reassuring to see something so familiar at A2...

Fossil fuels (coal, oil and natural gas) contain **nitrogen** and **sulphur** compounds as **impurities**. When these are burnt they produce **nitrogen oxides** and **sulphur dioxide**, which dissolve in rainwater and form weak nitric and sulphuric **acids**. This **acid rain** can be harmful to plants and animals:

- It **kills plants**, including trees, by damaging their **leaves** and by altering the pH of the **soil**.
- It **acidifies** lakes and rivers. The low pH kills animals and plants directly, but also causes **toxic aluminium ions** to be washed out of soil and into the water. On the other hand, **phosphate ions** (essential nutrients for plants) are **precipitated** as solid in acid conditions, and therefore sediment into the soil rather than remaining dissolved in the water.

Pollution Caused by Industry

Carbon Dioxide and Methane are Greenhouse Gases

This page is just for OCR, except for the box about oil.

Burning **fossil fuels** also produces **carbon dioxide**. Methane is produced by the **decay of waste** in landfill sites, but also from **farm animals**, **manure**, and from **mining** activities.

Carbon dioxide and methane both contribute to **global warming** — the greenhouse effect. Their presence in the Earth's atmosphere restricts heat from the Sun being reflected back out into space. Many scientists think that this could result in climate change, which could **disrupt ecosystems** and cause **flooding** (by melting the polar icecaps).

Efforts are now being made to **reduce emissions** of 'greenhouse gases'. Measures include:

1) Finding **alternative energy sources**, such as wind and solar power.
2) Increasing **taxation** on fuels.
3) Limiting the use of **landfill sites** (to reduce **methane** production).

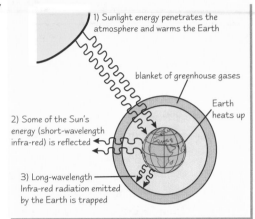

1) Sunlight energy penetrates the atmosphere and warms the Earth

blanket of greenhouse gases

Earth heats up

2) Some of the Sun's energy (short-wavelength infra-red) is reflected

3) Long-wavelength Infra-red radiation emitted by the Earth is trapped

Since the greenhouse effect is a **global** issue, **international co-operation** on the problem is essential. The **Kyoto Protocol** in 1997 was an attempt to get countries to agree to limit their production of greenhouse gases. But for various reasons, some of the big countries like the USA and Australia chose not to sign up.

> Another side-effect of using oil as a fossil fuel is that **oil spillages** sometimes occur from tankers at sea. Oil damages wildlife simply by its **physical effect** — it sticks to seabirds and other organisms, **smothering** them and preventing them from **moving freely**. It can also **poison** them if they try to remove the oil from their bodies. The oil can also harm **shore** organisms, especially **shellfish**.

 AQA B ONLY

CFCs Damage the Ozone Layer

CFCs (**chlorofluorocarbons**) were developed in the 1930s, mainly as components of **refrigerators**. From then until the 1970s, their use increased, especially as propellants in **aerosols**.

In the 1970s, it was discovered that CFCs damage the **ozone layer**. Ozone in the upper atmosphere protects us against **harmful UV rays** from the Sun. Ozone damage is therefore linked to increased rates of **skin cancer**.

In **1987**, an international agreement, the **Montreal Protocol**, was reached. All the major industrialised countries agreed to limit their production and use of CFCs. The agreement has since been tightened up, and CFC production has now largely **stopped**. However, CFCs stay in the atmosphere for a long time, and the hole in the ozone layer still remains.

Practice Questions

Q1   Most human beings contain traces of PCBs in their bodies, even though PCBs have not been used for thirty years. Explain why.

Q2   Name four heavy metal pollutants.

Q3   For each of the following pollutants, state one harmful effect that it has on the environment, and one measure that could be taken to control this pollution: a) sulphur dioxide   b) methane

Exam Questions

Q1   Suggest reasons for the following:
    a) The carbon dioxide concentration of the air is higher in London than on the Yorkshire Moors.   [2 marks]
    b) The concentration of carbon dioxide is lower in a tropical rainforest than in the Sahara desert.   [1 mark]
    c) The average concentration of carbon dioxide on Earth is higher now than it was 200 years ago.   [2 marks]

Heavy metal can also cause long-term problems with your hearing...

That's a different type though, I think. Anyway, these pages should come as a welcome bit of light relief — good old acid rain, the greenhouse effect and CFCs. You'll have been doing those every year since you were so high — they should be as familiar to you as your times tables. If they're not, then hang your head in shame and do something about it, quick.

Conservation of Resources

Once upon a time, people found out about useful things like glass and paper. So they thought "whoo-hoo" and started using them as fast as possible. Then someone said "erm, what happens when it all runs out...?"
These pages are for OCR (and are also useful for Edexcel).

We Need to **Conserve** Our Resources

Conservation is the **protection** and **maintenance** of natural resources. It's not the same as **preservation**, which is keeping things just as they are — natural systems are **dynamic**, with continual cycles and changes, so conservation needs **active management**. A healthy ecosystem is always fluctuating, and effective conservation has to take these natural changes into account. Conservation may also involve **repair** and **reclamation** after earlier mistakes.

Human life depends on the resources on Earth. These resources include:

1) **Food and water** — we rely on **agriculture** for food production, and on **water treatment** to recycle water. Water is naturally recycled in ecosystems, but **water treatment plants** make use of **micro-organisms** to provide safe and reliable drinking water for big population centres. Water recycling is described in the section below.

2) **Land to live on** — industrial activities, especially **mining**, can leave land derelict and unsuitable for habitation.

3) **Energy** — either **renewable** (e.g. wind power) or **non-renewable** (e.g. fossil fuels like coal and oil).

4) **Minerals** and other **raw materials**.

5) **Other species** — we must conserve other living things, from mammals down to microscopic prokaryotes. If we allow them to become **extinct**, we could lose **valuable resources** without even realising it. See pages 118-119.

Conservation of these resources is essential —

- for **economic** reasons — our **survival** depends on energy and other resources.
- for **ethical** reasons — we have a **duty** to protect our planet, and to pass these resources on to **future generations**.

Water Treatment is Used to *Recycle Water*

This diagram shows how nasty **sewage waste** is turned back into lovely clean **drinking water**.

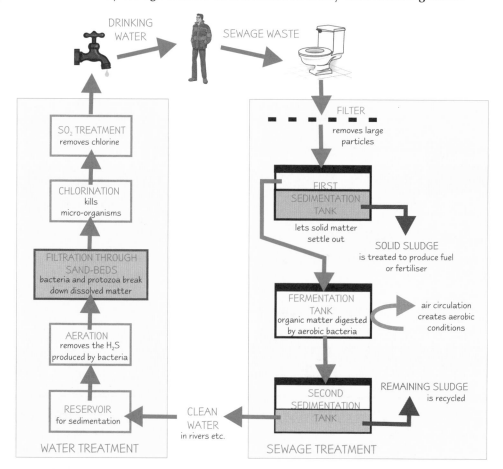

Conservation of Resources

Recycling Resources Means They Go on and on and on and on and...

Some resources cannot be renewed — **fossil fuels** are a good example. However, we can usually **recycle** materials. Doing this rather than throwing stuff away into landfill sites conserves our resources, and also reduces **methane** production from waste tips. For example —

1) **Paper** — recycling waste paper reduces the demand for **trees** to be cut down for paper production. However some people think that recycling paper is **not** environmentally friendly, because treating the old paper uses energy and produces toxic by-products. Trees for paper are actually a **sustainable** crop, because more can be grown to replace those taken, and so recycling is considered a waste of energy by some people.

2) **Glass** — recycling glass uses less **energy** than manufacturing new glass from scratch.

3) **Plastic** — recycled plastic also uses less **energy** than new production. And **pollution** is reduced — plastic production often creates toxic **by-products**.

The Government now encourages recycling by, for example, setting **recycling targets** for local authorities, with financial penalties if these targets are not met.

Active Conservation Can **Reclaim Land**

Some **mining** activities — for example **coal mining** and **china clay extraction** — produce large amounts of waste, which is often thrown up as **spoil heaps** near the mines. In recent years, efforts have been made to **reclaim** these tips, which can be dangerous as well as ugly. The tips are **landscaped** and then covered with **topsoil**, after which they can be built on, or used as farmland or as amenity areas. But it's essential to check first that the material is not polluted with **toxins**.

I remember when all this was spoil heaps...

Practice Questions

Q1 Define the term conservation. Explain how it is different to preservation.

Q2 Explain the difference between renewable and non-renewable energy sources.

Q3 Describe the role of micro-organisms in a sewage treatment works.

Q4 Explain the benefits of recycling: a) paper b) glass c) plastic.

Q5 How can land taken up by spoil heaps from mining be reclaimed for people to live on?

Exam Questions

Q1 Explain why throwing toxic waste materials down the drain may interfere with the way that a sewage treatment plant works. [2 marks]

Q2 Give one ethical reason and one economic reason why we need to conserve our natural resources. [2 marks]

So we clean water by chucking bacteria in it? I think I'll stick to orange juice...

Recycling is good. If you don't recycle stuff like tins, glass and paper then shame on you. It's really easy nowadays — if the council don't arrange to collect your recycling from you, your local supermarket will probably have the facilities to do it. It's good fun chucking all the bottles in so that they make a big smash. My ideal Saturday night out, is that.

Conservation of Species

We don't just conserve species to be nice. And it's not because they're cute or fluffy. There are lots of very serious reasons why we should maintain biodiversity, which you need to know all about. **These pages are for OCR and AQA B.**

Biological Conservation Maintains Biodiversity

Conserving other species seems the right thing to do. But these other organisms are also vital as part of the **ecosystem** (extinction of one species affects everything else). And they're a **resource** — living things are used by humans as sources of **food**, in **medicines**, and as tools for **biotechnology**. Once an organism is extinct, its genes are eliminated forever from the gene pool, and are no longer available to us. This is particularly significant now, because **DNA technology** means that **any** gene has potential for useful exploitation.

It's not because they're cute or fluffy. Not at all.

You can **classify** conservation efforts in different ways. Here are the five you need to know about:

1) **Biological** conservation — maintaining the diversity of living organisms **within habitats**.
2) **Environmental** conservation — conserving the **abiotic** (non-living) characteristics of ecosystems.
3) **Nature** conservation — the preservation of sites of **special scientific interest**.
4) **Species** conservation — the protection of **rare species**.
5) **Global** conservation — conserving the composition of the **atmosphere** and **oceans**.

Conserving Species Means Conserving Habitats

Some species are under threat because their natural **habitats** are shrinking. This happens for **two** main reasons:

- Human populations **expand**, so habitats are cleared to make way for homes etc.
- Over-exploitation of **resources**, e.g. through deforestation or mining.

Habitat conservation is important for its own sake, and also for the sake of the organisms that live there. By maintaining the **abiotic** characteristics of an ecosystem, the **biotic** component (the living organisms) can thrive.

In Britain, several measures are in place to promote habitat conservation:

1) **National Parks** — There are **fifteen** National Parks in Britain, ranging from Dartmoor to the Cairngorms. These regions are controlled by **Acts of Parliament** to conserve their natural beauty and to protect the organisms that inhabit them, but also to promote opportunities for the public to enjoy them.

2) **Sites of Special Scientific Interest** (**SSSIs**) — These are smaller areas (there are over **5000** SSSIs in Britain), often **privately owned**, which are particularly important because of their wildlife or geology. SSSIs are chosen by **English Nature**, by the **Countryside Council for Wales**, or by **Scottish National Heritage**. Once designated, there are **strict controls** on any proposed developments within them.

3) **Environmentally Sensitive Areas** (**ESAs**) — These areas are **nationally significant** places of ecological interest, where modern farming practices are threatening the environment. In ESAs, the government provides **financial incentives** to farmers to adopt environmentally friendly land management practices.

4) **Wildlife reserves** — Organisations such as the **Royal Society for the Protection of Birds** maintain and protect areas for **rare species** to live and breed safely.

Species can also be Conserved Outside their Natural Habitats

Even when habitats are threatened, some measures can be taken to conserve species:

- **Zoos** run captive **breeding programmes** for endangered species of animals. After successful breeding, the new generation can be **released** into the wild.
- **Botanical gardens** maintain **seed banks** — stores of the seed of threatened plants. They can be **germinated** later, and with careful storage they can be kept for an almost **unlimited** time.

Conservation of Species

A Case Study — African Elephants

There are only about **500 000** elephants in the whole of **Africa**. This may seem a lot, but it's down from about **4 million** in **1900**. If the numbers fall too far, **inbreeding** may weaken the population further.
Elephants are not just attractive animals, they're also an important part of the **forest/grassland ecosystem**. If their population crashes, this will affect many **other** species too.

Why has the elephant population fallen?

* The growing **human** population in Africa has led to elephant **habitats** being cleared.
* The **ivory trade** has meant elephants are **hunted** for their tusks.

So far, conservation has concentrated on limiting the **ivory trade**. Since 1990 this trade has been **illegal**, and poaching is now more **strictly policed**. Measures like this need **international co-operation**, and the ivory trade is included in the **Convention on International Trade in Endangered Species** (**CITES**).

Tropical Rainforests are Important for the Whole Biosphere

Tropical rainforests are important for **two** main reasons:

1) Their **biodiversity**, including many species found **nowhere else** on Earth.
2) Their effect on **global climate**. The rainforests absorb **carbon dioxide**, and therefore help prevent global warming.

In the past hundred years, over **50%** of the world's tropical rainforests have been destroyed, and this destruction still continues. **Why** has it happened?

* Clearing the forest for **agriculture**, **industry** or **living space**.
* Logging of trees for their **valuable wood**. Something to think about next time you sit on a mahogany loo seat!

Trees, of course, re-grow in time, so some logging activities are **sustainable**. But if over-exploited, the rainforest destruction becomes **irreversible**.

Conservation measures for rainforests include:

1) **Educating** local people about how to manage the land for a sustainable timber yield.
2) **International agreements** on managing forests, including a scheme to **certify** timber from well managed forests.
3) Developing **alternatives** to tropical woods — softwoods from managed woodlands elsewhere, or completely different materials like plastics.
4) Laying aside **conservation areas** as completely free from logging.

As with many conservation measures, controls need co-operation from **local people** and **international agreement**.

Practice Questions

Q1 Give one example of a conservation measure of each of these types of conservation:
    a) habitat conservation   b) nature conservation   c) species conservation
Q2 Explain the difference between a SSSI and an ESA.
Q3 Give two reasons why the conservation of tropical rainforests is important.

Exam Questions

Q1 Explain briefly why the environment has come more under threat over the past 100 years than in the previous 1000 years. [2 marks]

Q2 National parks are set up to protect wildlife, and also to provide opportunities for the public to enjoy them. Suggest how these two objectives might be in conflict with each other. [4 marks]

Q: How did a Chinese breeding programme help a giant panda get pregnant?

A: they showed her sex education videos. Seriously. Staff at the Wolong Giant Panda Protection Centre in south west China realised that the captive-born 4 year-old female, Hua Mei, had no knowledge of normal panda mating behaviour. So they showed her videos of mating pandas before a series of "blind dates" with males. What's even weirder is that it worked.

Adaptations to the Environment

Adapt to new conditions — or become extinct. Organisms that aren't well-suited to their environments can't survive. It's natural selection, and no species can avoid it. **These pages are for AQA B.**

The **Shape** and **Size** of Organisms are **Suited to their Environment**

Adaptations may be **structural**. For example, very hot (or cold) climates limit the **size** of organisms (see page 60). In hot conditions, large organisms have difficulty losing enough **heat**, because of their low **surface area: volume ratio**. And in cold conditions, small organisms may lose **too much** heat.

But the **shape** of an animal will also affect its **surface area: volume ratio**. This is why elephants have such **large ears** — essentially, they have radiators on the sides of their heads, to improve heat loss. The large **surface area** of the ears means that the animal can **lose heat** more easily (it has special **blood vessels** to help with this).

By contrast, small mammals that live in cold conditions have **compact rounded bodies**, to minimise the surface area. Compare the body shapes of the Arctic hare, which lives in very cold conditions, and the brown hare, which lives in a more temperate climate. The more compact body shape of the Arctic hare and its smaller ears help it to retain heat more easily.

Specialised Organisms Have Special Features

Xerophytes are plants that live in dry conditions (see page 60). They're **adapted** to cope with the lack of water. **Halophytes** are plants that tend to live in **salty conditions** (e.g. by the coast). They're **adapted** to tolerate high salt concentrations. These adaptations can be **structural** or **physiological**. For example —

1) **Glands** that **secrete salt**, removing it from the plant. **Cordgrasses** (*Spartina*) have these.

2) **Water stores**, so that the salt is **diluted** within the plant. Because of their water reserves, these plants look swollen and fleshy, so they're called **succulents**. *Salicornia*, the **glasswort**, is an example.

As well as structural adaptations, animals and plants might also have **physiological adaptations**. This means that the body or its metabolism works in a special way. Some examples were covered on pages 60-61.

1) The **kidneys** of desert mammals are adapted to produce a very **concentrated** urine, saving water.

2) Tropical plants like sugar cane have a special type of photosynthesis called **C4 photosynthesis** (see p. 61). In C4 photosynthesis **carbon dioxide** is fixed from the atmosphere to make a **4-carbon compound**, instead of the normal **3-carbon** compound of the **Calvin cycle** (see page 18). This way of fixing CO_2 is much quicker and more efficient than the normal pathway, especially when CO_2 concentrations are **low**. This is an advantage because tropical vegetation is very dense, and so **competition** for CO_2 is intense.

Animals can also have Behavioural Adaptations

Organisms may be adapted in their **behaviour** as well as in their structure and physiology. Any behaviour that improves survival chances is said to be **adaptive**.

Bird migration is an interesting example. Migration lets birds exploit food resources in **two completely different areas**, depending on the season. Also, nearer the poles the summer days are **longer** (giving more time for feeding), but nearer the equator the winter days are both **longer and warmer**.

In Britain, many of our summer birds (e.g. **swallows** and **house martins**) migrate to **Africa** once the weather gets colder. In the summer, food is plentiful (and competition is less intense) in **Britain**. In winter, when the weather gets bad, the birds go **south** to warmer climates.

Our winter birds, on the other hand, have migrated to Britain from even colder countries like **Scandinavia**. When these areas get covered with ice and snow, the migrants come south to Britain. Examples are **redwings** and **fieldfares**.

a house martin

Adaptations to the Environment

Changes in Day Length Trigger Migration in Birds

What triggers the migration behaviour of birds at the **right time**?
It's a **combination** of factors, but all of these things may be involved —

- Birds have an **innate annual cycle** that means that they tend to move south (for example) after six months, and then move north six months later. This type of **genetic programming** is called an **endogenous cycle**.
- Migration is also affected by **day length**, which therefore **reinforces** the endogenous cycle. Birds migrate south when the days become shorter than a certain **critical period**. This kind of response to day length is called **photoperiodism**.
- In many species, migration can also be triggered by **reduction in food supplies**.

Once the migrating behaviour has been triggered, birds take in **extra food** to build up their fat reserves. The migration itself depends on navigation by **vision**, **wind direction** and **magnetic fields**. **Learning** and past experience of the route may also play a part, and younger birds may rely on more experienced members of the flock to guide them.

Simple Organisms show Simple Adaptive Behaviours

Even very simple organisms show **adaptive behaviour**.

1) **Tactic responses** (**taxes**) are responses where the organism moves towards or away from a **directional stimulus**. For example, **woodlice** show a tactic response to light (**phototaxis**). They move **away** from the light, which keeps them concealed under stones during the day (where they're safe from predators and in damp conditions).

2) **Kinetic responses** (**kineses**) are slightly different. Here, the animal's movement is affected by the **intensity** of the stimulus. Woodlice show a kinetic response to **humidity**. In **high humidity** they move **slowly** and **turn** more often (so that they stay where they are). As the air gets **drier**, they move faster and faster and turn less often, so that they move into a **new area**. The effect of this response is that the animals move from drier air to more humid air, and then stay put. This improves the **survival** chances of the animals — it reduces their water loss, and it helps to keep them concealed.

Practice Questions

Q1 Define the terms xerophyte and halophyte.
Q2 Give two ways that halophytes can be specially adapted for their environment.
Q3 Describe three different factors that might cause a bird to migrate south.
Q4 Explain the difference between a taxis and a kinesis.

Exam Questions

Q1 The smallest mammal in Britain is the pygmy shrew. It eats more than its own body weight in food every day. Larger shrews eat about half their own body weight each day. Suggest why:
a) The smaller shrews need to eat relatively more food. [3 marks]
b) A mammal smaller than the pygmy shrew could not survive in Britain, but small insects can. [2 marks]

Q2 Maggots (juvenile flies) living in rubbish tips show a phototactic response. The youngest maggots move away from light, but older maggots (just before they begin to change into adults) move towards the light. Suggest why these different responses may both be adaptive. [2 marks]

You are the weakest organism... goodbye.

All these adaptations seem too clever to have happened just by chance, but remember that they evolve over millions of years. If a random trait appears due to mutation in a population and turns out to be an advantage, it'll then spread quickly. This is because if it helps an individual to survive, that individual will have time to reproduce more and pass the trait on.

Selective Breeding

Ignore this section if you're doing AQA or Edexcel. It's for OCR only.
Selective breeding is how we breed strains of plants and animals to our own specifications. Pretty useful, but confusing too.

Selective Breeding is Breeding Organisms with Particular Characteristics

Today all sorts of organisms are bred by humans because they provide us with food or because they have other useful features (e.g. wool). **Artificial selection** is when humans **breed** animals or plants with specific **characteristics**, hoping that their offspring will also show those characteristics.

Selective breeding works because:

1) The characteristic being selected is **genetically determined**, as opposed to **environmentally determined**.

2) There is **genetic variation** of the characteristic in the population. Artificial selection won't work if all the organisms have the same genes.

e.g. A plant could have big fruit either because of its genes or because of high quality soil. Artificial selection will only work if the big fruit is due to the genes and not the soil.

Artificial selection is a bit like natural selection, except that the **human** is doing the selection — not the environment. With artificial selection, organisms become **adapted** to become more useful to humans. They **don't** become adapted to their **environment**.

In artificial selection, organisms showing **extremes** of a characteristic are chosen for breeding (e.g. the cows which produce the **most** milk). This means that **directional selection** takes place from one generation to another (see page 53). **Natural selection** often shows directional selection too, but it's usually slower and improves fitness by changing many characteristics at the same time.

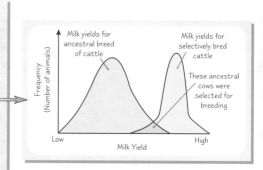

Heritability Means How Much a Characteristic is Determined by Genes

Artificially selecting particular characteristics isn't as straightforward as it might seem. The characteristics of an organism that make up its **phenotype** are determined partly by **genes** and partly by **environment**.

1) Genes and environment affect different kinds of **characteristics** to different **degrees**.

2) For example, **blood group** is determined entirely by genes, but **milk production** in cattle is determined by both genes and environment.

3) In order to find out whether artificial selection will work, it's important to be sure that the characteristics of interest are determined by a significant **genetic component** — the bigger the better, because they will be passed on to any offspring. This component is called **heritability**.

4) The heritability of a characteristic is the **proportion of the total phenotypic variation that is genetic** (not environmental). For example, the **heritability** of milk production is lower than that of blood group.

There's a nifty little **formula** you can use to work out **heritability**.

If **VP** is the total phenotypic variation, **VG** is the variation due to genes and **VE** is the variation due to the environment, then:

$$VP = VG + VE$$
$$\text{and } \textbf{heritability} = VG/VP$$

Environmental Variation can be Revealed by Breeding Pure-Bred Parents

Geneticists can find out the extent of **environmental variation** for a characteristic by breeding from parents that are **pure-bred** for that characteristic. This means using individuals that are **homozygous** for the characteristic.

Any **variation** seen in the offspring can therefore be put down to the **environment**, so giving us a measure of the **VE** component given above (since **VG** is zero). We can then measure the total phenotypic variation, **VP**, resulting from any cross (whether the parents are pure-bred or not) and use these values to work out the genetic component, **VG**, and – therefore – the **heritability**.

Selective Breeding

Plants are Selected for Food, Resistance and their Looks

The most common things that **plants** are selected for are:

1) **high yield** and **nutritional value** of food

2) **pest-** and **disease-resistance**

3) **tolerance** for harsh growing conditions, such as drought

4) **aesthetic features** for garden plants, such as flower colour or scent

The common (and unexciting) rose has been selected for its appearance and bred into loads of different (and beautiful) strains.

Plants can **reproduce** in different ways. Some of these ways are more suitable for selective breeding than others, so plant breeders have to think carefully about which **methods** to use. For example, taking **cuttings** of plants is a form of **asexual** reproduction, so can't be used for selective breeding (as all the resulting offspring are just **clones** of the parent). So this is what plant breeders normally do:

1) Individual plants with the desired **characteristic** are chosen.

2) A **pure-bred** (**homozygous**) variety is produced by **self-pollinating** these plants.

3) The different **pure-bred** plants are then **cross-pollinated** after several generations. This is called **hybridisation**.

4) Pure-bred plants have less **genetic variation**, so risk ending up homozygous for **susceptibility** to a disease. This problem is **prevented** in hybrid plants.

In plants that normally self-pollinate, pollen can be transferred by hand at this stage.

Selective Breeding is Harder with Large Animals

Selectively breeding plants is dead easy compared to selectively breeding **large animals** like cows (or even lions):

1) **Mating** can be **dangerous**, especially if the animals themselves are dangerous.

2) It takes a long time for large mammals to reach **sexual maturity**.

3) The **gestation period** (pregnancy) is often very long.

4) **Few offspring** are born at a time.

All this means that **breeding programmes** for large animals take a very **long time** and can be very **expensive**, as it costs a huge amount of money to keep large mammals and breed them.

Practice Questions

Q1 Name one characteristic that's largely genetically determined, and one that's largely environmentally determined.

Q2 Explain what is meant by the term 'heritability'. Show how it is determined, using VP, VE and VG.

Q3 What is meant by the term 'pure-bred'?

Q4 Give three examples of useful characteristics in plants.

Exam Question

Q1 An investigation was carried out on the heritability of height in a species of crop plant. The first stage involved propagating the plants asexually in order to produce a population of clones.

a) Why was it first necessary to produce a population of clones? [2 marks]

b) The variation in this asexual population was then calculated to be 0.3. The variation among the individuals in another population that was allowed to reproduce sexually was 0.8. Use these values to calculate the heritability of height in this species. [3 marks]

Sausage dogs didn't come from the wild...

You might think that selective breeding is a relatively new thing that we've developed with our knowledge of genetics... Well, you'd be wrong. We've been selectively breeding animals for yonks and yonks. All the different breeds of dog are just selectively bred strains which came from a general wolf-type dog back in the day. Even sausage dogs. Amazing eh?

Artificial Insemination and Transplantation

These pages are for OCR only.

This bit's all about the different techniques we've developed to make animal breeding more efficient. It's quite interesting, if you like thinking about cows and sheep getting down to it. Which, ahem, you don't, of course.

It's Important to Work Out **Which Animals** to **Breed** From

Sometimes it's impossible to select which individuals to breed from by only looking at **phenotypes**. For example, it could be that the characteristic we want is only expressed in one sex. Which **bulls** should we breed from to get female offspring which produce the most **milk**? Hmmm. Tricky one.

In this case, we can find out the **value** of the bull's **genes** by looking at the characteristics of his **offspring**. This is called **progeny testing**. It's used in both **plant** and **animal** breeding.

In other words, if he produces offspring with a high milk yield, then he must carry the genes for high milk yield, even if he can't show this by producing milk himself.

Artificial Insemination is Useful for Animal Breeding

As seen on page 123, there are practical **problems** in breeding animals — especially large mammals. **Artificial insemination** gives animal breeders more control over the **fertilisation**. **Semen** is collected and stored in **liquid nitrogen** (usually around -196°C). It is then warmed up and placed in the **vagina** or directly into the **uterus** through a very narrow tube called a **catheter**.

Artificial insemination is **great** because:

1) Breeders don't have to keep **males**, which are often more **aggressive** than females.
2) **Matings** (which can be traumatic) are avoided.
3) **Sperm** is cheaper and safer to **transport** than male animals.
4) **Selective breeding** and any **progeny testing** are **faster**, since semen can be used to inseminate many females at the same time.
5) It is possible to **test** for the genetic quality of sperm and **screen** for abnormalities.

However, there are **disadvantages** too:

1) Sperm could be **damaged** by the low temperatures in storage.
2) The convenience of fast fertilisation might encourage breeders to inseminate many females using one semen sample. This could lead to loss of **genetic diversity** by inbreeding (therefore increasing the risk of susceptibility to **disease**).

In Vitro Fertilisation Happens Outside the Female's Body

In vitro fertilisation (IVF) literally means 'in glass fertilisation'. This is a technique where **fertilisation** takes place in the laboratory, in a glass dish. It is an important technique used in research into animal breeding. Here's what happens:

1) The female mammal is treated with **hormones** such as FSH to stimulate production of many **oocytes** (eggs). This is called **superovulation**.
2) A **catheter** is used to **collect** oocytes from the mature follicles where they develop in the ovaries.
3) **Oocytes** and **sperm** are **mixed** in a glass dish containing sterile nutrient solution.
4) If successful fertilisation occurs, the resulting **zygote** divides by **mitosis** to form a ball of cells — the **embryo**.
5) The embryo is **implanted** into the uterus of a female mammal.

Artificial Insemination and Transplantation

Embryos can be *Transplanted* into *Surrogate Mothers*

An embryo can be **transplanted** into any female in the right stage of the **oestrous cycle** (something that can be controlled by **hormones**). It is even possible to **clone** early embryos so that numerous **genetically identical** embryos are produced, which are then transplanted into lots of different females.

Embryo cloning involves carefully **dividing** embryos into two, making artificial identical twins. This procedure, along with the **superovulation** technique, can massively increase the production of genetically desirable offspring.

Embryos can only be divided up to the 32-cell stage, before the cells have differentiated.

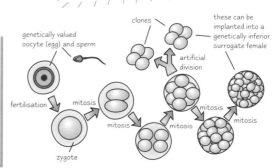

The advantages of this cloning and transplanting procedure are:

1) It **speeds up** the process of selective breeding.

2) A potentially **hazardous pregnancy** can be avoided in a valuable female by using less valuable **surrogates**. This has also been useful in the captive breeding of **endangered species**, where surrogates are even common domestic animals.

3) Just as with the sperm, it is possible to **screen embryos** for genetic abnormalities.

There are *Ethical* and *Social Issues* with *People* Using these Techniques

Artificial insemination and **IVF** are regularly used in **humans**. The **sperm** will normally come from the **male partner** in a couple who are having problems conceiving normally. However, sperm from a **donor** may be used if the male partner has problems with his sperm, or if a single woman wants a baby. Unfortunately, there a number of **complicated issues** surrounding this technique:

1) The child born might want to know the **identity** of their father, even though the donor would want to stay **anonymous**.

2) Many people think it's **wrong** for women to have the baby of someone they're not in a **relationship** with.

3) Sperm donors must be **screened** for genetic abnormalities and infections like HIV. This means that people with genetic disorders are told their genes are **unacceptable** and they're **prevented** this chance at reproduction. Saying which **genes** are **better** than others is a risky business.

Practice Questions

Q1 Give two advantages and two disadvantages of using artificial insemination in animals.

Q2 What is in vitro fertilisation?

Q3 Before in vitro fertilisation is attempted, animals are usually treated with a hormone to stimulate superovulation. Explain what 'superovulation' means and why it is necessary in this procedure.

Q4 Give one ethical consideration concerning the use of artificial insemination in humans.

Exam Questions

Q1 Egg production in poultry has been shown to be heritable enough to make the artificial selection of egg mass worthwhile. Describe a method of progeny testing which could be used to select the best male birds. [3 marks]

Q2 The bongo is a rare species of antelope that is currently being bred in some zoos. Embryo transplantation techniques have been used, whereby bongo embryos are allowed to complete their development in the more common eland antelope. Give two possible advantages of using this technique in this case. [2 marks]

AI — good film, shame about the ending...

Have a think about those ethical issues. One day we might all be screened and our genetic profiles kept on record. Then people with genetic disorders might be discriminated against — refused life insurance, not given jobs etc. There was a film about it a while ago called Gattacca — worth seeing. It had Uma Thurman for the boys and Ethan Hawke for the girls.

Genetic Diversity

These pages are for OCR only.

You've always been told you shouldn't make babies with your siblings. Now here's why...

Inbreeding Increases the Chance of Harmful Recessive Characteristics

All the genes and alleles in a population make up the **gene pool** of that population. The genes present in the gene pool can be affected by many factors. For example, **mutation** adds new genes to the gene pool and **selection** removes genes that are harmful in a specific environment. Gene pools also vary in terms of the **proportion** of **heterozygotes** and **homozygotes**, which is due to the amount of **inbreeding** in the population.

1) Continued breeding between **related** individuals is called **inbreeding**.

2) Related individuals are more likely to **share** the same kinds of alleles, because they've inherited these alleles from the same **ancestors**.

3) So inbreeding can **increase** the chance of these alleles coming together in **homozygous** combinations.

4) **Recessive** traits are only expressed as **homozygous** combinations, and some of them are **harmful**.

5) It can be especially bad if the harmful homozygous combinations occur at many gene **loci** — such individuals could have **lower fitness** (meaning that they produce **fewer offspring**). This effect is called **inbreeding depression**.

The opposite of inbreeding depression is called **hybrid vigour** (also known as **heterosis**). This happens when there is mating between unrelated individuals — it increases the number of **heterozygotes**, so making the expression of harmful recessive traits less likely.

It's Important to Conserve Genetic Diversity

There are good reasons for conserving **genetic diversity**. Genetic diversity in a population means that it can survive **environmental changes** because at least a few individuals in the population are likely to have the combination of alleles that will help them survive in the new conditions.

It may also be useful to conserve the genetic diversity of different organisms for the sake of **humankind** — different organisms could be used in **medicine** or **food production**, for example. Some people think we have a **moral obligation** to conserve genetic diversity too.

Conserving genetic diversity is really about conserving <u>genomes</u> — the total sets of genes in organisms.

Genetic diversity is being **lost** for all sorts of reasons:

1) Species are becoming **extinct** because of direct **exploitation** by humans, like hunting.

2) They're also becoming extinct by more **indirect** means, such as when **habitats** are altered for **cultivation**.

3) Genes are also lost during **selective breeding**, where certain types of alleles become more common at the **expense** of others.

Sorry, lads. It's too late for you.

Loss of genetic diversity is called **genetic erosion**. **Gene banks** are places where genetic diversity is preserved. They can be **protected** areas of habitat, or zoological or botanical gardens. Modern technology even lets us keep stocks of **gametes** or **seeds** in cold storage.

Genetic Diversity

Seed Banks are Important Gene Banks for Plants

The **seed** of a plant contains a tiny **embryo**, the result of cell division from a zygote after fertilisation. The embryo is in a kind of 'suspended animation' until the conditions are just right for **germination**. In this state many can survive **harsh conditions**, such as dehydration. Along with their small size, this makes them ideal for building up **gene banks** of plants.

The kinds of seeds that **can** survive a long time in cold dry storage are called **orthodox seeds**. However, the seeds of many tropical species are **recalcitrant** — this means they **can't** survive for long under such conditions.

recalcitrant seed (coconut)
orthodox seeds

Important seed banks are kept in many botanic gardens throughout the world. The Royal Botanic Gardens at Kew, for example, holds a seed bank at Wakehurst Place in Sussex.

Sperm Banks are Important Gene Banks for Animals

Like seeds, the storage of **sperm** doesn't take up much room. Semen is collected by a variety of techniques, and can then be stored at low temperatures in **liquid nitrogen** (just like in artificial insemination). **Eggs** and **embryos** of animals can be stored too, but their collection and storage is much harder and more traumatic.

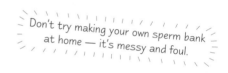

Don't try making your own sperm bank at home — it's messy and foul.

Gene banks aim to store a good range of **genetic diversity**. Ideally, this should **represent** the genetic diversity present in the **natural population**. This means that the sample of genetic material that is stored should be as **large** as possible.

Practice Questions

Q1 Explain what is meant by inbreeding depression.

Q2 Explain why breeding programmes for animals and plants should be designed to ensure that breeding takes place only between the most unrelated individuals within a species.

Q3 What is a gene bank?

Q4 Which are the better candidates for storage in a gene bank, recalcitrant or orthodox seeds?

Q5 How is sperm stored in a sperm bank?

Exam Questions

Q1 Many pedigree breeds of dog are known to suffer from an unusually high incidence of inherited disorders. For example, golden retrievers have a high incidence of bone dislocation. Explain why such inherited disorders are more common in pedigree breeds than in cross-breeds. [4 marks]

Q2 Give two characteristics of *both* plant seeds and animal sperm that make them good for storage in gene banks. [2 marks]

In Gene-sis, Adam and Eve started it all...

These have been a fairly easy couple of pages. OK, what have you learned? Don't have babies with your brother or sister. Don't hunt things to extinction. Do preserve lots of different types of habitat. Do collect and store seeds, eggs and sperm (erm... within reason). Now, that was a lark if you ask me. Just remember that it's all about maintaining genetic diversity.

Resistance

These two pages are just for OCR.

One day, we might live in a scary world full of ethically-dubious, disease-resistant designer babies. Which is why you should be grateful you only have to learn about resistance in crops and farm animals — not so many issues...

Disease **Resistance** is Controlled by **Genes**

Certain types of gene help organisms **resist** being harmed by chemicals or diseases. Resistance could be due to **single genes**, or **polygenes**.

An example of a **single gene** effect is the inheritance of **warfarin resistance** in rats. Warfarin is a **rodenticide** (rodent killer) that was widely used to control rats, until there was natural selection in favour of rats with a gene for **resistance** to it.

Warfarin resistance in rats is a **dominant** trait determined by a **single gene**.

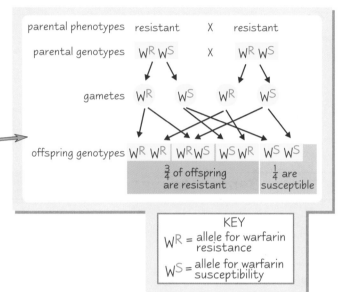

Polygenic inheritance of disease resistance occurs when lots of **interacting** genes determine a characteristic. This kind of inheritance is more **complex** and usually doesn't follow simple inheritance patterns like the one shown above.

A polygenic effect is seen in the inheritance of many types of disease resistance in **plants**.

Sometimes an interesting situation of **coevolution** occurs between the **host** and the disease-causing organism (**pathogen**), especially where **single gene** inheritance of resistance is taking place.

Resistance in the host depends on a **protein receptor** (determined by a single gene) intercepting the disease-causing **protein** of the pathogen (also determined by a single gene). A change in one may favour a change in the other by **selection**. This is called **gene-for-gene resistance**.

Selective Breeding Can Improve Disease Resistance

Breeders aim to produce organisms that have good **general resistance** to a wide variety of diseases. It's much better for a crop plant to have **some** resistance to a **lot** of different pathogens than for it to have total resistance to just one. In reality, after all, the plant will be exposed to a **wide range** of diseases when it is grown as a crop.

Here's how **disease resistance** is controlled in both **plants** and **animals**:

Daisy wondered whether there was something she wasn't resistant to.

1) Breeders **select** plants which show some **resistance** and just **breed** from these.

2) The resulting **offspring** are then **bred** to improve resistance. See p. 122-123 for more details on how they do this.

3) It's possible to breed for resistance because this trait often has high **heritability** in plants.

4) It's **not** easy to breed for resistance in **animals**, so breeders often resort to **vaccination** to control disease.

5) Vaccines aren't always available in **less economically developed countries**. So in these places, breeding animals to inherit resistance is more **important**.

Resistance

Bacteria Can Evolve **Resistance** to **Antibiotics**

Antibiotics are chemicals produced by microorganisms that **kill** or **inhibit** the growth of other **microorganisms**. They're important as medicine to control **bacterial infections**. However, bacteria can develop **resistance** to them:

1) Some bacteria carry **genes** that make them **resistant** to specific types of antibiotics. These have arisen by **random mutation**.

2) Due to **selection**, these genes become really **common** in populations of bacteria which are exposed to antibiotics — the resistant bacteria have a **selective advantage**. (The **antibiotic** is the agent of selection.)

3) Exposing bacteria to a certain type of antibiotic will **kill** the bacteria that **aren't** resistant, leaving the ones with the **resistant** genes behind.

4) These **thrive** in the absence of competition, and quickly **reproduce**, so soon the population is swamped with the resistance genes — an example of **natural selection**. (Despite this, antibiotics are often still an effective treatment...)

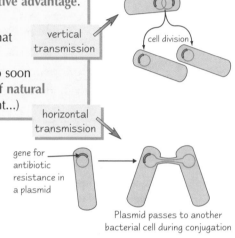

Bacteria can pass on the resistance genes to future generations of cells by **division** — this is called **vertical transmission** of the resistance.

Horizontal transmission is possible between neighbouring bacteria in the same population because they can join together to exchange **plasmids** (which often contain the antibiotic-resistance genes). Bacteria joining together like this is known as **conjugation** (see p. 199).

Insects Can Evolve **Resistance** to **Insecticides**

Bacteria evolve antibiotic resistance **quickly** because there are **loads** of them and they have a **short** generation time, which means we see many generations. The same can be said for many kinds of insect. Lots of insects carry genes that make them resistant to **insecticides** (the chemicals used to kill them).

The **spread** of **resistance** through a population is caused by **selection** of those with the resistance, just like in bacteria. Widespread use of the insecticide DDT has led to the evolution of **DDT resistance** in important pest species, such as *Anopheles* mosquitoes which carry malaria.

When bacteria or insects evolve **resistance** so that they can survive particular **concentrations** of antibiotic or pesticide, **higher** concentrations are needed to kill them. Unfortunately, this only makes them evolve even **more resistance**. Within the new resistant population, there is **selection** in favour of individuals resistant to even higher concentrations.

Practice Questions

Q1 Inheritance of resistance is sometimes polygenic. Explain what this means.

Q2 Explain the meaning of gene-for-gene resistance.

Q3 What is the difference between vertical and horizontal transmission of resistant genes in bacteria?

Q4 Explain why overuse of antibiotics could be a dangerous practice.

Exam Questions

Q1 Warfarin is a rodenticide. It has been used extensively to control numbers of wild rats. The first warfarin-resistant rats appeared in Britain in the 1950s and subsequently the trait spread quickly through the country. Warfarin resistance has been shown to be dominant and due to a single mutated gene.

   a) Suggest why a dominant resistant trait is likely to spread faster than a recessive one.   [3 marks]

   b) Explain why the spread of the warfarin-resistant trait is an example of selection.   [3 marks]

Q2 Explain how the characteristics of bacteria enable antibiotic-resistance genes to spread quickly in bacterial populations.   [4 marks]

Res-ist-ance is fu-tile...

You can learn a lesson for life on this page. Don't demand antibiotics from your doctor if you've just got a minor cough or achey toe. If you have antibiotics too often, all you're doing is killing off the weak bacteria so the big fat evil ones can take over. Then one day, the bacteria will be so strong that no antibiotic will be strong enough to kill them. Not a funny thought.

Human Genetics

These two pages are for OCR only.

These pages are about a few genetic disorders, how they're inherited and what they do. You'll also get to find out about genetic screening and pedigree analysis. Lucky you.

Cystic Fibrosis is a Recessive Genetic Disease

Cystic fibrosis is a condition which affects about 1 in 2500 babies — it's the most **common** major genetic disease in Britain. It's a **recessive** condition caused by **mutation** of a **gene:**

1) The gene responsible, **CFTR** (cystic fibrosis transmembrane regulatory protein) codes for a **channel protein** in cell membranes.

2) This protein allows the **facilitated diffusion** of **chloride** ions out of cells — **water** follows by osmosis.

3) Sufferers of cystic fibrosis have cells that **can't** let out chloride ions (and therefore as much water).

4) In parts of the body where **mucus** is produced by such cells, this makes the mucus **drier**. It clogs up **airways**, ducts of the **pancreas** and **reproductive** tracts, often leading to **death** from respiratory failure.

Huntington's Disease is a Dominant Genetic Disorder

Individuals with **Huntington's disease** suffer from involuntary **muscle movements** (choreas). It usually develops in **later life**. Their **mental ability** progressively worsens too. Scientists have identified the **gene** that causes the condition:

1) The **normal** form of the gene encodes for a protein called **huntingtin**, but we don't yet know what this protein does.

2) The **abnormal** form of the gene produces the **disease**.

3) It is caused by an expanding **triple nucleotide repeat**, CAG, in the gene. The number of repeats **increases** with meiosis and hence at each generation.

4) A **greater** number of repetitions means an **earlier** age of occurrence.

5) Unlike cystic fibrosis, it is a **dominant** condition.

Here's a picture of Korea, cos, erm, it sounds like chorea.

Laws of Inheritance Help us Understand Genetic Disorders

As a **recessive** condition, **cystic fibrosis** is inherited in a simple **Mendelian** way — it is determined by a **single** gene. This means that the only way two normal parents can produce a child with cystic fibrosis is if they are both **carriers** (**heterozygotes**) for the condition. In this case, there is a **1 in 4** chance of the baby being born with the disease at each pregnancy.

parental phenotypes	non-sufferer	X	non-sufferer		
parental genotypes	A a	X	A a		

gametes: A a   A a

offspring genotypes: AA | Aa | Aa | aa

offspring inheritances: $\frac{3}{4}$ chance of non-sufferer | $\frac{1}{4}$ chance of having cystic fibrosis

KEY
A = allele for non-suffering trait
a = allele for cystic fibrosis (recessive)

Because **Huntington's disease** is **dominant**, if a child is born with the disease, this means that at least one of the **parents** must have the disease too. Since the disease tends to develop later in life, this means that it is possible for sufferers to have children **before** their symptoms become apparent.

parental phenotypes: sufferer X non-sufferer

parental genotypes: B b X b b

gametes: B b   b b

offspring genotypes: B b | B b | b b | b b

offspring inheritances: $\frac{1}{2}$ chance of having Huntington's chorea | $\frac{1}{2}$ chance of non-sufferer

KEY
B = allele for Huntington's chorea (dominant)
b = allele for non-suffering trait

Human Genetics

Down's Syndrome is a Common Human Chromosome Mutation

About 1 in 700 babies are born with **Down's syndrome**. (The **likelihood** of having a baby with Down's syndrome increases with the **age** of the parents.) The symptoms are caused by **extra copies** of genes found on the **long arm** of **chromosome 21**. There are two possible causes of this:

1) In most cases, it occurs because an **extra chromosome 21** ends up in the zygote. This can happen because of **non-disjunction** of the chromosome 21 pair in the gametes. In other words, the two homologous chromosome 21s **don't split up** in meiosis I, or the two chromatids don't split up in meiosis II. Two copies of this chromosome end up in a gamete, and so fertilisation results in **trisomy** — **three** copies of the chromosome, instead of the normal two. This is **genetic**, but **not** hereditary.

2) In about 5% of cases, Down's syndrome is caused by a **translocation**. This is where a fragment of the long arm of chromosome 21 becomes **attached** to another chromosome. This is **hereditary**.

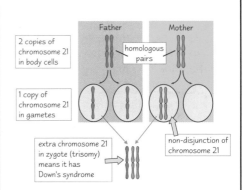

Pedigree Analysis and Genetic Screening Tell us About Inheritance

Potential carriers of genetic diseases often want to know their **risk** of having children with those diseases. There are two ways they can find out about this risk:

Pedigree analysis means examining a **family tree** to see how genetic disorders in certain members of the family can be inherited. Understanding the basic **laws** of genetics means that they can work out the **probability** of offspring inheriting a particular gene.

Genetic screening involves fairly complex techniques which can accurately tell whether there are **mutated** alleles present in unborn babies. Such techniques may make use of **genetic probes** (see page 107). Tests can be carried out on embryos **in vitro** (in glass, like a test-tube) or on foetuses **in utero** (in the uterus).

Two examples of **in utero** testing are:
- **amniocentesis** — taking samples of the amniotic fluid
- **chorionic villus sampling** — taking samples from the placenta

These tests are carried out along with **genetic counselling**.
Genetic counselling is about making sure that parents understand the **options** (such as whether or not to continue with the pregnancy) and **consequences** (such as whether the child born will need special care).

Practice Questions

Q1 Draw a diagram to illustrate a cross between two carrier parents for cystic fibrosis. What is the chance of their baby being born with the condition?

Q2 How does the inheritance of Huntington's disease differ from that of cystic fibrosis?

Q3 Explain how Down's syndrome arises by non-disjunction.

Q4 Briefly describe a method by which a fetus could be screened for a genetic abnormality.

Exam Questions

Q1 Huntington's disease is a dominant genetic disorder, which is eventually fatal.

a) Draw out a genetic cross diagram between two heterozygous sufferers of the condition. What is the probability of a child being born who will develop the condition? [4 marks]

b) Seeing as Huntington's disease is eventually fatal, suggest why the allele remains in human populations. [2 marks]

Q2 What is the advantage of genetic screening over pedigree analysis? [2 marks]

Something funny... blah blah blah... erm... not a funny page...

Genetic screening must be such a relief for people who think they're carriers of nasty things. It means that these people can have children without taking a huge risk. They can know in advance whether their child will have that nasty thing or not. But this is where we're getting into the realm of designer babies again... And the massive ethical debate that goes with it...

Human Genetics

These pages are OCR only.

Now you're going to learn some more about human genetics — gene therapy, transplantation and the like. This is the type of stuff that might be useful if you're trying to work out who can donate you a kidney if yours have worn out.

Gene Therapy is a Treatment for Genetic Disorders

The theory behind **gene therapy** is that if an allele **mutates** and fails to work properly, one that works can simply be **inserted** (as long as the faulty one is **recessive** and the working one is **dominant**). There are two main types of gene therapy:

- **Somatic cell therapy** involves inserting genes into **somatic** (body) cells.
- **Germ cell therapy** involves inserting genes into **germ cells** (gametes). Germ cell therapy isn't allowed.

Germ cell therapy may be hazardous. Inserted genes will be inherited by all cells of the offspring, having possible dangerous effects in an embryo.

In **somatic cell therapy**, normal functioning genes are added to **cells** that carry mutated ones. It **can't** be used to treat **dominant** conditions like Huntington's disease, because the dominant, faulty alleles would still have an effect. The technique is very **new** and has been developed in the UK to treat **cystic fibrosis** — a **recessive** condition.

1) Copies of the **normal** gene in loops of DNA (rather like plasmids) are **attached** to oily droplets called **liposomes**.

2) These are **sprayed** into the nose as an aerosol and are taken into the **abnormal** cells.

3) It is **not** a permanent cure because the cells lining the airways are constantly being **shed**.

4) Some people think that gene therapy is potentially **hazardous** because the genes could end up in **non-target cells** and have unforeseen effects.

Genetic Compatibility is Important in Transplantation

Many medical procedures involve taking **tissues** from one individual (the **donor**) and **transplanting** them into another individual (the **recipient**). This happens during blood transfusions (the blood being a liquid tissue), and during bone marrow or organ transplants.

1) Unless the donor is genetically identical, the immune system of the recipient treats foreign tissue as **non-self** and an **immune response** can be set up, with possibly **fatal** consequences.

2) The particular molecules of the donated tissue that are recognised as non-self are treated as **antigens** — they are usually **proteins**.

3) The recipient's **cytotoxic T-cells** (killer T-cells) recognise the non-self antigens on the transplanted tissue. The **receptors** on the T-cells are **complementary** in shape to the antigens, just as an enzyme is complementary to its substrate.

4) When the T-cell's receptors **bind** to non-self antigens on a donated cell, the T-cell treats the cell as it would a **pathogen** and tries to **destroy** it. This is what is meant by **rejection**.

5) **Close family members** are fairly likely to share the genes which code for antigens. This means that tissue donated by family members is more likely to be recognised as 'self'. It's therefore **less likely** to be **rejected** than that of strangers.

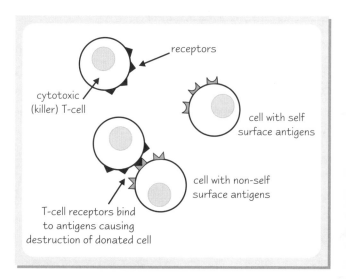

(Also see pages 194-197 for more on the immune response.)

Human Genetics

The **ABO Blood Group System** has a Simple Genetic Basis

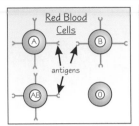

1) Blood group is determined by a gene at a **single locus**, which exists as the alleles I^A, I^B or I^O. (See page 47.)

2) This specific allele determines the type of **protein antigens** on the red blood cell surface. There are two types of red blood cell antigen — **A** and **B**.

3) These determine **blood groups** — a person with type A antigens will be blood group A, a person with type A **and** B antigens is blood group AB and so on... If you don't have any antigens on your red blood cells, you're blood group O.

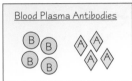

1) **Blood plasma** carries special **antibodies**. Again there are two types — A and B.

2) If blood antigens and antibodies of the **same type** mix, **agglutination** occurs — the antibodies bind to the antigens on the blood cells, forming a **clump**. This can be **fatal** because the clumps can **block arteries**.

3) So we **don't** have antigens and antibodies of the **same type** in our bodies. If we have type A antigens, we'll have type B antibodies and so on...

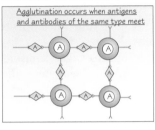

Most **blood transfusions** have to be from someone of the same blood group to prevent **agglutination** happening in the transfusion patient's blood. But there are two exceptions.

Blood groups	Red Blood Cell Antigens	Blood Plasma Antibodies
A	A	B
B	B	A
AB	A+B	none
O	none	A+B

People with **blood group O** (with **no antigens** on their red blood cells) can **donate** blood to anyone. Also, people with **blood group AB** (with **no antibodies** in their blood plasma) can **receive** blood from anyone. The reason for this is that it doesn't usually matter if donated blood contains the wrong **antibodies**, because there won't be a high enough concentration of them to cause serious agglutination. BUT, donated blood can't contain the wrong **antigens**, because the large amount of **antibodies** in the patient's blood will attack them, causing agglutination.

MHC Genes Tend to be Inherited **Together**

The **major histocompatibility complex (MHC)** is a cluster of **genes** bunched tightly together on part of chromosome 6. These genes cause many of the immune responses behind **rejection** of transplanted tissues and organs.

Rejection is less likely if the donor and recipient are **genetically similar**. But, there are potentially loads of possible **combinations** of alleles in the MHC, so it's **unlikely** that an **exact match** between donor and recipient will be found.

However, it's possible to find an **approximate match**. Because they're **inherited together** (they make up a **linkage group** of genes), genetic similarity at one locus is more likely to mean similarity at another. So **family members** are more likely than strangers to have similar MHCs. This means that **transplanted** organs or tissues are less likely to be **rejected** if they are donated by members of the recipient's **family**.

Practice Questions

Q1 Briefly describe the theory behind gene therapy.

Q2 Explain why somebody with blood group B could not donate their blood to somebody with blood group A.

Q3 What is the major histocompatibility complex?

Exam Questions

Q1 A mother has blood group O and a father has blood group AB. Would the father be able to act as a blood donor if his own son is the recipient? Explain your answer fully. [5 marks]

Q2 Close family members are genetically similar in terms of the major histocompatibility complex of chromosome 6. Explain why close family members are often good matches for organ transplantation. [5 marks]

Bloody good bit on blood groups here... (hee hee, I wasn't swearing...)

This is where you're allowed to get jealous of identical twins. Not only do they get the pleasure of pretending to be each other (which must be so much fun), but they have a constant source of genetically identical organs, bone marrow and blood. So should they ever need a transplant, they've got a really high chance of it working fine first time. Lucky so and sos.

The Liver

These pages are for OCR. Edexcel only need to know how the liver regulates glucose levels. You've already seen some of the liver's functions on pages 64-65. This stuff goes into a bit more detail, but don't panic — it's not complicated.

The **Liver's Structure** is related to its **Function**

The liver has to keep a **constant internal environment** in the body, so that the body cells can function properly. The **hepatic artery** supplies the liver with **oxygenated blood** and **deoxygenated blood** is taken away by the **hepatic vein**. The **hepatic portal vein** brings the liver **products** of **digestion** from the **ileum**.

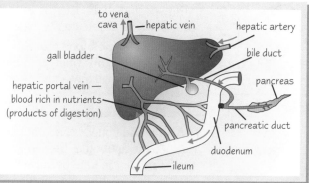

The liver is made up of cells called **hepatocytes** arranged in rows called **lobules**. Between the lobules there are **blood spaces** called **sinusoids** and small **channels** for **bile** to flow along called **bile canuliculi**. Bile **emulsifies fats** in the small intestine (see page 135).

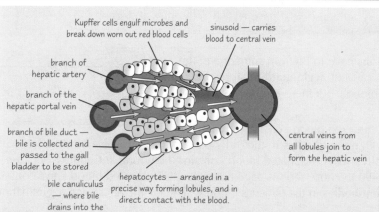

The **Liver** helps **Regulate Blood Glucose Levels**

Body cells mostly respire **glucose**. Blood glucose concentration in the blood is kept steady by two hormones, **insulin** and **glucagon**, in a **negative feedback mechanism** (see page 86 for the full details). The liver is responsible for **removing** glucose from the blood, or breaking down stored substances to release glucose.

When there's a **shortage** of blood **glucose**, the **liver** can:

1) **Break down glycogen** into **glucose**.
2) Convert other sugars (e.g. **fructose**) into **glucose**.
3) Convert **lipids** into **glycogen** then **glucose**.
4) Convert **glycerol** and **amino acids** into **glucose**.

The **Liver** controls **Fat Metabolism**

Triglycerides

1) When there's too much **glucose** in the blood, the **liver** converts it to **triglycerides**. Lipid compounds like **triglycerides** aren't soluble in water, so they're **associated** with **proteins** by the **liver** to make **water soluble lipoproteins**, so that they can be **transported** around the body in the blood.

2) The triglycerides are **transported** to the **adipose tissue** and **stored**. Most adipose tissue is stored under the skin (**subcutaneous fat**) and used for insulation and protection of vital organs.

3) When they're needed again, triglycerides are **hydrolysed** by lipase enzymes into **fatty acids** and **glycerol**, releasing large amounts of **energy**.

Cholesterol

1) This is a lipid found in all animal **cell membranes**. It's also needed to make **steroid hormones** and **vitamin D**.

2) The **liver** can make **cholesterol** if it needs to, or **remove excess** cholesterol in **bile**.

3) If you have high cholesterol levels it can lead to **gall stones**, which **block** the **bile duct**, **atherosclerosis** (narrowing of arteries), **coronary heart disease** or a **stroke**.

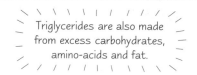

Triglycerides are also made from excess carbohydrates, amino-acids and fat.

The Liver

The *Liver* makes **Bile**

Bile is made by the **hepatocytes** from **water**, **bile salts**, **bile pigments**, **cholesterol** and **phospholipids** and is stored in the gall bladder. It **emulsifies fat** and provides lipase enzymes with the **optimum pH** by **neutralising stomach acid** (see page 99). When **fats** reach the **small intestine**, they trigger the release of **two hormones** into the blood:

1)   **Secretin**, which stimulates **bile production**.

2)   **Choleocystokinin-pancreozymin** (**CCK-PZ**) which stimulates the **release of bile** into the **intestine**.

Two of the main components of bile are **bile pigments** and **bile salts**:

Bile pigments The pigments **bilirubin** and **biliverdin** are produced from the breakdown of **haemoglobin**, and excreted in bile. If these pigments aren't **excreted**, it can make your skin look yellow (**jaundice**) — often a sign of **liver problems**.	**Bile salts** Bile salts **emulsify fats** in the small intestine. Too **little** bile salt **concentrates** the **cholesterol** in bile, which may **precipitate out** in the **gall bladder** or **bile duct** as **gall stones**.

The *Liver* synthesises *Plasma Proteins*

Plasma (blood) proteins are synthesised from **amino acids** in the liver. There are **three** different types of plasma protein:

1)   **Albumins** — these help keep a **balance** of **fluids** inside and outside of the **blood vessels**.
They also act as **transport molecules** for **fatty acids**, **bile pigments**, **salts** and some **steroid hormones**.

2)   **Globulins** — these transport **lipids**, some **vitamins** and **iron**.

3)   **Fibrinogen** — this is converted to threads of **fibrin**, which form a **mesh** to trap blood cells, helping **blood** to **clot**.

The *Liver* removes **Harmful Substances** *from the* **Blood**

The **hepatocytes** transform harmful substances, like **alcohol**, **drugs**, **spent hormones** and **toxins**, into **less toxic compounds** so they can be **easily excreted** from the body, e.g. ammonia is excreted by the kidneys.

The **liver** metabolises **ethanol** in **alcoholic drinks** into **ethanal**. If you drink a lot of alcohol over a long period of time, **enzyme activity decreases** in the hepatocytes, the cells die, and **scar tissue** (**collagen**) blocks blood flow to the liver (**cirrhosis**). **Albumin** production is **inhibited**, so fluid **leaks** into surrounding tissues, causing **swelling**. **Nitrogenous compounds**, e.g. ammonia, **build up** and start to **poison** the **brain**.

Practice Questions

Q1   List 3 ways in which glucose is produced by the liver when in short supply.

Q2   What are the functions of the plasma proteins?

Exam Questions

Q1   Outline the role of the liver in the control of blood glucose.   [8 marks]

Q2   Jaundice is frequently an indication of liver damage. In this condition, the skin takes on an abnormal yellow colouration because bile pigments are circulating in the blood.

a) Explain why bile pigments are circulating in the blood of a person suffering from liver damage.   [2 marks]

b) Suggest why a person suffering from liver damage is often prescribed smaller doses of any medication required for other complaints.   [2 marks]

Liver and bacon casserole, anyone...

*I don't know why anyone would want to eat liver. It's got the most revolting wibbly-wobbly texture. Eurghh.
Thank goodness they've stopped serving it for school dinners. Still, you've got to admit that it's a really
useful organ with lots of handy functions and not liking liver is no excuse for not learning what's on this page.*

The Skeleton and Joints

These pages are for OCR and Edexcel only. Read these pages for proof that skeletons aren't that scary.

Compact Bone *is made of cells called* Osteocytes

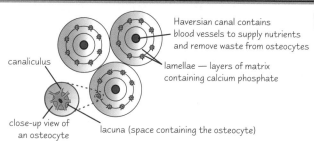

Haversian canal contains blood vessels to supply nutrients and remove waste from osteocytes

lamellae — layers of matrix containing calcium phosphate

canaliculus

lacuna (space containing the osteocyte)

close-up view of an osteocyte

Osteocytes (bone cells) are embedded in a **matrix** of **collagen** and inorganic salts, like **calcium phosphate**. The collagen makes the bone **strong**. The osteocytes arrange themselves in **concentric rings** called **Haversian systems**, so the matrix is laid down in layers (**lamellae**) around each Haversian canal. The spaces that contain the osteocytes are called **lacunae**. The **narrow channels** crossing the lamellae are called **canaliculi**.

The matrix is just the stuff in between the cells in bone and cartilage. Keanu Reeves spent 3 films finding out the answer to that question. Duh.

The Limbs *of all* Mammals *are Similar*

The limbs of all mammals are designed in the same basic form of the pentadactyl limb — part of the evidence that all mammals have a common ancestor. The pentadactyl limb has three parts — a hand / foot with five digits (fingers or toes), a lower limb containing two bones, and an upper limb containing one bone.
The limbs are adapted to different functions in different species — bats have limbs adapted into a wing shape, and seals have flippers to move them through the water.

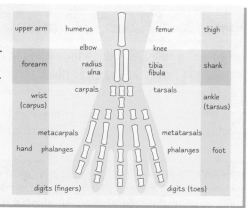

upper arm	humerus		femur	thigh
	elbow		knee	
forearm	radius ulna		tibia fibula	shank
wrist (carpus)	carpals		tarsals	ankle (tarsus)
	metacarpals		metatarsals	
hand phalanges			phalanges	foot

digits (fingers)   digits (toes)

The Structure *of* Vertebrae *is related to their* Function

Thoracic and **lumbar vertebrae** make up part of the **vertebral column**, which **supports** and **protects** the **spinal cord**. Thoracic vertebrae are in the upper part of the spine, and lumbar vertebrae are in the lower part. The parts of each vertebra are **adapted** for different functions.

The **thoracic** vertebrae are the **least flexible** because they can only move forward and sideways slightly. They **support the ribs**.

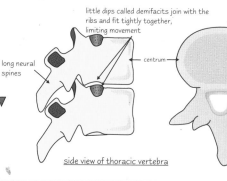

little dips called demifacits join with the ribs and fit tightly together, limiting movement

large centrum to give greater flexibility between lumbar vertebrae

long neural spines

centrum

transverse processes (neural spines) are long and wide, pointing downwards

extra processes keep vertebrae in position when under stress

side view of thoracic vertebra

cross-section of lumbar vertebra

The **lumbar vertebrae** provide **rigid support**, but they also let you **bend**, move sideways and **rotate** your upper body.

Muscles *work in* Antagonistic Pairs

You need **muscles** to move your bones.
Muscles are attached to bones by **inelastic tendons**. The tendons can't stretch, so when a **muscle contracts** it **shortens** and **pulls** the bone.
A different muscle returns the bone to its original position.
Muscles are usually found in **antagonistic pairs** — in the arm, when the **biceps contracts**, it pulls the lower arm upwards. When the **triceps contracts** the lower arm goes back to its original position. This is a **lever action**.

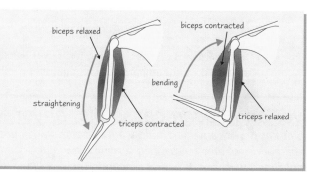

biceps relaxed   biceps contracted

bending

straightening

triceps contracted   triceps relaxed

The Skeleton and Joints

Synovial Joints occur at the Elbow, Knee, Shoulder and Hip

Some joints hold bones tightly together and allow **very little movement**, such as the bones of the skull, and others allow **free movement**, e.g. synovial joints.

Synovial joints act as **levers** with **antagonistic muscles** and occur where **two bones meet**.

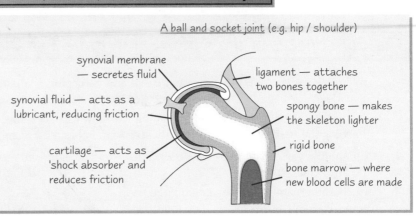

A ball and socket joint (e.g. hip / shoulder)

synovial membrane — secretes fluid

synovial fluid — acts as a lubricant, reducing friction

cartilage — acts as 'shock absorber' and reduces friction

ligament — attaches two bones together

spongy bone — makes the skeleton lighter

rigid bone

bone marrow — where new blood cells are made

Cartilage is a type of Skeletal Tissue

The simplest form of cartilage is **hyaline cartilage**. It's made of living cells called **chondrocytes** that secrete a matrix called **chondrin**. It's **strong**, but **flexible** and **elastic**. It **reduces friction** at joints, allows limited **movement**, provides **support**, e.g. in the trachea, and is responsible for **growth** in long bones.

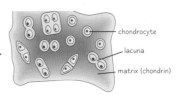

chondrocyte

lacuna

matrix (chondrin)

Osteoporosis and Osteoarthritis are effects of Ageing

As you get older, you can start to feel the effects of ageing in your bones. **Osteoporosis** and **osteoarthritis** are two common problems in **older people**:

1) **Osteoporosis** is a loss of **calcium** from **bone tissue**, which makes bones weaker, **more brittle** and **more likely to fracture**. This often affects **women** more than men — calcium needed by the body comes from the bones, and it's controlled by a hormone called **parathormone**. **Oestrogen inhibits parathormone** but during the **menopause** (see page 161), as oestrogen levels **decrease**, **more calcium** is **removed** from the bones. Osteoporosis can be treated with calcium supplements or, for women during the menopause, Hormone Replacement Therapy (HRT). It can be prevented by getting lots of calcium in your diet and doing weight-bearing exercise (e.g stair climbing).

2) **Osteoarthritis** is where the **cartilage deteriorates** and wears away, and there's **less synovial fluid** in the joint. The bones can get **damaged** or deformed as they **rub** against ligaments and other bones. This can lead to painful, swollen joints. It can be treated with drugs for the pain and swelling, or in severe cases, **joint replacement** — that's why a lot of older people have **hip replacements**.

Practice Questions

Q1 Name three functions of the skeleton.

Q2 How is the structure of thoracic and lumbar vertebrae related to their function?

Q3 Give one difference between osteoporosis and osteoathritis.

Exam Question

Q1 a) Explain why osteoporosis-related bone fractures are significantly more common in women than men by the time they reach the age of 70. [2 marks]

b) Explain how osteoarthritis causes joint pain. [2 marks]

Make sure you've learnt the bare bones of this section...

This stuff isn't very difficult, but there's a lot of info crammed on to these two pages. It might be stuff that you haven't really covered in detail before, so make sure that you know it. You need to be able to recognise the structures of compact bone and hyaline cartilage under a light microscope, so learn the main features of each. Ah go on, it'll be worth it.

The Mammalian Ear

These pages are for OCR only. *Your ears are always being bombarded with sound. They do a pretty good job of passing on all the messages they get. Try reading this page out loud and let your ears help with your revision...*

The **Ear** consists of the **Outer Ear**, **Middle Ear** and **Inner Ear**

1) The **outer ear** contains the **pinna**, **ear canal** and **eardrum**. The **pinna** (the outside bit of your ear) is made of cartilage. It **collects** and **focuses** sound waves into the **ear canal**.

   The **ear canal** is lined with skin containing **hair** and **wax-secreting cells**, which stop foreign objects reaching the **eardrum**. The **eardrum** (tympanic membrane) **transmits vibrations** to the **middle ear**.

2) The **middle ear** contains the **ear bones** and the **Eustachian tube**.

   The **hammer, anvil** and **stirrup** are three tiny bones in an air filled cavity. They transmit **vibrations** to the **inner ear**.

   The **Eustachian tube** connects the middle ear to the back of the nose. It **equalises** the **pressure** between the **middle ear** and the **air outside** (when your ears 'pop' when you change altitude in an aeroplane, you're equalising the pressure in your middle ear).

3) The **inner ear** contains the **cochlea** and the **semicircular canals**.

   The **cochlea** is a **coiled tube** divided into **three** fluid-filled canals — the **vestibular**, **median** and **tympanic canals**. The fluid (perilymph) is kept in by the membranes of the **oval** and **round** windows.

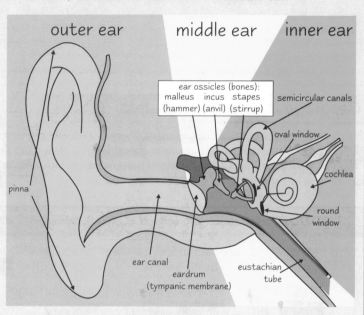

Sound travels as *Waves*

Sound is produced by the **vibration** of **air particles** and travels into your ear as waves.

1) **Sound waves** are collected by the **pinna** and channelled through the **ear canal**, making the **eardrum** vibrate.

2) Vibrations from the eardrum make the **malleus**, **incus** and **stapes vibrate** against the **oval window**. These act as a **lever system** to **increase** the force of vibrations from the eardrum. It's also increased because the **surface area** of the **eardrum** is **larger** than that of the **oval window**.

3) The sound waves enter the inner ear as pressure waves that move the **perilymph** through the **inner ear** and **across** the **membrane** of the **middle canal**.

4) **Sensory hair cells** (**mechanoreceptors**) on the membrane are **stimulated** when the fluid **passes** over them. The **hair cells** are attached to the **nerve cells** of the **auditory nerve**, which sends the **nerve impulses** to the **brain** for processing — it tells you what you're hearing.

5) The **round window** bulges **outwards** to **release** the **energy** of the **pressure waves** so the ear drum doesn't get damaged.

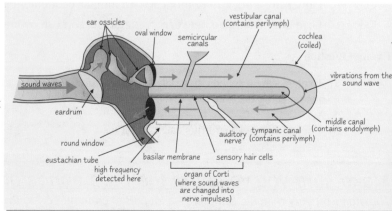

The Mammalian Ear

Semicircular Canals help you keep your Balance

The bony, semicircular canals have **three loops** of **canals** filled with a fluid called **endolymph**. Each has an **ampulla** which connects to the **vestibule** and then the **cochlea**. The three canals detect the **direction** and **rate of change** of the position of your **head**. **Receptors** in the ampullae detect fluid movement, and send messages to the brain as your head changes position. If you spin round fast, the fluid in the canals tends to **stay behind** in its **original position**, which can make you feel dizzy.

Two cavities called the **utricle** and **saccule** contain receptors that give information on the **position** of the head and body, so they help keep you **upright.**

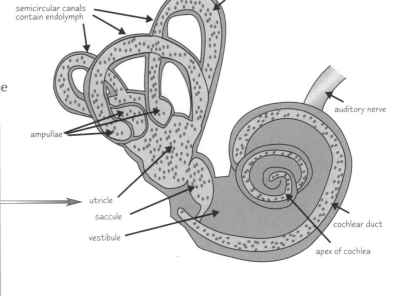

semicircular ducts

semicircular canals contain endolymph

auditory nerve

ampullae

utricle

saccule

vestibule

cochlear duct

apex of cochlea

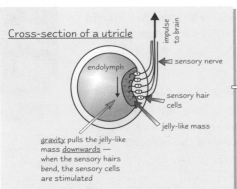

Cross-section of a utricle

impulse to brain

endolymph

sensory nerve

sensory hair cells

jelly-like mass

<u>gravity</u> pulls the jelly-like mass <u>downwards</u> — when the sensory hairs bend, the sensory cells are stimulated

Practice Questions

Q1 How are bacteria prevented from entering the middle ear?

Q2 Outline the role of the ear ossicles.

Q3 Describe the part played by the Eustachian tube.

Q4 Explain the significance of the inner ear being fluid-filled.

Q5 How is balance achieved?

Exam Question

Q1 Ménière's disease is a disorder of the inner ear characterised by fluctuating hearing loss, attacks of vertigo (a sensation of movement when standing still), and tinnitus (ringing in the ears). It is thought to be caused by under or over-production of endolymph (fluid) in the cochlear duct.

Treatment for the condition may be medical or surgical. Medical treatment aims to reduce the volume of endolymph. This may be achieved by administering diuretics, which increase the volume of urine produced.

a) Name the structure in the cochlea responsible for the perception of sound waves. [1 mark]

b) Suggest how an increase in fluid may cause hearing loss. [3 marks]

c) Suggest how treatment with diuretics may improve this condition. [2 marks]

d) Suggest what may occur in the ear to cause the sensation of movement when standing still. [1 mark]

Bet you didn't realise there was a blue mutant snail living inside your ear...

The ear isn't all that complicated really. There's something amazing about how those vibrations end up in our brains as words and songs and the teachers' voices droning on and on in the background and lawnmowers and pneumatic drills and screaming babies and people scraping their nails down blackboards and aaaargh, pass me those earplugs before I go mad.

SECTION TEN — MAMMALIAN PHYSIOLOGY AND BEHAVIOUR

Simple Behaviour Patterns

These pages are for OCR, AQA A and AQA B. If you're doing AQA A you only need the bit on taxes and kineses.
You've already covered some behaviour patterns in the core content so these pages shouldn't be too hard.
There are a few classic case studies you need to learn, but then you should be fine.

Behaviour *is an Organism's* Response *to* Changes *in its Environment*

Behaviour lets an organism **respond** and **survive**. Behaviour can either be **innate** or **learned**, but most behaviour relies on a blend of the two. Both your **genes** and your **environment** play a part in influencing your behaviour and it's sometimes hard to decide what is innate and what is learned. One example is **human speech** — almost all humans are born with the **innate ability** to speak, but if a child is born **deaf**, or is brought up in **isolation** and doesn't hear people **speaking**, then it won't learn to speak. Human babies have to **learn** language.

Innate Behaviour *is* Inherited *and* Instinctive

Innate behaviour is **instinctive**. Animals can respond in the **right way** to the stimulus **straight away**, even though they've never done it before, e.g. newborn mammals have an instinct to suckle from their mothers. Instinctive behaviour can be a fairly simple **reflex**, or a set of **complicated** behaviour, like a courtship ritual.

Bradley thought that the piglets' suckling instinct was getting rather out of control

> **Reflex actions** are simple innate behaviours, where a stimulus produces a fairly simple response, like sneezing, **salivation**, coughing, and blinking. They often protect us from **dangerous stimuli**. Although reflexes are **involuntary** (automatic) actions, they can sometimes be **modified**, e.g. humans usually learn to **control** their bladder **sphincter** when they're quite young (see page 81).

> Taxes and **kineses** are **reflexes** which allow animals to move away from **unpleasant stimuli** (see p.121).
> 1)   A **taxis** is a **directional movement** in response to a **stimulus**,
>   e.g. earthworms show **negative phototaxis** — they move away from light.
> 2)   A **kinesis** is a change in the rate of movement, depending on the intensity of a stimulus,
>   e.g. **sea anemones** wave their tentacles more when stimulated by **chemicals** emitted by their prey.

Innate responses are often stereotyped. **Stereotyped responses** are controlled by **sequences** of behaviour called **Fixed Action Patterns** (**FAP**), e.g. nest building, courtship. These are **automatic** and the **same type** of **response** is always given to a stimulus, with **fixed patterns** of **coordinated** movements (see pages 164-165 for more on this).

Learned Behaviour *is* Modified *in response to* Experience

Learned behaviour isn't **innate** — you have to learn it, obviously. It lets animals **respond** to **changing conditions**. Animals can learn to avoid predators and harmful food, and to find food or a suitable mate. Some important examples of learned behaviour are **habituation**, **classical conditioning**, **operant conditioning** and **insight behaviour**:

Habituation

> If you keep on giving an animal a stimulus that isn't **beneficial** or **harmful** to it, it quickly learns **not** to respond to it. This is called **habituation**. That's why you can often **sleep through** loud and familiar noises, like traffic, but might wake up instantly at a quiet but **unfamiliar** noise. By **ignoring** non-threatening and non-rewarding stimuli, animals can spend their time and energy more **efficiently**.

Classical Conditioning

> **Classical conditioning** happens when an animal **learns** (passively) to associate a 'neutral stimulus' with an important one, e.g. a dog associates a bell ringing with the arrival of food. The response is **involuntary**, **temporary** and **reinforced** by **repetition**.

> #### Ivan Pavlov — Classical Conditioning in Dogs
> Pavlov studied the behaviour of dogs and noticed that they would **salivate** (drool) every time they saw or smelt food. He began to ring a **bell** just before each time the dogs were given their food. After a while he found that the dogs salivated when the bell was rung even if he didn't give them their food.

Simple Behaviour Patterns

Operant Conditioning

Operant conditioning or **'trial and error learning'** is where an animal **learns actively** to **associate** an **action** with a **reward** or a **punishment**. This happens in **humans** when children are rewarded or punished for **specific behaviour**.

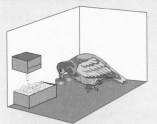

Burrhus Skinner — Operant Behaviour in Pigeons and Rats

Skinner trained **rats** and **pigeons** to obtain a **food reward** with a **'Skinner box'** that he invented. The animal had a choice of buttons to press. When the animal pressed a particular lever or button, it was **rewarded** with food. He found that pigeons and rats used a system of **trial and error** to learn which button to press to get the reward.

Insight Behaviour *is the most* Complex *form of* Learned *behaviour*

This type of behaviour involves **solving a problem** by looking at it, thinking about it, and using **previous experiences** to help solve it:

Wolfgang Kohler — Insight Behaviour in Chimpanzees

People used to think that only humans made and used tools. **Kohler** worked with **chimpanzees**, putting them in a play area with, for example, a bunch of **bananas out of reach**. To get the bananas, the chimp would have to **use an object** as a **tool**. The chimp had different length sticks and wooden boxes. They **used** sticks as **tools** to pull in the bananas, clubs to get them from above, longer sticks to climb up and even **piled up** the **boxes** to climb up them.

Kohler concluded that chimps showed **insight behaviour**.

Imprinting *involves* Innate *and* Learned *Behaviour*

Most behaviour is a **combination** of innate and learned behaviour, e.g:

Baby ducks learn to recognise and follow the **first moving object** that they see during a **critical period** soon after they **hatch**. This is called **imprinting**. The first thing they see is usually their **mother**, but if they see a human, they can become imprinted on the human instead. Although they have an **innate instinct** to follow the first moving object they see, they have to **learn** what their 'mother' looks like, because they have no **innate experience** of what an adult duck looks like.

Practice Questions

Q1 Describe the difference between 'taxis' and 'kinesis'.

Q2 A cow touches an electric fence and gets a shock. From then on, it avoids the fence. What kind of conditioning is this?

Q3 Draw a table of differences between 'classical conditioning' and 'operant conditioning'.

Exam Questions

Q1 Describe a type of behaviour that appears to be a combination of genes and environment. [3 marks]

Q2 Suggest how operant conditioning could be used in dog training. [3 marks]

Oh bee-have...

It's a bit weird to think that a lot of our behaviour is instinctive and the rest of it we've probably been trained to do by our parents. I'm starting to wonder if we ever do anything just because we want to — after all, school is just one long system of rewards for some behaviours and punishments for others... might as well be a pigeon in a box, I reckon.

Growth and Development

This section is for OCR and AQA B only. It's not a bad one — the stuff on reproduction should be pretty familiar. I was going to say you do it every year, but realised that could be taken the wrong way... ***These two pages are for OCR and AQA B.***

Cell Division and Enlargement lead to Growth

Growth in organisms usually involves an increase in **size**, an increase in **mass** and an increase in the **number of cells**. It's how a single-celled **zygote** eventually becomes a fully grown **adult**. It involves an **irreversible** increase in mass which happens mainly through **cell division** (**mitosis**) and some **cell enlargement**.

Growth can be **measured** in different ways for different organisms:

1) **Microbes** — growth of **populations** over time can be measured by:
   a) **Culturing** them in a growth medium, e.g. on agar plates, and counting the number of **visible colonies**. Every single bacterium can produce a visible colony, so **serial dilutions** (see page 181) might be needed.
   b) Counting **individual microbes** on a microscope slide or **haemocytometer** (see page 181). **Replication** is important (taking lots of repeat readings), as there could be an **uneven distribution**.
   c) **Turbidity** (cloudiness) readings can be done by measuring how much **light** passes through a **bacterial suspension** at various stages in the growth of the population using a **colorimeter**.
   d) Measuring the concentration of **metabolic by-products** produced by the microbe at various stages in the growth of the population, e.g. the amount of **carbon dioxide** produced by **yeast**.

2) **Plants** — taking **dry mass** measurements is a common method — the sample is dried out (often in a hot oven) and **weighed** at intervals until no more mass is lost. This means all the mass is due to the plant, and not to any extra water it might have absorbed. Obviously this **kills** the organism, so you have to measure **future growth** by taking **random samples** from a big population of plants, which must be the **same age** and growing under **identical conditions**. An easier way is to measure changes in **length**, and this doesn't kill the plant.

3) **Animals** — **mass** and **length** are common measurements for both the whole organism and for its various body parts. **Dry mass** measurements can be done but they have one pretty obvious drawback...

Growth **rates** in all these organisms can be measured in two ways:

1) **Absolute growth rate** is the amount of growth made by an organism **in a given period of time**. For example, if a root grew by 9 mm over 24 hours, its absolute growth rate would be **9 mm per day**. A curve showing the **total** amount of **growth** would look different to one showing absolute growth rate. It would increase steadily, because organisms don't tend to shrink, and it'd probably be sigmoid in shape. But a **curve of absolute growth rate** can increase and decrease, because an organism can grow more quickly for a while and then grow more slowly.

2) **Relative growth rate** relates the **absolute growth rate** to an organism's **size**, e.g. if a 10 g organism and a 20 g organism both grow 1 g in a day, the 10 g organism has **twice** the relative growth rate. Graphs of relative growth rate often show a **decrease** over time.

Typical Growth Curves

The classic **growth curve** is **sigmoid** or **s-shaped**. It has four clearly defined **phases** (although the sigmoid shape is just the first three), and can be seen by growing bacteria in a **closed environment** — see page 24 for details.

In a more natural **open environment**, growth curves often flatten out at the **stationary phase** and then **fluctuate** for long periods of time, depending on **limiting factors** such as food supply, predators, etc. A natural population won't often enter the final **death phase**.

In **humans**, different growth curves are produced depending on what's measured — e.g. **absolute mass** or **height** over time, **relative growth rates** of certain parts (e.g. the brain or reproductive organs), etc. Similar growth curves can often be produced in **plants** by measuring **dry mass**.

Note the <u>sharp increase</u> in first few years, then a <u>growth spurt</u> at puberty.

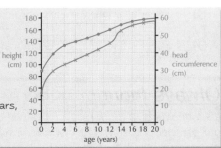

Growth and Development

OCR ONLY

Development is a Series of Changes involving Cell Differentiation

Growth basically means getting bigger, but **development** is more complicated. It involves producing **specialised** cells and arranging them into **tissues** and **organs**. The cells have to become **adapted** for the functions they'll carry out. This process is known as **differentiation**. Differentiation can be seen in **root** and **shoot tips** (apices), behind the areas where growth is happening by **cell division** and **elongation**. The new cells made by cell division have to undergo changes to turn them into, for example, a specialised leaf cell or xylem cell.

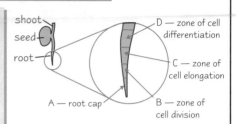

AQA B ONLY

Puberty is the onset of Sexual Maturity

Puberty is the stage between childhood and adulthood. It's started by the release of a hormone called **GnRH** from the **hypothalamus**, which stimulates the **pituitary gland** to release other hormones:

1) In females **FSH** and **LH** are released, which causes the ovaries to release **oestrogen** and start the **menstrual cycle**.

2) In males **FSH** and **LH** are released, which causes the testes to release **testosterone** and start to produce **sperm**.

These **sex hormones** cause the development of the **sexual characteristics**:
A **long** pre-puberty stage in humans means there's a long time for the **brain** to develop before maturity. There is also a long time for **learning** and developing **complex behaviour**. This has been important in **evolutionary terms**, as humans have been able to use their intelligence to survive.

Female	Male
Menstruation	Penis and testes grow
Breasts grow	Muscles develop
Pubic hair	Pubic and body hair
Hips widen	Larynx enlarges so voice deepens

AQA B ONLY

Ageing is associated with less Efficient Functioning of the Body

The effects of **ageing** are sadly inevitable, and include:

1) Changes in **physiological functions** — organs begin to function **less efficiently**. For example, there's a decline in **cardio-vascular** efficiency and capacity with old age.

2) **Tissue degeneration** — structural proteins become **harder** and **less elastic**.

3) Accuracy of **DNA replications** declines, so some **genetically abnormal** cells start to be formed during **mitosis**. These cells won't be able to do their jobs properly. They might even die or become **cancerous**.

4) Decline in the effectiveness of the **immune system** — making many diseases **more serious** in later life.

Practice Questions

Q1 Explain the difference between absolute growth rate and relative growth rate.

Q2 Give three ways in which human growth could be measured.

Q3 Explain the difference between growth and development.

Exam Questions

Q1 Describe how various hormones trigger the changes seen in boys at puberty. [5 marks]

Q2 From the graph shown on the right, find:
a) the age (in years) at which rate of growth is highest. [1 mark]
b) the rate of growth at 4 years of age. [1 mark]
c) when growth has stopped. [1 mark]

Q3 Many people become less mobile and experience more health problems in their old age. Briefly explain why, in terms of changes within the body. [8 marks]

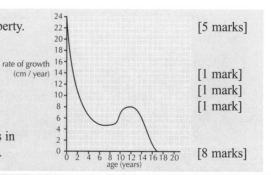

I always thought puberty was associated with immaturity...

Actually, that kind of comment isn't going to wind you lot up any more, is it? Most of you are safely past all that, and sighing in disgust at the teenaged antics of the fifteen year olds in the corridor. You're probably nodding smugly in agreement, but don't get too pleased with yourselves — just round the corner waits ageing, and that's even less fun.

Asexual Reproduction

Asexual reproduction seems pretty weird to us mammals. Imagine if we went round with babies budding off us, or if your leg fell off and turned into your kid. Still, it'd be a boring world if we were all the same. **These two pages are for OCR.**

Asexual Reproduction *produces* Genetically Identical Offspring *(clones)*

Asexual reproduction happens in a **wide range** of different organisms using lots of different **mechanisms**. You need to know one example from each of the **five kingdoms**:

1. Prokaryotes (bacteria)

In these single-celled (unicellular) organisms, a process of **binary fission** (splitting in two) usually occurs to produce clones. The DNA in the parent cell undergoes **replication** before it splits in two.

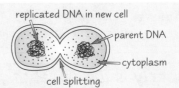

replicated DNA in new cell
parent DNA
cytoplasm
cell splitting

2. Protoctists

These include unicellular **protozoa** and **plant-like algae**.
Simple protoctists such as the **amoeba** also reproduce by **binary fission**.

3. Fungi

Some **unicellular** fungi, such as **yeast**, reproduce by **budding**. The new yeast cells bud out from the parent's body and then **detach** from it. More **complex** fungi such as **mushrooms** reproduce using large numbers of microscopic **spores**. The above ground part of the mushroom is the **fruiting body**, which produces spores in the **gills** underneath the cap to be carried away by the wind.

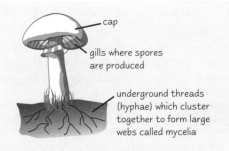

cap
gills where spores are produced
underground threads (hyphae) which cluster together to form large webs called mycelia

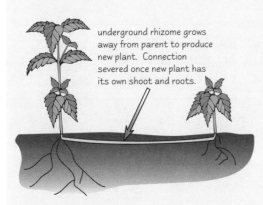

underground rhizome grows away from parent to produce new plant. Connection severed once new plant has its own shoot and roots.

4. Plants

Many plants reproduce asexually. New individuals grow out from a part of the parent plant. This is known as **vegetative growth** or **vegetative propagation**. Examples of how new plants arise include:

- **Tubers** (swollen underground stems) in **potatoes** that can survive the winter in the soil.
- **Runners** (horizontal stems growing on top of the soil) in **strawberries**. They're produced throughout the growing season.
- **Bulbs** (underground shoots surrounded by fleshy leaves) in **daffodils**.
- **Corms** (swollen underground stems) in **crocuses**.
- **Rhizomes** (horizontal stems growing under the soil) in **mint**.

Plants can also be produced by <u>artificial propagation</u>, which is asexual reproduction manipulated by humans. See page 146.

5. Animals

Asexual reproduction in animals is **less common** and is usually only seen in **simpler** animals. Examples include:

- **Regeneration** of separated body parts (also known as **fragmentation**) which is done by **sponges**, **flatworms** and **starfish**.
- **Parthenogenesis** — development of young from **unfertilised eggs**. Seen in insects like **ants**, **bees** and **aphids**.
- **Budding** in *Hydra* (a **Cnidarian**, like jellyfish and corals) where offspring are produced from **buds** grown on the parent.

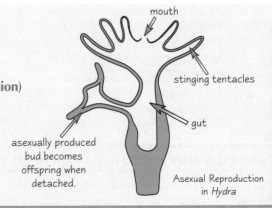

mouth
stinging tentacles
gut
asexually produced bud becomes offspring when detached.
Asexual Reproduction in *Hydra*

Asexual Reproduction

Asexual Reproduction has Advantages and Disadvantages over Sexual Reproduction

ADVANTAGES	DISADVANTAGES
No need to find a mate — therefore faster	One parent leads to lack of variation
Usually no gamete production — saves time and energy	Huge numbers of potential offspring can be wasted, e.g. spores
Dispersal and colonisation of offspring can be easy, e.g. spores	Limited dispersal, e.g. tubers, can lead to overcrowding
Identical clones maintain 'good' strains	'Bad' strains can't improve and may become extinct

These advantages and disadvantages have **evolutionary** consequences. If there's no **genetic variation** then there can't be any '**survival of the fittest**'. This means that a group of clones or even whole species could be **wiped out** by one particular pathogen or by a particular combination of environmental changes.

Sexual reproduction gives rise to different **strains** and **varieties**. This encourages **resistance** to pathogens to develop. If there's an **epidemic** of a particular pathogen in a population, any resistant individuals will survive. They'll go on to reproduce and the resistance will become **widespread** in the population.

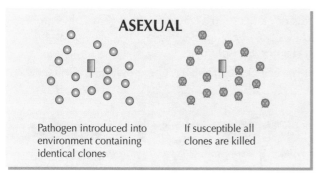

ASEXUAL

Pathogen introduced into environment containing identical clones

If susceptible all clones are killed

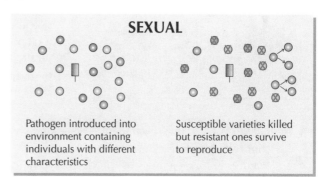

SEXUAL

Pathogen introduced into environment containing individuals with different characteristics

Susceptible varieties killed but resistant ones survive to reproduce

Practice Questions

Q1   What is meant by binary fission?

Q2   Give four examples of plants that reproduce asexually.

Q3   Name two organisms that reproduce by budding.

Q4   What advantages are there in reproducing by producing spores?

Exam Question

Q1   a) Match each of the organisms below to its reproductive technique.

Organism	Reproductive Technique
Hydra	Tubers
Potato	Parthenogenesis
Aphid	Sexual
Cod	Binary fission
Amoeba	Budding

[5 marks]

    b) Explain how potato plants carry out asexual reproduction.   [5 marks]

<u>Stick to the pros and cons in the box above please — no getting creative...</u>

Asexual reproduction goes on more than a lot of people think. Actually, most people don't think about it at all — they tend to dwell on the other kind, for some reason. But now you've got to start thinking about it, because for the exam you need to know at least one example from each of those five kingdoms. Quick, what about the Protoctista? Fungi? OK I'll stop now.

Artificial Propagation and Cloning

Lots of plants reproduce asexually naturally. But we humans have come up with cunning ways to meddle with this so that we can make loads of pretty or useful plants for our own use. **These two pages are for OCR only**.

Artificial Propagation is Asexual Plant Reproduction Manipulated by humans

Knowledge of plant growth and development has been used commercially to develop methods of **artificial propagation**. This is a quick and easy way to **increase** plant numbers. It lets commercial growers produce lots of **identical clones** of varieties with desirable features, so it's particularly important in the **horticultural industry**.

These are the main methods of artificial propagation of plants:

1. **LAYERING** — This is a technique often used with plants that produce **horizontal runners**, like **strawberries**. The runners are pegged down onto the soil or into a pot, encouraging the **buds** on the runners to **root**. When the new plants have rooted they can be cut from the parent and grown elsewhere.

    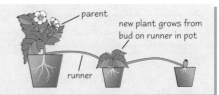

2. **CUTTINGS** — Bits of some plants (often the **shoots** and **leafy parts**) can be cut off and made to **root** when put in soil or compost. Hormone-based **rooting powder** can be used to encourage root growth. Examples include:

    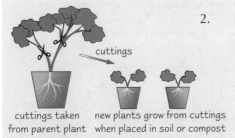

    a) The young leafy tips of plants such as **lavender** and **geranium** can be cut off and placed in a pot of compost or even a test tube of water until they root and can be transplanted.

    b) The shoots of fruit bushes like **currants** can be grown in a similar way.

3. **GRAFTING** — This is where the **upper section** of a **shoot** (the **scion**) is removed from a plant with desirable features (e.g. apple tree that produces lots of fruit), and grafted onto the **lower section** of **shoot and roots** (the **stock**) of another variety which has a better root system. This gives a plant with the best features of both original plants. It's often used to produce **fruit trees** and **roses**.

    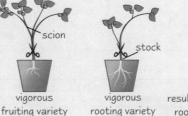
    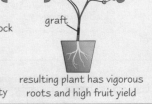

4. **BUDDING** — Plants like **roses** have newly growing **axillary buds** which are found between the side shoots and the main stem. These can be **removed** and grown in a special medium that **promotes growth** and allows **root development**.

5. **TISSUE CULTURE** — (also known as MICROPROPAGATION)

    Tissue culture involves producing **clones** from various plant parts. It's done in laboratories and greenhouses under closely monitored conditions. Some tissue called an **explant** is removed from the parent (stock) plant and grown in a culture medium containing **sugars**, **nutrients**, **minerals** (such as N,P and K) and **plant hormones** at a **constant temperature**. It's done in **sterile** conditions to prevent growth of microbes that could harm the plants.

    Different tissues can be used, but rapidly growing **meristems** are the most common. These can be treated with plant hormones such as **cytokinins** to promote **shoot** growth, and **auxins** to stimulate **root** growth. **Non-meristematic tissue** can be used to produce a mass of **undifferentiated** cells called a **callus**. This can then be treated with **hormones** to give normal plants.

    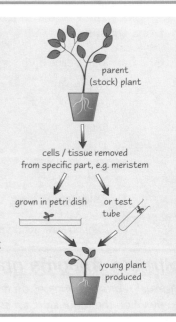

    As the tissues produce shoots and roots, they can be moved to **potting compost** to grow into mature plants. This is done on a **huge scale** producing large numbers of clones. Lots of different plants e.g. **raspberries**, **houseplants** and **orchids** are produced using tissue culture.

Artificial Propagation and Cloning

Tissue Culture has Numerous Advantages and Disadvantages

The development of **tissue culture** has meant new plants can be produced on an **industrial scale**.
A worldwide industry now exists to supply growers with huge numbers of **genetically identical plants**.
But the technique has **disadvantages**, as well as **advantages**:

Advantages

- Desirable features can be **reproduced exactly**.
- Only a **small amount** of tissue is needed.
- New plants are produced very **quickly**.
- Only **one parent** is needed to provide the stock plant.
- Sterile conditions prevent plant **diseases**.
- No seed **germination problems**.
- The plants produced are often **stronger** than seedlings.
- New plants can be produced **all year** round.
- New plants in the early stages are small and easy to **transport**.
- **Genetically engineered** plants are easily propagated.
- Clones can be used to synthesise lots of the same **chemicals** and **metabolites** for use in industry.

Disadvantages

- If clones are susceptible to a disease it could wipe out the **whole population**.
- **Fewer** new plants are produced compared with sowing seeds.
- Sterile conditions are hard to set up and maintain — **contamination** happens easily.
- Labour can be **expensive**.

Desirable features, such as attractive flowers, can be reproduced exactly

Practice Questions

Q1 Name four methods of artificial propagation.
Q2 What are the scion and the stock of a plant?
Q3 Which hormones are important in micropropagation?
Q4 Give five advantages and two disadvantages of producing new plants from cuttings rather than seeds.

Exam Question

Q1 Match each of the terms below with its correct definition:

Term	Definition
cytokinin	Tissue culture used to produce plant clones.
grafting	Hormone used to promote root growth.
auxin	Technique of staking horizontal runners.
micropropagation	Procedure to join two parts of different plants.
layering	Hormone which promotes cell division.

[5 marks]

Mmmm, imagine grafting Brad Pitt onto David Beckham...

Or Colin Farrell onto Jonny Wilkinson. Or for the boys, perhaps Kylie Minogue onto Holly Valance. I could go on for hours like this, but I'm just distracting you from the really exciting business of layering, grafting and tissue culture. Which can make you a very nice flower with good roots, and then copy it lots of times. So hurrah. Well done.

Plant Pollination

Some plants can reproduce sexually, which is quite clever of them considering they can't move. They've had to evolve other ways of getting their sex cells together, even going so far as to use slugs. Ew. **These two pages are for OCR.**

*Pollination is the **Transfer of Pollen** between Flowers*

Pollination is the transfer of pollen from an **anther** (male) to a **stigma** (female) of flowers of the **same species**.
It happens **before** fertilisation — don't get pollination and fertilisation mixed up. There are **two types** of pollination:

1) **Self-pollination** — pollen is transferred to the stigma of the **same flower** (or another flower on the same plant). This is possible because most flowers are **hermaphrodites** — they have both female **and** male sex organs.

2) **Cross-pollination** — pollen is transferred to the stigma of a **different flower** (on another plant) of that species.

Self-pollination is not as effective as cross-pollination at generating **genetic variation**, which can be a disadvantage (see p.145) Some plants have evolved ways to make sure **cross-pollination** is more likely than self-pollination:

1) **Protandry** — **anthers** mature before **stigmas**, so self-pollination can't happen, e.g. **wood-sage**.

2) **Protogyny** — **stigmas** mature before **anthers**, e.g. **ribwort plantain**.

3) **Dioecious** species — some species have evolved so that they have either all **male parts** or all **female parts**, making self-pollination impossible, e.g. **holly**.

*Most Flowers use either **Wind** or **Insects** for Sexual Reproduction*

So some plants can reproduce **sexually**, but as they can't move they need cunning ways of getting the **male gametes** (in pollen) to the **female bits** of other flowers. Most use either **insects** or the **wind** to do this, and have certain **adaptations** depending on which method they use.

Insect-pollinated flowers have adaptations to attract insects, such as **brightly coloured petals**, **nectar** and a strong **scent**. Insects land on the flower and **pollen grains** get stuck to them. They then deposit the pollen onto other flowers they land on. Flowers often have intricate structures to make sure that pollen is both **picked up** and later **deposited** by the insect.

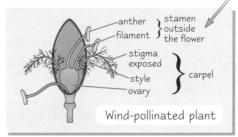

The male gametes are found in the <u>pollen grains</u>. They're made in the <u>anther</u>. The female gametes are found in the <u>ovules</u>. They're made in the carpel, which later becomes the <u>ovary</u>.

Each carpel consists of an ovary and ovules. It later becomes the ovary.

Insect-pollinated plant

Grasses are **wind-pollinated**. They don't need to attract insects, so they tend to be **dull** and **unscented**. Often they don't have petals, leaving the anthers and stigmas **exposed** to the wind. The wind blows pollen grains from the **anther** of one plant to the **stigma** of another plant. This means that **large quantities** of **light** pollen have to be made, to **increase** the chances of pollination happening.

Wind-pollinated plant

The main features of a **typical wind-pollinated** and a **typical insect-pollinated** flower are **summarised** in the table:

FEATURE	INSECT-POLLINATED	WIND-POLLINATED
petals	large, brightly coloured, shiny	inconspicuous, dull and small
sepals	present beneath the petals	small or absent
stamen (male)	sturdy and found inside flower	flimsy with large anthers and long filaments hanging outside flower
stigma (female)	sticky and enclosed within flower	long feathery stigmas give a big surface area to catch pollen
pollen grains	large and sticky	light and dry
scent	present	absent
nectar	produced in nectaries to attract insects	none produced

Examples of <u>wind-pollinated</u> flowers include grasses, ragweed and maize. There are some plants that are even pollinated by <u>water</u> e.g. Canadian pondweed. Examples of <u>insect-pollinated</u> flowers include buttercups, sweet peas, poppies, foxgloves and orchids. There are a few other animals that also carry out pollination e.g. <u>humming birds</u>, <u>moths</u> and even <u>slugs</u>.

Plant Pollination

Pollen Grains contain the Male Gametes and are made in the Anthers

Inside an **anther** there are four compartments called **pollen sacs** where the pollen is made. Each of the four pollen sacs has a thick wall with an inner layer called the **tapetum**. At the centre of each pollen sac are **diploid pollen mother cells** (with two sets of chromosomes) which each divide by **meiosis** to make four **haploid pollen cells** (with one set of chromosomes each) called a **tetrad**.

Compare this process to <u>spermatogenesis</u> in the testes of male animals (p.153).

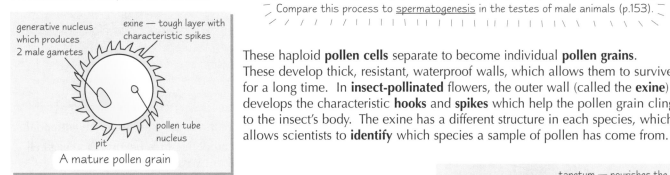

generative nucleus which produces 2 male gametes
exine — tough layer with characteristic spikes
pollen tube nucleus
pit
A mature pollen grain

These haploid **pollen cells** separate to become individual **pollen grains**. These develop thick, resistant, waterproof walls, which allows them to survive for a long time. In **insect-pollinated** flowers, the outer wall (called the **exine**) develops the characteristic **hooks** and **spikes** which help the pollen grain cling to the insect's body. The exine has a different structure in each species, which allows scientists to **identify** which species a sample of pollen has come from.

The **haploid nucleus** in each **pollen grain** divides in two by **mitosis** to form a **generative nucleus** and a **pollen tube nucleus** (see p.150). It's the **generative nucleus** that eventually divides to form the two **male gametes** used to **fertilise** the female gametes. When the pollen grains are ripe, the **anthers** split open (**dehiscence**). This **releases** the pollen grains from the pollen sacs and anther, ready to be transferred to a stigma in **pollination**.

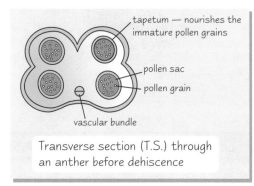

tapetum — nourishes the immature pollen grains
pollen sac
pollen grain
vascular bundle

Transverse section (T.S.) through an anther before dehiscence

Practice Questions

Q1 Explain the difference between self-pollination and cross-pollination.
Q3 Why is cross-pollination usually considered preferable to self-pollination?
Q2 Describe five features of a typical insect pollinated flower.

Exam Questions

Q1 a) State whether the cells at each of the stages i) to iv) in the diagram are haploid or diploid.

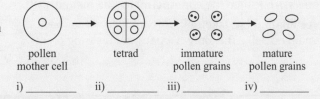

pollen mother cell
tetrad
immature pollen grains
mature pollen grains

i) _____ ii) _____ iii) _____ iv) _____ [4 marks]

b) Gardeners are advised to grow maize (corn on the cob) plants in blocks rather than in rows, to aid pollination. Suggest why this might be. [2 marks]

Q2 a) Define cross-pollination and self-pollination. [2 marks]
b) Outline three methods that plants have evolved to avoid self-pollination. [6 marks]
c) Suggest why it might have been an evolutionary advantage to avoid self-pollination. [2 marks]

Cross-pollination — it's certainly getting on my nerves...

Easy — insect-pollinated flowers have to look and smell nice to get the insects' attention, and they produce nectar to entice them in. Wind-pollinated flowers have sticky-out bits so the wind can blow the pollen away, and feathery parts to catch it. Spend a bit more time on the development of plant male gametes though. It might take a while to get it clear in your mind.

Sexual Reproduction in Plants

Once plants have managed to get their gametes sorted out, they can move on to fertilisation. It's a bit more complicated than the good old sperm and egg though — sorry. **These two pages are for OCR.**

The **Female Gametes** Develop inside the **Ovule**

The **carpels** contain the **ovules**, which in turn contain **embryo sac mother cells**. These mother cells are **diploid** and divide by **meiosis** into **four haploid cells**, but only one cell survives to become the **embryo sac**.

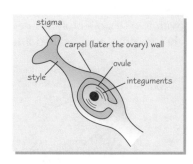

The embryo sac develops and its nucleus undergoes **mitosis** three times to form **eight haploid nuclei**. These are arranged so that eventually there are two **polar nuclei** at the centre and **three nuclei** at each end of the embryo sac.

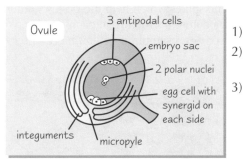

1) The **three nuclei** at one end of the embryo sac form the **antipodal** cells.

2) At the other end the **largest** cell of the three becomes the **egg cell** (**female gamete**) while the other two are called **synergids**.

3) The two **polar nuclei** fuse to become a **diploid polar nucleus**.

Integuments are just protective outer layers. They will eventually become the tough seed coat around the developing plant embryo (see bottom section).

Double Fertilisation occurs inside the **Embryo Sac**

1) If a compatible **pollen grain** lands on the **stigma** of a flower, the grain absorbs water and splits open.

2) A **pollen tube** grows from it down the **style**. There are **three nuclei** in the pollen tube — one **tube nucleus** at the tube's tip and two **male gamete nuclei** behind it. The tube nucleus makes **digesting enzymes**. These are **released** by the tube so that they digest surrounding cells, making a way through for the pollen tube.

3) When the tube reaches the **ovary**, it grows through the **micropyle** (a tiny hole in the ovule wall) and into the **embryo sac** in the **ovule**.

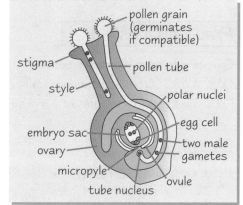

4) In the embryo sac, the **tube nucleus** disintegrates and the tip of the pollen tube bursts, releasing the two **male gamete nuclei**. One of them fuses with the **egg nucleus** to give a **diploid zygote**. This divides by **mitosis** to become the **embryo** of the seed.

5) The **second** male nucleus fuses with the **polar nuclei** at the centre of the **embryo sac**. This produces a **triploid nucleus**, which becomes a **food store** (called the **endosperm**) for the mature seed.

6) So a **double fertilisation** has taken place (**two** male nuclei have fused with female nuclei). This **only** happens in flowering plants.

The Embryo develops in the **Seed** and the Ovary becomes the **Fruit**

Once fertilised, the ovule is known as the **seed**. The **zygote** divides to form the **embryo**, which has a tiny root called a **radicle** and a shoot called a **plumule**. It also usually has two **cotyledons** (seed leaves) which become either the first leaves or a food store.

The **endosperm nucleus** surrounds the embryo and nourishes it. In some seeds such as wheat (a **monocot** — only one cotyledon) the endosperm remains to act as a food store. The **integuments** around the ovule become the tough, waterproof **seed coat** or **testa**.

The **ovary** develops into the **fruit** after fertilisation. It contains the seeds and has a wall (called the **pericarp**) which is designed to help with seed dispersal. It might become swollen and fleshy to attract **animals**, or simply dry up and split open to scatter its seeds.

seed (previously the ovule)

fruit (swollen ovary)

Sexual Reproduction in Plants

Germination occurs when the Seed Breaks its Dormancy

Seeds may remain **dormant** for a long time until **germination** is triggered. This happens due to a combination of environmental factors including availability of **water and oxygen**, a suitable **temperature**, and often **light**.

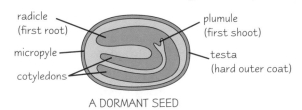

radicle (first root)
plumule (first shoot)
micropyle
testa (hard outer coat)
cotyledons

A DORMANT SEED

You might be surprised to know that light is needed for the germination of the seeds of many species. This helps ensure that the seeds only germinate when they're near enough to the surface, and when they're not in an area that's too heavily shaded.

Water is taken in (**imbibition**) and **enzymes** can begin to break down food stores for the embryo to use. Oxygen is needed for **respiration** and a suitable temperature is required for the **enzyme-based reactions**. **Hormones** such as **gibberellins**, **ABA** (**abscissic acid**) and **cytokinins** play a key role in controlling germination:

1) **Gibberellins** break seed **dormancy** and bring about the production of new **enzymes** which are involved in mobilising food stores for the growing embryo.
2) **ABA prevents** germination and must be broken down before germination can occur.
3) **Cytokinins** can **speed up** germination.

The **plumule** grows upwards into the light so **photosynthesis** can start, and the **radicle** grows downwards into the soil to absorb **water** and **minerals**.

Plumules are positively phototropic (grow towards light) and negatively geotropic (away from gravity). Radicles are positively geotropic and negatively phototropic.

In **dicotyledonous** seeds (two cotyledons) —
- if the cotyledons are carried **above** ground, it's known as **epigeal** germination.
- if they remain **below** ground, it's called **hypogeal** germination.

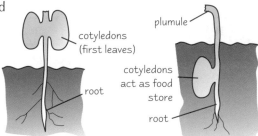

cotyledons (first leaves)
root
epigeal germination

plumule
cotyledons act as food store
root
hypogeal germination

Practice Questions

Q1 How many haploid and diploid nuclei are there in a mature embryo sac?
Q2 What is the function of the polar nuclei in the ovule?
Q3 What is the difference between epigeal and hypogeal germination?

Exam Questions

Q1 a) The diagram shows the events leading to fertilisation in a flowering plant. Label parts A-D.

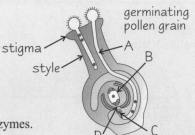

germinating pollen grain
stigma
style
A
B
C
D

[4 marks]

b) Explain why the pollen tube needs to secrete digestive enzymes. [2 marks]

Q2 What do each of the following structures become after fertilisation:
a) integuments   b) ovule   c) ovary? [3 marks]

I preferred avocados when I wasn't aware that they were ovaries...

The plants are back to their clever tricks again. This time they're solving the problem of how to disperse their offspring when they're stuck in one place. One popular method is to turn the ovary into something delicious, so that a passing animal decides to eat it. Then it poos out the seeds somewhere else, giving them a wonderful start in life. Thanks mum.

Human Reproductive Organs

If this book falls open at these two pages all the time, you ought to be ashamed of yourself. You're an A2 student now, and far too mature to be entertained by diagrams of willies. My copy always opens at the page on Glycolysis. Honest. **These two pages are for OCR.**

Ova *are produced in the Female's* Ovaries

The female reproductive system produces and develops **ova** (female gametes). The eggs released from the ovaries are actually called **secondary oocytes** rather than ova, because they haven't yet completed the second meiotic division (this only happens during fertilisation). The oocytes are delivered to the **uterus** via the **fallopian tubes (oviducts)**.

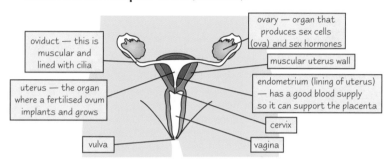

oviduct — this is muscular and lined with cilia

ovary — organ that produces sex cells (ova) and sex hormones

muscular uterus wall

uterus — the organ where a fertilised ovum implants and grows

endometrium (lining of uterus) — has a good blood supply so it can support the placenta

cervix

vulva

vagina

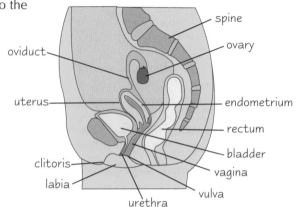

spine

ovary

oviduct

uterus

endometrium

rectum

bladder

vagina

clitoris

labia

vulva

urethra

Sperm *are Produced in the Male's* Testes

The male reproductive system produces **sperm** (male gametes) and delivers them to the female's vagina during **sexual intercourse**.

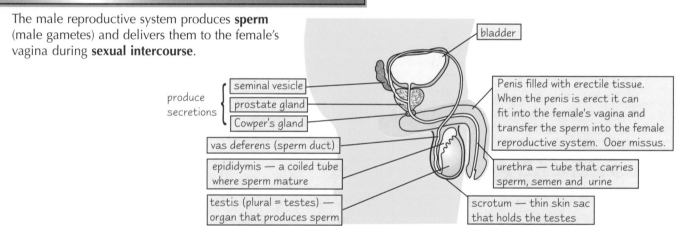

bladder

produce secretions

seminal vesicle

prostate gland

Cowper's gland

vas deferens (sperm duct)

epididymis — a coiled tube where sperm mature

testis (plural = testes) — organ that produces sperm

Penis filled with erectile tissue. When the penis is erect it can fit into the female's vagina and transfer the sperm into the female reproductive system. Ooer missus.

urethra — tube that carries sperm, semen and urine

scrotum — thin skin sac that holds the testes

The Male *Gamete is called a* Sperm*, the* Female *Gamete is called an* Ovum

Thousands of sperm are made every **second** in the testes of mature males.

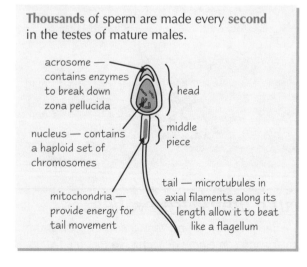

acrosome — contains enzymes to break down zona pellucida

head

nucleus — contains a haploid set of chromosomes

middle piece

mitochondria — provide energy for tail movement

tail — microtubules in axial filaments along its length allow it to beat like a flagellum

Usually only **one** secondary oocyte is released halfway through the menstrual cycle each **month**.

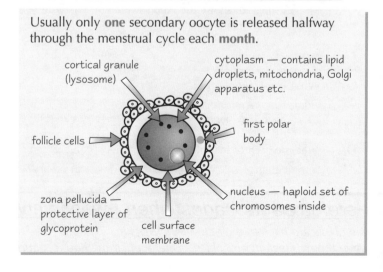

cortical granule (lysosome)

cytoplasm — contains lipid droplets, mitochondria, Golgi apparatus etc.

follicle cells

first polar body

zona pellucida — protective layer of glycoprotein

cell surface membrane

nucleus — haploid set of chromosomes inside

Human Reproductive Organs

The Production of Gametes (ova and sperm) is called Gametogenesis

In the ovaries of the **female**, the ova are made by **oogenesis**. This starts when the female is still a **foetus**, as follows:

1) The cells around the outside of the **ovaries** divide by mitosis to make **oogonia**.
2) The oogonia move towards the middle of the ovary and become **primary oocytes**.
3) **Follicle cells** surround the primary oocytes, which are then called **primary follicles**.
4) At **puberty**, sex hormones like **FSH** stimulate some of these primary follicles to divide by **meiosis**, producing **secondary oocytes**.
5) Each month, one of these secondary oocytes develops inside a **Graafian (mature) follicle**. The follicle travels to the surface of the ovary and bursts, releasing the oocyte into the oviduct (**ovulation**).

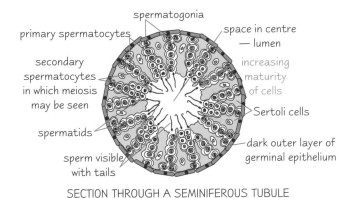

SECTION THROUGH A SEMINIFEROUS TUBULE

Mammalian **testes** are made up of **seminiferous tubules**. **Sperm** are produced in the walls of the seminiferous tubules (see diagram) from puberty onwards — a process called **spermatogenesis**:

1) Diploid cells in the **germinal epithelium** of the tubules divide repeatedly by mitosis producing **spermatogonia**.
2) These grow into **primary spermatocytes**.
3) These divide by meiosis to produce haploid **secondary spermatocytes**.
4) These divide once more to produce **spermatids** which mature into **sperm**.

Gametogenesis is controlled by Hormones

Spermatogenesis and **oogenesis** are controlled by two hormones, **LH** and **FSH**, that are released from the **pituitary gland**. These hormones are called **gonadotrophins** and their release from the pituitary is controlled by the release of another hormone, called **GnRH**, from the **hypothalamus**. The processes are examples of **negative feedback mechanisms**:

Spermatogenesis happens constantly —

1) **LH** stimulates the production of **testosterone** in the testes, which in turn stimulates **sperm production**.
2) **FSH** stimulates **Sertoli cells** in the testes, which help with sperm production.
3) **Testosterone** regulates **LH** release by **negative feedback** (see p.160 for more on negative feedback).

Oogenesis is part of the **menstrual cycle** and is linked to **uterus development**.

This is only part of the menstrual cycle, which is shown fully on page 160.

1) **FSH** stimulates a **primary follicle** in an ovary to begin dividing (see above).
2) The follicle secretes **oestrogen** which inhibits **LH** and **FSH** (negative feedback again).
3) But a **surge** in oestrogen levels half way through the menstrual cycle now **stimulates** LH and FSH production.
4) The **LH** increase stimulates development of the **secondary oocyte** inside a **Graafian Follicle**, and causes **ovulation**.

Practice Questions

Q1 What is the function of: a) the uterus   b) the urethra   c) the epididymis.
Q2 What type of molecule does the acrosome of a sperm contain?
Q3 Name three hormones involved in oogenesis.

Exam Question

Q1 Define gametogenesis and describe the process in a human male.   [10 marks]

Get this page learnt — you may well be tested on the testes...

Lots of diagrams on these pages. Don't skim over them, they're not just there to look pretty. You've really got to learn all the different bits and what they do, so practise by covering up the labels around the sides and checking you can remember what they say. Don't be embarrassed — staring at pictures of reproductive organs is OK when it's educational.

Human Reproduction

Here's a funny bit — a description of sex in scientific language. **Page 155 is for OCR only, this page is for AQA B too.**

The **Male Gametes** are Transferred to the **Female** during **Sexual Intercourse**

Sperm that are made in the testes collect in the **epididymis** and are stored in the **vas deferens** (see diagram on p.152). In humans, **internal fertilisation** occurs (secondary oocytes and sperm meet inside the female) so a process is required to deliver the sperm to an oocyte in an oviduct. This process is **sexual intercourse** or copulation.

In the male the penis becomes **erect** through physical or psychological stimulation. **Nerve impulses** sent to the penis cause the arteries to **dilate**, filling the spongy **erectile tissue** with blood. During intercourse, **contraction** of the muscles around the **seminal vesicles**, **vas deferens**, **prostate gland** and **urethra** eventually result in the release of **semen**. This is known as **ejaculation**. The semen usually contains several **million** sperm in an **alkaline** fluid.

Sperm are usually released at the top end of the vagina near the **cervix** during intercourse. The alkaline semen protects the sperm from the **acidic conditions** in the vagina, and allows some of them to pass through the cervix into the **uterus**.

Capacitation Prepares the Sperm to **Fertilise** an Oocyte

In the first few hours after ejaculation, the sperm undergo a series of changes to get them ready to fertilise an oocyte. This process is called **capacitation**. It involves the removal of some of the **outer proteins** on the sperm, and the **reorganisation** of its plasma membrane. Capacitation makes the sperm more **mobile** and prepares its membranes for the **acrosome reaction** needed to enter the oocyte. The sperm is now ready to **fertilise** an oocyte, as outlined below:

1) The sperm's **acrosome** releases enzymes to digest through the **follicle cells**.

2) **Membrane receptors** on the sperm then bind to receptors on the **zona pellucida**, causing more enzymes to be released from the acrosome.

3) The sperm **enters** the secondary oocyte, which then undergoes the **second division of meiosis** to give the **ovum** and a **second polar body**.

4) As soon as the sperm penetrates the cell surface membrane, **cortical granules** are released by the ovum. These cause the **zona pellucida** to form a barrier called the **fertilisation membrane**, to prevent any other sperm entering.

5) The sperm's nucleus **fuses** with that of the ovum to make a diploid **zygote** (23 chromosomes from the haploid sperm, 23 from the haploid ovum).

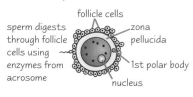
sperm digests through follicle cells using enzymes from acrosome — follicle cells — zona pellucida — 1st polar body — nucleus

sperm penetrates zona pellucida to fuse with surface membrane — cell surface membrane — cortical granules combine with zona pellucida to form barrier

sperm enters egg to fuse with nucleus — 2nd polar body released

Contraception Prevents **Pregnancy**

Contraception often involves **preventing fertilisation**, but some forms can involve the **death** of the young embryo. This means that there are **ethical** as well as biological considerations. In biological terms the various methods are:

1) **Tubal ligation** — the woman's **oviducts** are cut and tied so that ova released can't meet any sperm (sterilisation).

2) **Vasectomy** — the **vas deferens** are cut and tied so no sperm are released when the man ejaculates (sterilisation).

3) **Condom** — a thin rubber sheath is placed over the **penis** to stop sperm entering the vagina.

4) **Diaphragm** — a rubber cap covers the **cervix** to stop sperm getting through and reaching the uterus and oviducts.

5) **Female condom** — a rubber sheath lines the **vagina**, again to stop sperm getting through the cervix.

6) **Coil (IUD)** — inserted through the cervix and into the uterus by a doctor. Prevents **implantation**.

7) **Pill** — taken orally each day. Contains **hormones** designed to stop **ovulation** so no ova are released (see p.162).

8) **Emergency pill** — taken after **unprotected sex**. Contains hormones that prevent **fertilisation** and **implantation**.

9) **Hormone-based injections / patches** — work like the Pill to stop **ovulation**, but the effect is more **long-term**.

Newer methods include the **male pill** and **contraceptive sprays**. There are also natural methods like the **rhythm method** (avoiding sex around the time of **ovulation** each month) and **celibacy** (not having sex — well, that one's cheating slightly).

> **Ethical** considerations include:
>
> 1) Some **religions** (e.g. Roman Catholic) don't allow the use of contraceptives. This can lead to unwanted pregnancies.
>
> 2) **Age-related** issues — at what age should contraception be discussed, and do parents need to be involved?
>
> 3) Contraception involving prevention of **implantation** could be seen as allowing the death of a **potential human life**.

Human Reproduction

Abortion is the Removal of an Unwanted Foetus

The loss of the embryo or foetus can be a **natural** event that ends pregnancy (a **miscarriage**), but since 1968, **surgical removal** of the foetus or the use of **drugs** to kill it has been made legal in the UK. Common methods include:

1) **Vacuum aspiration** with a pump to remove the foetus.
2) **Dilation and curettage (D&C)** in which the cervix is dilated and the uterus scraped to remove the foetus.
3) **Contractions induced** to expel the foetus (premature labour), by injecting **prostaglandins** into the amniotic fluid.
4) **Abortion pill** — a drug using **anti-progesterone** and **prostaglandins** to induce a miscarriage.

Potential **medical** problems with abortions include the risk of **infection**, putting **future pregnancies** at risk and **infertility**.

Ethical issues are complicated and are often based around the woman's rights compared to the foetus' rights. Pressure groups known as **Pro-choice** and **Pro-life** exist at the two ends of the spectrum. Some of the common arguments for and against abortion are summarised in the table:

ARGUMENTS IN FAVOUR OF ABORTION	ARGUMENTS AGAINST ABORTION
May be needed to protect the woman's health	Destroying a life / potential life
Foetuses with serious defects that wouldn't survive long outside the womb could be aborted at an early stage	The 24 week limit in the UK is very arbitrary
The baby may not be wanted, e.g. if the mother is very young	Women who do want babies may abort healthy foetuses for other reasons, e.g. if they're the 'wrong' sex
The planet is already overpopulated, and abortion is one way to limit the number of babies being born	The father's rights to their potential child might be ignored
The foetus wouldn't survive outside the womb, so is only a potential life	Some women can be emotionally traumatised by the experience

If you can think of any other arguments for or against abortion, great — remember them for the exam.

IVF is Fertilisation Outside the Body

In vitro fertilisation (see page 166) is a way of producing babies for couples that are having **problems** conceiving naturally. Secondary oocytes are fertilised outside the body in a **laboratory** before being reintroduced to the mother's uterus. **Ethical** considerations include:

1) The production of **unwanted embryos** — several will be fertilised, but not all of them will be used.
2) **Psychological** and **financial** repercussions for the parents — it can be expensive, and success isn't guaranteed.
3) The risk of **multiple births**, which can be risky for both the mother and the babies.

Practice Questions

Q1 Why is semen alkaline?
Q2 What is meant by 'capacitation'?
Q3 How does the egg prevent 2 sperms from fertilising it?
Q4 Outline the pros and cons of the various types of contraception.

Exam Questions

Q1 A sperm passes through the following parts on its journey from epididymis to oviduct:
<u>vagina, uterus, vas deferens, urethra, cervix</u>. Place them in the correct order:

i) Epididymis, ii) ...., iii) ...., iv) ...., v) ...., vi) ...., vii) oviduct [5 marks]

Q2 Discuss the ethical issues involved in the medical procedure of abortion. [10 marks]

Let's talk about delivery of sperm to an oocyte in an oviduct, bay-bee...

OK it's not as catchy, but it's more scientifically accurate and that's what counts in this book. What a hassle sex is — people who don't want babies getting pregnant, people who do want babies not getting pregnant. Plus you can get all kinds of nasty diseases. Personally when I start getting broody I'm just going to grow a baby at the end of my arm like a strawberry plant.

Embryo Development

The placenta develops alongside the foetus, and leaves just behind it too. It's there that substances are passed from the mum's blood to the foetus's and vice versa. **Page 156 is for OCR and AQA B, and page 157 is just for OCR.**

The **Placenta** allows Transfer of Materials between Mother and Foetus

The **placenta** starts to form shortly after the **blastocyst** (the mass of cells formed from the zygote) implants in the uterus wall. The blastocyst develops into an **embryo** and starts to form a layer called the **chorion** which grows into the **endometrium** (lining) of the uterus. Tiny projections called **chorionic villi** are produced with capillaries inside containing the foetus's blood. Spaces develop around the villi in the endometrium called **sinuses**, which become filled with the mother's blood. The villi themselves are filled with **microvilli** and **mitochondria** to aid transport of substances between the mother and the embryo, which soon develops into a **foetus**. The two blood supplies mustn't mix as there are differences in blood **pressure**, and possibly blood **groups**, but they flow very **close** to each other to allow easy transport of materials.

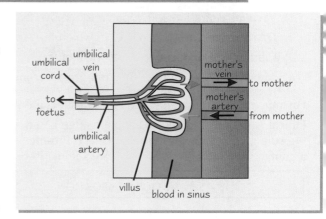

The **Structure** of the Placenta helps its role in **Exchange of Materials**

The placenta is involved in the **transfer** of the following substances:

1) **Oxygen** from mother to foetus — **oxygenated** blood enters the sinuses from the mother's arteries and flows next to the foetus's blood in the **umbilical vein**. It diffuses along a **concentration gradient**, from the higher concentration in the mother's blood to the lower concentration in the foetus's blood. Foetal **haemoglobin** has a **higher affinity** for oxygen than the haemoglobin in the mother's blood, which means that foetal haemoglobin combines with oxygen at the same partial pressure that maternal haemoglobin releases it.

2) **Carbon dioxide** from foetus to mother — carbon dioxide **diffuses** out of the umbilical artery in the opposite direction to oxygen as the foetal blood flows past the mother's blood.

3) **Nutrients** from mother to foetus — the developing foetus needs the same nutrients as the mother, including **glucose**, **fatty acids**, **glycerol**, **amino acids**, **vitamins** and **minerals**. Glucose is passed to the foetus by **facilitated diffusion**, amino acids are transferred by **active transport**, and vitamins and minerals move across to the foetus by a **combination** of diffusion and active transport.

4) **Water** from mother to foetus — water passes into foetal blood by **osmosis**.

5) **Urea** from foetus to mother — this waste material **diffuses** into the mother's blood to be excreted.

6) **Antibodies** from mother to foetus — some antibodies can pass across the placenta from mother to foetus, giving the foetus **temporary passive immunity** to some diseases when it's born.

As well as being a **transport organ**, the placenta also acts as an **endocrine gland** during pregnancy. It secretes **oestrogen**, **progesterone** and **HCG**, which amongst other things help to maintain the **endometrium** and develop the uterus and breasts.

The placenta also acts as a **barrier** against harmful bacteria and most viruses. It can't stop the **rubella virus** or **HIV** though, and it's not a barrier to harmful substances and drugs such as **alcohol**, **nicotine** and **heroin** which can harm the foetus.

Transport method	Example
diffusion	O_2, CO_2, urea
facilitated diffusion	glucose
active transport	amino acids
osmosis	water

Embryo Development

The **Amnion** surrounds and **Protects** the developing Foetus

The **amnion** is a tough membrane surrounding the foetus. It's filled with **amniotic fluid**, which is similar to **serum** in **blood plasma** and contains proteins, carbohydrates, lipids and dissolved salts. It forms a protective, shock-absorbing cushion which **protects** the foetus from physical damage and allows it to **move**.

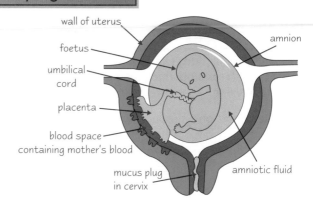

The foetus both swallows and urinates into the amniotic fluid during pregnancy. Eww.

The **Mother's Actions** during Pregnancy Directly Affect the **Foetus**

Nowadays a lot of information, advice and antenatal care is given to mothers to help them have a **healthy pregnancy** for themselves and their baby. The factors given in the table can have a big impact on **foetal development**:

FACTOR	RECOMMENDATIONS	PROBLEMS
Diet	Healthy eating guidelines are produced for pregnant mothers. These include a slightly increased intake in energy, and a higher intake of protein and certain vitamins and minerals, e.g. folic acid and calcium.	i) A lack of iron can cause anaemia ii) If folic acid is lacking, there's a bigger risk of defects in the foetal neural tube (which eventually becomes the whole CNS) iii) Slow growth of foetus if there's a lack of protein iv) Problems with the foetal skeleton if there's a lack of calcium
Smoking	Mother should not smoke	i) Babies born underweight and premature, or miscarried ii) Increased risk of cot death and respiratory disease in the child The main causes of these problems are that carbon monoxide affects haemoglobin's ability to carry oxygen, and nicotine narrows the blood vessels.
Alcohol	Don't drink more than the low recommended level for pregnant women, and preferably none at all.	Excess consumption can lead to impaired mental and physical development, heart problems and learning difficulties in the baby. This is know as foetal alcohol syndrome.
Drugs	Stick to recommended doses of prescribed drugs. Avoid all illegal drugs.	Drug abuse can cause drug dependency in babies, underweight and premature babies, and birth defects.

Other factors that can affect foetal development include **access to healthcare, exercise, stress**, etc.

Practice Questions

Q1 Why is it important that a pregnant woman's blood doesn't actually mix with that of her foetus?

Q2 Name two viruses that can pass across the placenta from mother to foetus.

Q3 What is the amnion?

Q4 Give three problems that might be caused if a woman doesn't follow the recommendations concerning her diet while she is pregnant.

Exam Question

Q1 a) List the substances that pass from the mother to the foetus and vice versa via the placenta. Explain how each substance moves from one bloodstream to the other. [12 marks]

b) Give two other functions of the placenta, besides transport. [2 marks]

So whether we remember it or not, we've all tasted our own wee...

The placenta's important. It lets the blood of mum and foetus get close together without mixing, so that stuff can pass across. There are six main things that are exchanged across it (try writing down all six now without looking — people always seem to forget one). Plus it secretes some very important hormones and acts as a barrier to stop harmful stuff crossing.

Controlling Growth in Plants

Back to plants now, just briefly. Plants rely on a combination of plant growth substances to control the different stages of their growth and development, like germination, flowering and fruit production. ***These two pages are for OCR.***

Day Length *determines when Plants produce* Flowers

Plants usually **flower** only at certain times of the year. The timing of this flowering is triggered by changes in **day length** (**photoperiod**). Some plants only produce flowers when day length is **long** and exceeds a certain **critical value**. These are called **long-day plants** (LDPs), and tend to flower in **spring** and **summer**, e.g. roses.

Other plants only produce flowers when day length is **shorter** than their critical value. These are called **short-day plants** (SDPs). They tend to flower in **autumn** and **winter**, e.g. poinsettia (see photograph).

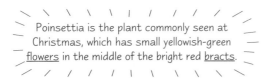
Poinsettia is the plant commonly seen at Christmas, which has small yellowish-green flowers in the middle of the bright red bracts.

Studies have shown that it's not actually the length of the **days** that determines when plants flower, but the length of the **nights**. If the dark periods are **interrupted** by short bursts of light, the plant is fooled into thinking it's a short night and so the timing of the flowering process is **disturbed**. So LDPs respond to **short** periods of darkness, and SDPs need **long** dark periods to flower. This mechanism is known as **photoperiodism**.

1) Photoperiodism depends on a plant **pigment** that can detect light and darkness.
2) The pigment involved is called **phytochrome**. It has two forms, **Pr** and **Pfr**.
3) **Pr** is converted to Pfr when it absorbs **red** light (wavelength of around **660 nm**).
4) **Pfr** is converted to Pr when it absorbs **far-red** light (wavelength of around **730 nm**).
5) **Pfr** is also converted **slowly** into Pr in **the dark**.
6) In the **daytime** sunlight contains more **red** light than far-red light. The Pr is converted to Pfr, and so Pfr is the **main form**.
7) At **night** Pfr is turned back into Pr, and so Pr is the **main form**.

- The build up of **Pfr** stimulates **LDPs** to flower.
- It has the **opposite** effect in **SDPs** though, where Pfr **inhibits** flower production. They can only flower when **nights** are long and all the Pfr is converted back to **Pr**.
- Other factors such as **gibberellins** (plant growth hormones) and **temperature** also have an influence on photoperiodism.

Plant Growth Regulators *are important in the* Maturing *of* Fruits

The fruit industry has developed rapidly since the use of **plant growth regulators** to control fruit formation and ripening became common practice. Plant growth regulators used for this include:

1) **Ethene** — this is synthesised in plants and is easily **passed** to other plants as a gas. Ethene stimulates the enzyme **cellulase** to break down cellulose **cell walls** in fruit and speed up **ripening**. Some fruit, e.g. bananas, are now harvested and transported **before** they're ripe because they're less likely to be damaged. They're then treated with ethene to **ripen** them just in time to hit the shelves in your local supermarket.
2) **Auxins** — these can trigger the development of fruit even if **fertilisation** hasn't happened. This is called **parthenocarpy** and it's used to produce seedless grapes.
3) **Gibberellins** — these can have a variety of effects, but often work in a similar way to **auxins**.

Controlling Growth in Plants

Seed Dormancy *helps prevent Unsuccessful Germination*

Seeds usually remain **dormant** for a while after they're dispersed, until conditions are suitable for **germination**. Plants often produce their seeds at the **end** of the growing season (i.e. autumn) when the conditions aren't very good for growth, so the seeds usually stay dormant over the **winter** or perhaps for as long as **several years**. Seed dormancy therefore makes **successful germination** more likely.

It's the relative concentrations of various **growth regulators** that control seed dormancy. **ABA** (**abscisic acid**) is the main **inhibitor** of germination. It's the **decrease** in concentration of ABA which ends seed dormancy, along with an **increase** in the concentration of **gibberellins** secreted by the embryo. Gibberellins stimulate the production of the enzyme **amylase** in some seeds. This breaks down stored starch into **maltose** to provide substrate for the embryo to grow.

> The main factors affecting germination are:
>
> 1) **Water** — needed as a **solvent** and **transport medium**.
>
> 2) **Oxygen** — needed for the **aerobic respiration** of food stores.
>
> 3) **Temperature** — lots of seeds need a period of **low temperature** before germination can happen, e.g. apple seeds. This helps to make sure they don't germinate in the **autumn**. A **higher temperature** helps once germination has started, because many **enzymes** work best at higher temperatures.
>
> 4) **Light** — lots of species need **light** for their seeds to germinate, e.g. lettuce. Light increases Pfr levels (see previous page) which **stimulates germination** in the presence of water and oxygen. This means that seeds stay dormant if they're buried **too deeply** underground or if there isn't going to be enough light where they are for **photosynthesis**.
>
> 5) **Physical damage** — some seeds need to have their **seed coat** (**testa**) damaged before germination can occur, probably because the tough seed coat doesn't let through enough water when intact. This damage might happen due to the seed being partly **digested** by animals.

Experiments *can be done to Investigate the Factors affecting* Germination

The main factors affecting germination (see box above) can easily be **investigated** using **cress seeds** grown on cotton wool in a petri dish. The seeds can be exposed to the different factors in **varying amounts** to see what effect they have on germination. A **control** with **warm** conditions and adequate **water**, **oxygen** and **light** should be used. Individual dishes of seeds can then be tested by **changing** one of these conditions. It's important to **replicate** the experiment with a **few dishes** under each set of conditions. Measurements of **shoot** and **root** length or **percentage germination** can then be calculated, and the **average** over different dishes for a particular set of conditions found.

Practice Questions

Q1 At what time of year will an LDP tend to flower?

Q2 What is photoperiodism?

Q3 Which form of phytochrome builds up in daylight hours?

Q4 Which enzyme does ethene stimulate that speeds up the ripening of fruit?

Q5 Name four factors that affect seed germination.

Exam Questions

Q1 In an experiment to investigate seed germination it was found that germination rates increased with temperature between 10 and 40 °C, but dropped rapidly at higher temperatures. Explain these findings. [6 marks]

Q2 Explain why the reduction in levels of the pigment Pfr is important to short-day plants. [3 marks]

Seeds have the right idea — refuse to come out until it's warm and sunny...

So that's how they make nice grapes and tangerines that aren't full of annoying seeds. Plants normally wait until fertilisation has happened and the seeds are ready before they produce the auxins that trigger the development of fruit. But if you can extract auxins or produce synthetic ones, you can set off fruit development before any seeds are made. Good stuff.

Hormonal Control in Humans

Sorry lads — these pages are pretty much devoted to the inner workings of the ladies. Now one of them is bound to star going on about how easy you have it in comparison. Got a point, though... **These two pages are for OCR and AQA B**. **If you're doing Edexcel you need to know the bottom section on this page.**

The **Human Menstrual Cycle** is Controlled by **Hormones**

The human **menstrual cycle** lasts about **28 days**. It involves the development of a **follicle** in the ovary, the release of a **secondary oocyte**, and thickening of the **uterus lining** so a fertilised ovum can **implant**. If there's no fertilisation, this lining breaks down and leaves the body through the **vagina**. This is known as **menstruation**, and it marks the end of one cycle and the start of another.

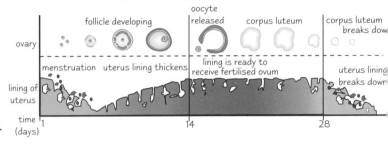

The menstrual cycle is controlled by four **hormones** — **follicle-stimulating hormone (FSH)**, **luteinising hormone (LH)**, **oestrogen** and **progesterone**. They're either produced by the **pituitary gland** or by the **ovaries**:

Hormones released by the anterior pituitary —

1) **FSH** is released into the bloodstream at the **start** of the cycle and is carried to the **ovaries**. It stimulates the development of one or more **follicles**, which in turn secrete **oestrogen**.

2) **LH** is released into the bloodstream around **day 12**. It causes **release** of the secondary oocyte (**ovulation**). When the oocyte bursts out, it leaves its **primary follicle** behind. LH helps the follicle turn into a **corpus luteum**, which is needed later.

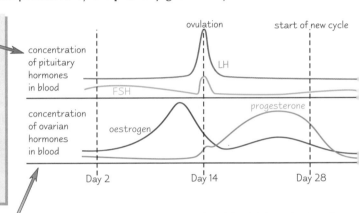

Hormones released by the ovaries —

1) **Oestrogen** is produced by the **developing follicle**. It causes the **lining** of the uterus to **thicken**. It also **inhibits** the release of **FSH**. This stops any **more** follicles maturing. But then a **peak** in oestrogen levels starts a **surge** in FSH and LH production, which triggers **ovulation**.

2) **Progesterone** is released by the **corpus luteum** after ovulation. Progesterone keeps the uterus lining **thick**, ready for implantation if fertilisation occurs. It also **inhibits** release of FSH and LH. If no embryo implants, the corpus luteum **dies**, so progesterone production stops and **FSH inhibition** stops. This means the cycle starts again, with development of a **new follicle**.

The secondary oocyte is released around <u>day 14</u> *of the menstrual cycle, and must be fertilised within* <u>24 hours</u>. *If sexual intercourse leads to fertilisation, the fertilised ovum moves down the oviduct to the uterus where it* <u>implants</u> *in the wall. This takes up to* <u>3 days</u>.

Negative Feedback Loops regulate Hormone Concentrations

The stages of the menstrual cycle are carefully controlled by hormones, and **negative feedback loops** exist to keep the system in balance. In the menstrual cycle, **GnRH** stimulates the pituitary to release **LH** and **FSH**, which then stimulate release of **oestrogen** and **progesterone**. But when levels of oestrogen and progesterone get **too high**, they **inhibit** the release of GnRH, so the release of **all the others** stops too. Then when levels **fall** again, the hypothalamus stops being inhibited and everything gets going again. This is what's meant by **negative feedback**.

Positive feedback loops also exist, e.g. uterus contractions that are caused by <u>oxytocin</u> stimulate <u>more</u> oxytocin to be released from the <u>posterior pituitary</u> during labour.

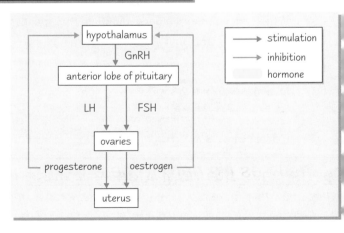

Hormonal Control in Humans

Hormones also control Pregnancy, Birth and Lactation

During pregnancy (at 12 weeks) the **placenta** takes over from the corpus luteum in secreting progesterone and oestrogen.

1) **Progesterone** levels increase during pregnancy, which helps to maintain the **uterus lining** (**endometrium**). It also helps develop the uterus and breasts and **inhibits** contractions of the uterus. Levels then **fall** before birth.

2) **Oestrogen** levels rise during pregnancy too. Like progesterone, it helps maintain the **uterus** and **endometrium**. It also **inhibits FSH** so that no more oocytes are released during pregnancy. Oestrogen levels **carry on rising** at the end of the pregnancy and this overides the effects of **progesterone**, allowing the uterus to start **contractions**.

3) **Human chorionic gonadotrophin** (**HCG**) is secreted by the **embryo** and placenta. It stimulates production of **oestrogen** and **progesterone** by causing the **corpus luteum** to develop. It also triggers the production of **testosterone** in a male embryo. HCG levels begin to **fall** after a few months of pregnancy.

4) **Oxytocin** is released by the **posterior pituitary gland** and stimulates **contractions** of the uterus before birth. Increasing oxytocin levels increase the **number** and **force** of the contractions and allow the baby to be born. Oxytocin also causes **milk** to be **released** when the baby is feeding.

5) **Prolactin** is secreted from the **anterior pituitary gland** and stimulates **milk production** (**lactation**) after birth.

OCR ONLY

Hormones are involved in PMT, the Menopause and HRT

1) **Pre-menstrual tension** (**PMT**) (also known as **pre-menstrual syndrome**, **PMS**) is experienced by many women in the days before **menstruation**, i.e. at the **end** of the menstrual cycle. It includes various physical and psychological symptoms, like **irritability**, **depression**, **mood swings**, **bloating** due to water retention and **breast tenderness**. It's thought to be caused by changes in **oestrogen** and **progesterone** levels, which can affect **neurotransmitters** in the brain controlling mood and emotion.

2) **Menopause** — This is the time when a woman's menstrual cycle **stops** and she can't have any more children, usually between the ages of **45 and 55**. It's also known as 'the change', and it's due to impaired functioning of the **ovaries** which start responding less to **FSH**. This means that **less** oestrogen and progesterone are produced. Symptoms include **hot flushes**, **fatigue**, **mood swings** and the development of **osteoporosis** (loss of **bone mass**) caused by loss of **calcium** from the bones (see p.137).

3) **Hormone replacement therapy** (**HRT**) uses **oestrogen** to reduce **calcium loss** and so reduce the chances of developing **osteoporosis**. It can also lessen the other unpleasant symptoms of the menopause, but there are problems associated with HRT including a slight increase in the risk of **heart disease** and certain **cancers**.

Practice Questions

Q1 Which hormones in the menstrual cycle are released by: a) the pituitary gland   b) the ovaries?

Q2 Give an example of a negative feedback loop in the menstrual cycle.

Q3 Give three common symptoms of PMT.

Q4 What causes the menopause?

Exam Question

Q1 a) The graph on the right shows the levels of 4 hormones during pregnancy — HCG, oestrogen, prolactin and progesterone. Identify hormones A, B and C.

[3 marks]

b) Where is the hormone prolactin produced and what is its main effect?

[2 marks]

Sometimes it's hard to be a woman...

Aargh, so many hormones on this page! It's making me feel weepy and crave chocolate just reading about them all. The menstrual cycle is probably familiar to you from GCSEs, and there are four main hormones that control it — FSH, LH, oestrogen and progesterone. Then HCG, oxytocin and prolactin come into play during pregnancy and birth. Phew.

Hormonal Control in Humans

I'm sure you'll be devastated to hear that these are the last two pages in this section. They're about controlling growth, with a bit on contraceptives and pregnancy testing thrown in for variety. **These two pages are for OCR and AQA B.**

AQA B ONLY

The Hormones **Oestrogen** and **Progesterone** are used for **Contraception**

Hormone-based **contraceptives** can be divided into the following categories:

1) The **combined pill** is taken orally and is a complex of synthetic hormones, usually **progesterone** and **oestrogen**. It mimics the hormone levels of **early pregnancy** and so inhibits the release of **FSH** (follicle stimulating hormone) and **LH** (luteinising hormone). This stops any **secondary oocytes** being released (ovulation).

2) The **mini pill** just contains **progesterone**. It's used by women who have a higher than normal risk of developing certain **health problems**, e.g. **thrombosis**, perhaps due to past problems or their family history. This is because the combined pill also slightly **increases** the risk. The mini pill doesn't always prevent **ovulation**, but it prevents **fertilisation** and **implantation** by changing the consistency of the **mucus** in the uterus to stop sperm swimming, and by making the **lining** of the uterus unsuitable for implantation.

3) The **morning-after pill** is taken after sex as an **emergency** measure (within a few days), e.g. if a condom has broken. It contains **oestrogen** and **progesterone**, which prevent **implantation**.

4) **Injections** and **implants** containing **progesterone** are also possible. They can stop **ovulation** for a few **months**.

AQA B ONLY

The Hormone **HCG** is used for **Pregnancy Testing**

The hormone **human chorionic gonadotrophin** (**HCG**) is released by the **embryo** (and placenta) and excreted in the pregnant woman's **urine**. Pregnancy testing kits contain a strip which has coloured **HCG antibodies** attached to it, and this is dipped into the urine sample. If HCG is present in the urine, it **attaches** to the antibodies and a coloured line appears on the strip to show that the woman is pregnant. A **control line** also appears whether the woman is pregnant or not, to show that the test's worked. (See page 187 for more about how pregnancy testing kits actually work.)

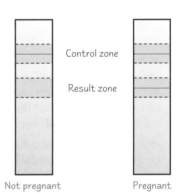

Control zone

Result zone

Not pregnant   Pregnant

Thyroxine secretion is controlled by the **Hypothalamus** and **Pituitary Gland**

Hormones help control our **growth** and **development**. The hormones involved are produced by the **hypothalamus** and the **pituitary gland**, which are located at the base of the brain.

The hypothalamus is a regulatory centre that secretes **thyrotrophin releasing hormone** (**TRH**). TRH affects the **anterior lobe** of the **pituitary gland**, causing it to release **thyroid stimulating hormone** (**TSH**) into the bloodstream. TSH in turn affects a gland near the trachea in the neck called the **thyroid gland**. It stimulates the thyroid gland to release another hormone called **thyroxine**, which increases **metabolism**.

The whole system has a **negative feedback loop** so that levels of these hormones are kept in check and growth is regulated:

Thyroxine **inhibits** the production of **TRH** and **TSH**, so if the metabolic rate gets too **high** due to too much thyroxine, less TRH and TSH are released. Thyroxine levels then **fall** and so does the **metabolic rate**. If the metabolic rate falls too **much**, the **low** thyroxine concentration means that **more** TRH and TSH are released. This **increases** the thyroxine levels again.

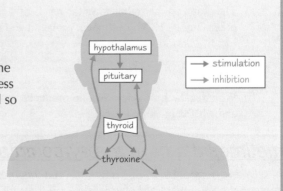

hypothalamus

pituitary

thyroid

thyroxine

→ stimulation
→ inhibition

Hormonal Control in Humans

Thyroxine controls Growth and Development

1) The **thyroid gland** consists of a network of **blood capillaries** with **follicles** between them. The **follicle walls** are a single layer of cells that secrete **thyroglobulin** into the **lumen** at the centre of the follicle, where it's **stored**.

2) Thyroglobulin is a large protein-based molecule that contains **iodine**. It's taken up by the cells when **thyroxine** is being made. Enzymes convert the thyroglobulin into the smaller molecule thyroxine, which can enter the **blood capillaries**.

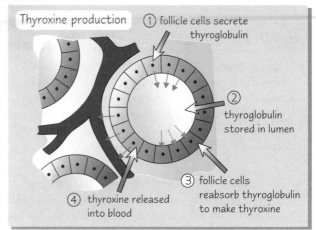

Thyroxine production
① follicle cells secrete thyroglobulin
② thyroglobulin stored in lumen
③ follicle cells reabsorb thyroglobulin to make thyroxine
④ thyroxine released into blood

3) Thyroxine affects nearly **all** the cells in the body by increasing their **metabolic rate**. It increases the rate of **cellular respiration**, and the rate of protein, carbohydrate and fat **metabolism**. This leads to **growth**, and also to the release of more **heat** and **energy** by the cells.

4) Thyroxine is particularly important in the growth of **teeth** and **bones** in children, as well as in **brain development**. Women who don't get enough iodine when they're pregnant can't make enough thyroxine, and can have babies whose **nervous** and **skeletal systems** aren't properly developed. Adults suffering from a lack of iodine in the diet develop a **goitre** (enlarged thyroid gland). Adults with an underactive thyroid lack energy, become depressed and may gain weight.

Growth throughout the body is also promoted by a non-specific hormone simply called **growth hormone** (**GH**). It's produced by the **anterior lobe** of the **pituitary gland**, and its release is controlled by the **hypothalamus**. It stimulates **protein production**, and it's important that the correct amounts of GH are produced if a normal **body size** is to be reached.

The development of **secondary sexual characteristics** is caused by **testosterone** and **oestrogen** produced by the **testes** and **ovaries** respectively. See page 143 for more on how these characteristics are controlled by hormones during puberty.

Practice Questions

Q1 Which hormones are used to produce the combined contraceptive pill?

Q2 Name the hormone that's detected by a pregnancy test.

Q3 Where in the body are the thyroid, hypothalamus and pituitary gland?

Exam Question

Q1 Match each of the glands listed below with its secretion:

Gland	Hormone
Hypothalamus	Testosterone
Pituitary	Thyroxine
Thyroid	Oestrogen
Ovaries	TSH
Testes	TRH

[5 marks]

Come on kids — let's play count the hormones!

There are ten different hormones mentioned on this page. If you find them all, you don't win a prize! Anyway, that's the end of this section so you can have a break from hormones now after this final extravaganza. And here are the hormones, just to prove there really were ten: oestrogen, progesterone, FSH, LH, HCG, TRH, TSH, thyroxine, GH, testosterone.

Courtship and Territory

This section is for AQA B, but it's also got a few bits for Edexcel. *In the animal kingdom, males court females to try and persuade the females to mate with them. They might do this by doing silly dances, bringing the female presents of food or even fighting other males. So not that different to humans, then.* **These two pages are for AQA B only.**

Courtship involves Showing Off

Reproduction is vital for the survival of any species. In order to reproduce successfully, adults must choose a mate that is the **right species** and that is as **strong** as possible. This will give any young produced the greatest chance of **survival**. In most species it's the **male** that displays courtship behaviour to attract a female, although in some cases it's the other way round. To be chosen, the male must prove to a female that he's fit and virile, and this is where courtship behaviour comes in. Each species produces a **unique courtship signal**, so there's a huge **variety** of courtship behaviour seen in the animal kingdom. A courtship ritual usually includes one or more of the following:

1) **Sounds** — e.g. courtship **songs** in many bird species.
2) **Visual signals** — e.g. displaying colourful **plumage**.
3) **Behaviour** — e.g. performing courtship **dances**, bringing **gifts** of food. Some animals build **nests** or defend a **territory**.

Male fireflies attract females by literally 'flashing' at them. They emit flashes of light, but each species has a slightly different frequency of flashing.

Male frigate birds inflate the sacs on their chests so that they look bigger and more impressive to females.

There are several reasons for the elaborate and prolonged courtship displays of some species:

1) The female needs to be sure that the male is of the **right species**. Mating between different species is a waste, as fertile offspring aren't produced. For this reason, even **closely related** species have **different** courtship rituals.

2) The male needs to be sure that the female is **sexually receptive** at the time. If she isn't, the male will know because she **won't respond** to the courtship behaviour.

3) The **quality** of the display may make a female more likely to choose a particular male. She wants to mate with a male who will pass on alleles to her young that will help them to **survive** — or that will help them to **breed** successfully in the future (see the section below on **sexual selection**).

Sign Stimuli are often Important in Courtship Displays

A **sign stimulus** is an **external signal** that triggers a particular response in another individual of the same species. The response is **innate** (genetically inherited), not learned, and occurs in **all** individuals of the species. For example, male **sticklebacks** attack other males, but court females. The **sign stimulus** triggering each behaviour is the abdomen of the other fish. If it's **red** (as in a male), the fish **attacks**. If it's **grey and swollen** (like a female with eggs) the male **courts**.

Experiments show that this happens even with models that don't look **anything** like a stickleback — it's only the colour and shape of the **abdomen** that's important.

This sort of behaviour often results from **innate releaser mechanisms**. This is a mechanism in the brain, which automatically triggers a specific behaviour or **fixed action pattern** in response to the sign stimulus.

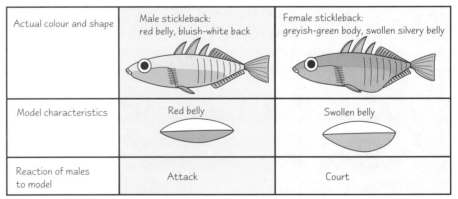

Actual colour and shape	Male stickleback: red belly, bluish-white back	Female stickleback: greyish-green body, swollen silvery belly
Model characteristics	Red belly	Swollen belly
Reaction of males to model	Attack	Court

Natural Selection can be Sexual Selection

Natural selection isn't just 'survival of the fittest', it's really '**breeding of the fittest**'. Some traits are seen in animals not because they help them to **survive**, but because they help them to **breed**. This is known as **sexual selection**. For example, **sexual selection** has resulted in male **peacocks** developing large and spectacular tail feathers because they **attract female peacocks**. A tail like the peacock's probably **reduces** its survival potential, as it's very visible to predators and slows the animal down. But it increases its **reproductive potential**, and so the trait has survived.

Courtship and Territory

It *Always Helps* if you **Smell Nice**

1) Courtship behaviour is influenced by a **complex interaction** of different **hormones**, with the **sex hormones** (not surprisingly) being particularly important.

2) Other hormones can also play a part in mating behaviour. For example, some species form long-term '**pair bonds**' — rarely, a male and a female may even stay together **for life**. One species that tends to form monogamous relationships for life is the **prairie vole**. Research on prairie voles has linked this behaviour with two **brain hormones**, oxytocin (in the female) and vasopressin (in the male).

3) Some female animals secrete another type of chemical which helps them to attract a mate. These are called **pheromones**. A pheromone is defined as **a chemical substance produced by one animal that affects the behaviour of another**. Different types of pheromone have **different functions**, but one important type of pheromone is that produced by a female to alert males of her species in the area to the fact that she is **sexually receptive**. The pheromone then attracts the male and can also act as a **sign stimulus** to start mating behaviour. The advantage of pheromones is that they act over **long distances** and can attract males that the female isn't even aware of. They will also affect **all** the males in the area, allowing the female to select the 'best' male as her mate.

> One of the best known pheromones is produced by the **female silk moth**. It can attract males from as far away as **10km** and the male can detect even a **single molecule** of it.

Territoriality is Having your **Own Patch**

Territoriality is a common behaviour in many different types of animal, including fish, birds and mammals. A **territory** is an area that is defended by an individual or group, which other members of the same species are **not permitted to enter**. Not all species have territories, but in those that do, territories have a **variety** of functions. They allow animals to **control access** to critical **resources**, such as nesting sites or food. And in many species, territories allow one sex (usually the **male**) to defend an area to which the other is attracted for **mating**.

In such species, having a desirable territory can be a **big advantage** in terms of breeding success. If a male can establish his own territory and keep other males out, it helps to convince the female that he's a **suitable mate**. A male that successfully defends a desirable territory is likely to be **strong** and **healthy**, and therefore an ideal mate. He will also have access to the **resources** needed to provide for any young produced. Males who have **not** been able to successfully find and defend a territory are **unlikely** to find a mate. Mating with a male that didn't have his own territory would be a **waste** for the female of such a species, as her young would be **unlikely to survive**.

Lions defend territories as a group, whereas leopards have their own individual territories.

Practice Questions

Q1 What is a sign stimulus?
Q2 Give an example of sexual selection.
Q3 What is a pheromone?

Exam Questions

Q1	State three advantages of courtship behaviour for the organisms involved.	[3 marks]
Q2	(a) Explain the advantage to a female of producing pheromones.	[2 marks]
	(b) Suggest why it is usually the females of a species that produce the pheromones, rather than the males.	[1 mark]
Q3	Suggest how territoriality may have evolved in a species by natural selection.	[5 marks]

Hmmm, lady frigate birds certainly have strange taste...

In most species, the female chooses from several males who all want to breed with her. This is because it's usually the female that invests the most time and energy in the offspring. Males mate with as many females as possible, but the female only has one set of young at a time to pass on her genes. In species where only males care for young (e.g. the seahorse), they also choose the mates.

Infertility and Pregnancy

In most women, conception happens naturally and a successful pregnancy follows. But sometimes there are problems, so modern science has developed ways to help potential mothers have babies if they want to. **These pages are for AQA B.**

Sometimes **Conception** can't happen without **Help**

Infertility can be caused by problems in **either** the male **or** the female **reproductive system**. This section deals with **female** infertility, which is most commonly caused by the following conditions:

1) **Abnormal ovulation**, or irregular release of a secondary oocyte from the ovary. Normally one oocyte is released **each month** under the direction of several **hormones**. But if any of these hormones aren't functioning properly, ovulation will happen **irregularly** or **not at all**.

2) A **blockage** in one or both of the woman's **fallopian tubes**. This means that the oocytes can't get to the uterus and sperm can't reach the oocytes.

Abnormal Ovulation is usually Treated using **Drugs**

The development of oocytes and their release in ovulation is controlled by hormones called **gonadotrophins**, which are produced in the **pituitary gland**. If these hormones are at low levels, oocytes might not be released. Treatment for this condition may involve the injection of extracted or synthetic **gonadotrophins**, or taking **drugs** which stimulate their production (e.g. **clomiphene**).

Blocked Fallopian Tubes Require **In Vitro** **Fertilisation**

If a **blockage** is stopping the oocyte and sperm meeting, or the oocyte getting to the uterus, then *in vitro* **fertilisation** (**IVF**) often gives the best chance of conceiving a child. This is where sperm and oocytes are taken from the parents (or possibly from a donor) and fertilisation happens in a **laboratory** instead of inside the woman. The embryos can then be **implanted** back into her uterus once fertilisation has happened. The full procedure is shown on the right.

This is a **complex** and **expensive** process and there's **no guarantee** of success. The technique involves the use of **drugs** and minor **surgical procedures**, so it carries more **risk** for the mums and babies than a natural pregnancy does. Because it's quite likely that **none** of the embryos will develop, until recently **several** embryos were implanted back into the uterus to increase the chances of at least one developing into a baby. But the high number of **multiple births** (twins and triplets) associated with IVF has meant that **fewer embryos** are now implanted back into the uterus. Multiple births are **more risky** for both the mum and the babies.

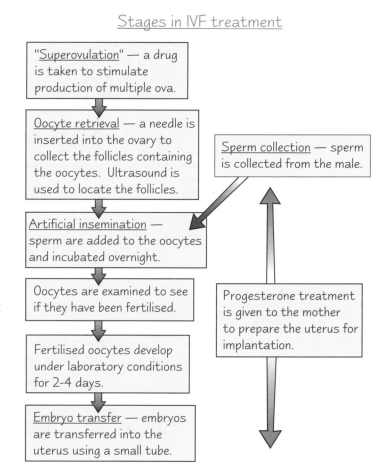

Stages in IVF treatment

"Superovulation" — a drug is taken to stimulate production of multiple ova.

Oocyte retrieval — a needle is inserted into the ovary to collect the follicles containing the oocytes. Ultrasound is used to locate the follicles.

Sperm collection — sperm is collected from the male.

Artificial insemination — sperm are added to the oocytes and incubated overnight.

Oocytes are examined to see if they have been fertilised.

Progesterone treatment is given to the mother to prepare the uterus for implantation.

Fertilised oocytes develop under laboratory conditions for 2-4 days.

Embryo transfer — embryos are transferred into the uterus using a small tube.

Infertility and Pregnancy

A Woman's body *Functions Differently* during *Pregnancy*

Pages 156-7 described how a baby develops inside the mother's uterus. While this happens, the **mother's body** has to **adjust** to the demands of pregnancy. Some of the main changes are:

1) The mother puts on **weight**. This might seem obvious, but it's **not** just the weight of the baby. The **uterus wall** and the **breasts** increase in size. The **placenta** adds some extra weight. There will be an increase in the amount of **blood** and **body fluids**, and **fat** will be stored up ready to provide extra energy for **breast feeding**. **How much** weight a woman puts on depends on her **height** and **starting weight**. This is calculated as her **BMI** (Body Mass Index). The graph shows **recommended weight gain** for different BMIs.

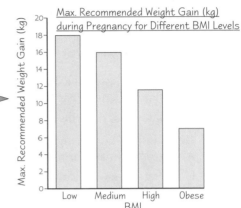

2) **Plasma volume** and **red blood cell mass** increase. This is to meet the **metabolic demands** of pregnancy and breast feeding. The effects are shown in the graph.

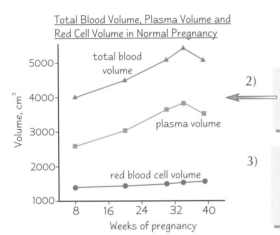

3) **Cardiac output** increases. The mother's **heart rate** goes up and so does the **stroke volume** (how much blood is pumped out with each beat). Again, this helps to meet the extra **energy demands** on the mother's body.

4) **Kidney function** alters. The mother's body has to deal with the **baby's wastes** in addition to her own. To cope with this, her kidneys filter fluid at an **increased rate**.

Pregnancy puts quite a strain on a woman's body, especially the **heart** and the **kidneys**. A healthy woman will be able to deal with these extra demands, but this is one reason why **regular medical checks** during pregnancy are essential.

Practice Questions

Q1 What are the two main causes of infertility in women?

Q2 Name a drug that can be used to stimulate ovum development in infertile women.

Q3 What is in vitro fertilisation (IVF)?

Exam Questions

Q1 Describe two treatments available for the correction of abnormal ovulation in infertile women, and explain how they work. [4 marks]

Q2 Explain why it is normal for a pregnant woman to put on weight during pregnancy. [6 marks]

Q3 IVF treatment allows women who cannot conceive naturally to have children. Describe the stages in this treatment. [7 marks]

Sounds a lot of effort for something that poos and vomits on you...

Still, a lot of people seem to want them, and so I suppose it's nice that science has come up with ways to help out. Don't worry too much about learning those graphs above, it's just the facts that you really need to know. The diagram showing the stages in IVF treatment is important too. Try drawing it from memory, then have a look and see what you forgot.

Population Sizes

These pages are for AQA B.

Pages 24 and 25 dealt with the way in which a population grows in a laboratory culture. In the real world, however, things are a little more complicated. Especially when we're talking about more complex organisms than bacteria...

Population Change *involves lots of 'Rates'*

If you want to **predict** what will happen to a population, you need to know how many individuals are being **added** to it and how many are **leaving** it. Changes in population size are caused by the following factors:

1) **Birth rate** — how many individuals are being **born** (per unit of population) in a **given time**.
   Sometimes referred to as **natality rate** (which is probably better, because plants, for example, aren't 'born').
   Birth rate = **no. of births / no. in population**

2) **Death rate** — how many individuals are **dying** (per unit of population) over a **given time**.
   Also referred to as **mortality rate**.
   Death rate = **no. of deaths / no. in population**

3) **Immigration** — the number of individuals **joining** the population from somewhere else in a **given time**.

4) **Emigration** — the number of individuals that are **leaving** the population to move elsewhere in a **given time**.

> **Population growth = (no. individuals born + no. of immigrants)**
> **– (no. of individuals dying + no. of emigrants)**

Birth rate and **death rate** can be used to predict population growth as they usually remain **fairly constant**, but it's hard to predict **immigration** to and **emigration** from a population without **long-term data**. This kind of data can be used to plot **population growth curves**, which you need to be able to **interpret**. A typical population growth curve is shown on the right.

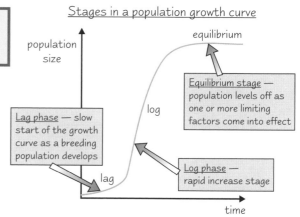

Population Data *can be used to plot* Survival Curves *and* Age Pyramids

If data are available about **death rates** at **different ages** in a population, a **survival curve** can be plotted. This shows how the death rate in a population changes with **time**, i.e. whether the death rate increases, decreases or stays the same as the population gets **older**. Data like this can also be used to find the **life expectancy** of the population — this is the average age at which individuals in the population die.

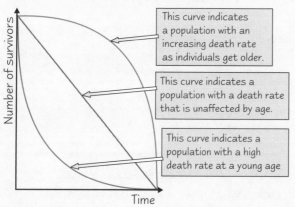

Useful information is also provided by plotting an **age pyramid** of the population. The height of the pyramid represents all the ages present in the population, with the youngest at the bottom and the oldest at the top. The **width** of the pyramid represents the **number of individuals of that age** in the population. The shape of the pyramid indicates **future trends** in the population.

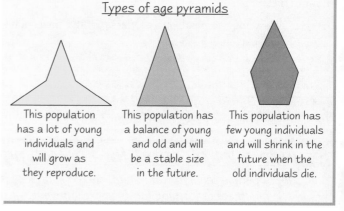

Population Sizes

The **Demographic Transition Model** shows Trends in **Human Populations**

The **demographic transition model** of human populations indicates trends in **birth** and **death rates** over a **long period** of time, as the population becomes more **developed**. It consists of **four** stages:

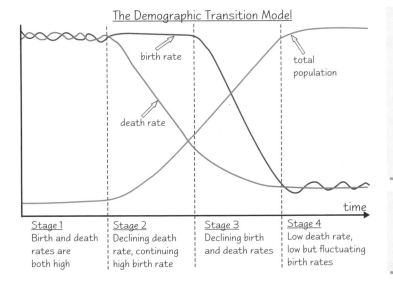

The Demographic Transition Model

birth rate

death rate

total population

time

Stage 1	Stage 2	Stage 3	Stage 4
Birth and death rates are both high	Declining death rate, continuing high birth rate	Declining birth and death rates	Low death rate, low but fluctuating birth rates

Stage 1 is when the level of development is **low**. There are **high** birth and death rates. The **high birth rate** is related to the absence of **family planning**, high **infant mortality**, and the need for lots of children to earn money for the family and to take care of their parents when they're old. The **high death rate** is due to **poor sanitation** and lack of **clean water** (which both cause **disease**) and shortage of **food**. The population is **stable**, as birth and death rates **balance**.

Stage 2 occurs when sanitation, water supply, food supply and medical care **improve**. There's still a **high birth rate**, but the **death rate** drops so the population **grows rapidly**.

Stage 3 is linked with an increase in **family planning** and the fact that, with **low infant mortality**, there's no need to have lots of children. The **birth rate falls** and, with the death rate remaining **low**, the population growth rate **slows**.

In **stage 4** the birth rate and death rate are both **low**, and the population becomes **stable** again.

Practice Questions

Q1 State the four factors that influence the size of a population over time.

Q2 What is meant by birth rate?

Q3 What is shown by an age pyramid?

Exam Questions

Q1 Look at the age pyramid on the right and explain, with reasons, what you would expect to happen to the size of the population in the future.

Years of age
85
80
75
70
65
60
55
50
45
40
35
30
25
20
15
10
male   5   female
0
10   5   0 0   5   10
Numbers (hundreds)

[5 marks]

Q2

Number of survivors

low density

high density

Time

The graph on the left shows the survival curves of a deer population when the population is at high density and at low density.
Suggest reasons for the difference between the two curves.

[4 marks]

Just the thing to cheer you up — lots of graphs about death...

... no wonder you kids enjoy doing biology revision so much. You've got to learn the basic shapes of these curves and pyramids, but understand that they won't always look the same — it depends what's going on in the individual populations. You need to be able to look at a curve or pyramid and use it to say what's happening in the population it represents.

Infectious Disease and Immunity

AQA B ONLY

These pages are for AQA B, and Edexcel need to learn the section about natural immunity on page 170.
Well isn't this a nice varied section — from courtship through pregnancy and population growth and on to infectious disease, all within eight pages. What on earth will be next — boomerang trajectories? Cutlery maintenance? Who knows.

Coughs and Sneezes Spread Diseases...

Any micro-organism that causes disease is called a **pathogen**. For a pathogen to affect you, it has to actually get **inside your body** (or at least into your skin). Your body has a number of entry points, and for every one there are some microbes that use it to get in.

- **Droplet infection** occurs when you cough or sneeze out droplets containing pathogens, which then get breathed in by other people. **Chickenpox**, **influenza** and **TB** are examples of diseases spread in this way.

- Microbes can also get transferred by **contact**. This can be **direct** (one person touching another), or **indirect** (a person touching something that's been touched by another person). Examples include **conjunctivitis**, **athletes foot** and **cold sores**. Transmission is more likely if the skin is broken, as it's a lot easier for the pathogen to get in.

- Disease can also be passed on via **food** and **drink**. **Food poisoning** is the obvious example, and **cholera** is another.

- Other ways that pathogens can enter the body are **sexual transmission** (e.g. HIV) and via **carriers** (**vectors**) — usually insects that bite humans (e.g. **malaria** is spread by *Anopheles* mosquitoes).

You will be Naturally Immune to some Diseases

To understand **immunity**, you need to know about **antibodies** and **antigens**. Antibodies are chemicals that **destroy microbes**, and they're produced by a type of **white blood cell** called a **B-lymphocyte**.
The B-lymphocytes 'recognise' the disease microbe by the pattern of chemicals on the pathogen's surface. These chemicals are called **antigens**. There are many different types of B-lymphocytes, each producing antibodies against a **different antigen**.

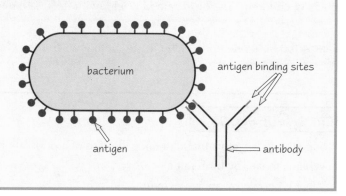

1) If a B-lymphocyte recognises a pathogen of 'its type', the cell starts **dividing rapidly** by mitosis (**clonal expansion**) to give lots of **plasma cells** and **memory cells** of that type. The plasma cells then start producing lots of the type of **antibody** needed to fight the disease.

2) When the pathogen has been destroyed, the **memory cells** remain in the body to counteract the infection if it returns.

3) This gives your body an **immunological memory**, and means it can react much **faster** if that pathogen invades the body again — probably before you start to show any symptoms. You will have become **naturally immune** to that disease. This type of natural immunity, when you make your own antibodies, is called **active immunity**.

The other main type of immunity is **passive immunity**, which is **temporary**. Babies are **naturally passively immune** to a number of diseases when they're born, because some of their **mother's** antibodies have come across the placenta or in her breast milk. They have no '**memory cells**' though, and after a while the antibodies **break down** and the immunity goes. The passive immunity protects the baby in its first few months, and gives it a chance to build up its own **active immunity** to common pathogens.

AQA B ONLY

Pathogens can Outwit the Immune System

There are some diseases that you **can't** become permanently immune to, either naturally or after **vaccination** (see next page). The **common cold** and **flu** are examples. The micro-organisms avoid the immune response because they constantly evolve into new strains with **different antigens**. When the new strains get into the body, the B-lymphocytes can no longer recognise them and the build up of antibodies has to start all over again.

Infectious Disease and Immunity

Vaccines help out our Immune Systems

While your B-lymphocytes are busy dividing to build up their numbers to deal with a pathogen, you suffer from the disease. **Vaccination** can help avoid this. It lets your body increase its level of the B-lymphocyte and antibody needed to fight a particular disease, **without** the microbe being present and increasing in numbers at the same time. This means you get the **immunity** without getting any **symptoms**. Vaccines come in various types:

1) **Dead pathogens** (e.g. **whooping cough**). The pathogen is **killed** but the antigens remain on its surface, so the body undergoes an immune response without the micro-organism causing any damage.

2) **Non-virulent strains** of a pathogen (e.g. **rubella**). Some strains of a pathogen stimulate **antibody production** but don't actually cause the disease.

3) **Modified toxins** (e.g. **diptheria**). Toxins (poisons) produced by pathogens can act as **antigens**. The toxin can be treated with **heat** or **chemicals** so that it doesn't produce any symptoms, but still triggers an immune response.

4) **Isolated antigens** (e.g. **influenza**). Sometimes, the antigen can be **separated** from the pathogen and injected to trigger an immune response.

5) **Genetically engineered antigens** (e.g. **hepatitis B**). The hepatitis B antigens have been **isolated** and can be made by **genetic engineering**. They're injected without the virus, so the disease can't develop.

Vaccines give you **active artificial immunity**. It's **active** immunity because your body makes its own antibodies, and will produce memory cells to help prevent the disease in the future. And it's called **artificial** immunity because you haven't actually **caught** the disease in the natural way — your body has been 'tricked' into reacting against a harmless pathogen.

Sometimes people need to be given '**instant immunity**' (vaccination takes a while to be effective). For example, if you get a bad cut you're often immunised against **tetanus**. This involves injecting the patient with the **tetanus antibodies** themselves, rather than with a harmless pathogen. It gives you **artificial** immunity **immediately**, but it's also **passive** immunity because your B-lymphocytes haven't been involved in producing the antibodies and there will be no memory cells. The immunity will wear off when the antibodies break down.

> Vaccines not only protect the people vaccinated, but, because they reduce the **occurrence** of the disease, those **not** vaccinated are also less likely to catch it. This is called the **herd immunity effect**.

Practice Questions

Q1 What is a pathogen?

Q2 What type of blood cell produces antibodies?

Q3 What is meant by the 'herd immunity effect'?

Exam Question

Q1 `The graph on the right shows the number of cases of polio reported each year since 1991 in the Indian province of Tamil Nadu. A vaccination programme was in place throughout this period.

(a) The same vaccine programme was in place over the years 1991-1993, yet the cases of polio dropped dramatically. Suggest a reason for this.   [1 mark]

(b) There are now very few cases of polio in Tamil Nadu. Explain why the vaccination programme should **not** be discontinued.   [1 mark]

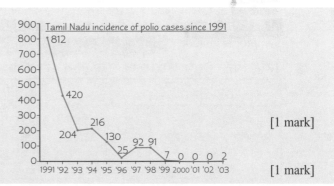

Immunity — another good idea from God / random mutations / aliens...

Delete as appropriate, according to your own belief system. I'm a big believer in political correctness, me. Personally I think it was the aliens, but I'm not saying that's right, it's just my opinion. It's also my opinion that you need to get this page learnt, as thoroughly and as quickly as you can. See, some of my opinions are pretty sensible. Others — not so much.

Effect of Lifestyle on Health

Page 172 is for AQA B and Edexcel, page 173 is for AQA B only.
There's one basic rule in life — anything you enjoy is likely to be bad for your health. With the exception of exercise, but who really prefers jogging to pizza? Most things are OK in moderation, but if you overdo the fun, you pay for it. Boo.

You Have to Watch Your Diet

As far as **diet** goes, the main problems are:

- eating **too much**
- having too much **fat** or **salt** in your diet.

Here are some of the nasty things that can happen (even the words are pretty nasty, but the effects are worse):

Atherosclerosis. This is the 'furring up' of the **coronary artery** by fats — mostly **cholesterol**. The fatty deposit inside the artery wall is called an **atheroma**, and it's caused by a diet that's too high in animal fats, which the body turns into cholesterol. This **restricts the blood flow to the heart** and can result in a **myocardial infarction** (heart attack, to you and me).

normal cross-section of artery

fatty material deposited in vessel wall

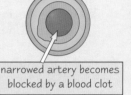
narrowed artery becomes blocked by a blood clot

Atherosclerosis makes the formation of **clots** more likely, and if these block a blood vessel in the brain, a **stroke** results. When the blood supply to a part of the brain is cut off, that part gets damaged. The medical term for a stroke is a **cerebrovascular accident**.

Being overweight or eating too much fat or salt can result in **high blood pressure**. Of course, the medical term isn't 'high blood pressure' but **hypertension**. Hypertension makes your heart work harder and can cause heart attacks and strokes. It also increases the risk of developing an **aneurysm** (a bulge in an artery) which can be **fatal** if it bursts.

There can be Nasty Consequences if you Don't Take Care of your Lungs

Here are some nasty diseases caused by mistreatment of your **lungs**.

1) **Lung cancer.** The main cause of lung cancer is **smoking** and you're also more likely to get it if you live or work in an atmosphere where other people smoke. **Tumours** grow in the epithelial tissue of the lungs and damage them, and tumour cells travel around the body in the blood and set up **secondary tumours** (**metastases**) elsewhere.

2) **Chronic bronchitis** is another disease that results from **smoking**, and it can also be caused by **air pollution**. Irritation causes the bronchioles in the lungs to become **inflamed**, which can block some of them. This makes breathing more difficult. The person coughs — which is an unsuccessful attempt to remove the blockage — and the **alveoli** (air sacs) in the lungs can **tear** as a result.

3) **Emphysema** is caused by smoking and air pollution too. It often follows if you've had chronic bronchitis for a long time. **Macrophages** digest the particles inhaled and produce **elastase**. The elastase digests the **elastin** in the alveoli wall, so they can't recoil any more. They burst, reducing the surface for gas exchange. The lungs **enlarge** but get **less efficient**. The person will be breathless and may even need to be given **oxygen** to help them breathe.

4) **Pneumoconiosis** is fibrosis and scarring of the lungs due to long-term breathing of dusts like coal, silica or asbestos. It's seen mainly in people who work in dusty environments, like coalmines or quarries.

5) **Tuberculosis (TB)** is an **infectious** disease caused by a bacterium, unlike the other four. Symptoms include a bad cough, fever and unexplained weight loss.

TREATMENT OF LIFESTYLE-RELATED HEALTH PROBLEMS

Disease	Treatment
atherosclerosis	surgical removal of section of artery, angioplasty (unblocking of arteries using a needle)
hypertension	low sodium diet and drug treatment (e.g. diuretics, beta blockers, calcium channel blockers), low fat diet and exercise programme
coronary heart disease	surgery, low fat diet, exercise programme, drug treatment (e.g. aspirin, statins, nitrates)
chronic bronchitis	drug treatment (e.g. bronchodilators, corticosteroids), oxygen therapy, lung transplant
tuberculosis (TB)	drug treatment (long-term course of several antibiotics)
pneumoconiosis	no treatment, except to avoid further inhalation
lung cancer	radiotherapy, chemotherapy, surgery

Effect of Lifestyle on Health

Holidays *may be* **Good for You**, *but excessive* **Sunbathing Isn't**

Everyone likes sunlight (except vampires), but actually it's quite nasty. It contains damaging **ultra-violet radiation**. This radiation can cause **skin cancers** of various sorts, because it damages the **DNA** in the cells it hits. This can cause them to divide uncontrollably and become **malignant**. Ultra-violet (UV) radiation comes in various types:

1) **UVA** is what causes you to **tan** — it stimulates production of **melanin** (a protective brown pigment) in the skin. It penetrates deep into your skin and prolonged exposure can cause **skin cancer**.

2) **UVB** is what causes **sunburn**. It also causes **skin cancer**.

3) **UVC** doesn't usually get through the atmosphere — it's filtered out by the **ozone layer**. If it does get through, where the ozone layer is thin or absent (e.g. due to **CFCs**, see p.115), it's very damaging to the skin.

Most skin cancers are fairly easy to treat and not too dangerous, particularly if they're caught early, but there's one very serious type called **malignant melanoma** which kills about **1600** people each year in the U.K. One reason it's so serious is that it can **spread** very quickly to other parts of the body if it's not treated at an early stage.

Protection against UV rays involves:

1) not exposing yourself to the sun for **long periods** (wear loose clothing and a sun hat instead of exposing your skin).
2) avoiding the **midday sun** (from around 11 a.m. to 3 p.m).
3) not using **sunbeds**.
4) using **sunscreen** of at least factor 15.

She looks happy now,
but a few years down the line...
...she'll realise how stupid that hat looked.

Practice Questions

Q1 What is an atheroma?
Q2 What is a cerebrovascular accident?
Q3 Name three lung diseases that can be caused by smoking.
Q4 Give one type of treatment for each of the diseases you named in your answer to Q3.
Q5 Which components of sunlight lead to skin cancer?

Exam Questions

Q1 Explain the health problems that can be caused by a diet too high in saturated fats. [6 marks]

Q2 Explain why sunbathing can be dangerous, and state ways in which these problems can be reduced. [10 marks]

Hmmm. I think I'd have preferred the cutlery maintenance...

...but at least it wasn't boomerang trajectories. A much trickier subject than many people think. These two pages aren't that bad really, but there are quite a few medical terms you have to learn, especially in that first section. Doctors like to sound clever, so they use words like 'myocardial infarction' instead of the much simpler 'heart attack', and you'll have to too.

Screening Programmes

Screening programmes are used to check whether people in high risk groups are developing diseases before they start to feel ill. This allows treatment to begin really early if a problem's found, which increases its chances of success. **These two pages are for AQA B, and the last section on genetic screening is also for OCR.**

Screening can Detect some Diseases Before any Symptoms are Seen

Some serious diseases have symptoms that can go **unnoticed** for a while, as they don't cause sickness, pain or discomfort. To detect these diseases early enough to allow for **successful treatment**, a variety of **screening programmes** have been introduced. A screening programme looks at as many people as possible, but especially those in some sort of **high risk category** (e.g. because of their age or because they have a family history of a disease), to detect those that have an early stage of a disease.

X-rays can be used to Detect Breast Cancer

A **breast x-ray** or **mammogram** is currently used to detect **breast cancer** in its earliest stages. Early detection can improve the chances of **successful treatment** and recovery. Breast x-ray screening aims to show changes which are **too small** to be felt by the patient and which can't be found in any other way. A **tumour** will show up on the x-ray as a **shadow**. Screening is normally carried out on women **over 50**, and is recommended once every **three years** Radiation is **dangerous** to some extent (it can damage cells and lead to cancer), which is one reason why screening isn't done more often. It's done on women over 50 because they're **more at risk**, and in younger women the breast tissue tends to be more **dense** and x-rays can't pick up tumours very easily. It's estimated that the screening programme currently saves around **300 lives** per year in the U.K.

Ultrasound is a Very Safe Screening Technique

Both **x-rays** and **endoscopy** (see next page) carry a small **risk** for the patient. But the use of **ultrasound** to build up images of the body has no known side-effects or dangers. For this reason it's routinely used to screen **developing foetuses** during pregnancy.

Ultrasound is emitted from a **transducer** which is placed on the mother's abdomen and moved to "look at" different parts of the foetus. **Repetitive ultrasound beams** scan the foetus in thin slices and are **reflected back** onto the same transducer. The information obtained from different reflections are used to build up a picture (a **sonogram**) on a monitor screen. Sonograms are used to determine **age**, **size** and **growth** in the foetus, and to screen for any **problems** with its development.

Because of the **safety** of the procedure, ultrasound is being developed to screen for other conditions such as **blood clots**, **ovarian cancer**, and **breast cancer**. For breast cancer, it may be **more reliable** than x-rays for use in women **under 50**, as ultrasound can pick up irregularities in **denser** breast tissue than x-rays can (see above).

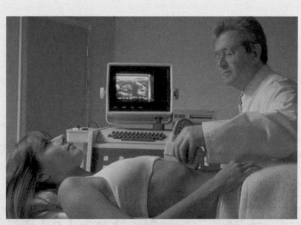

A doctor uses ultrasound to build up a sonogram on the monitor

Screening Programmes

Endoscopy Looks Inside the Body

1) **Endoscopy** involves passing small **optic fibres** into the body which are attached to a **camera** so a doctor can actually look inside your body for physical changes.

2) A miniature instrument can also be used to **snip off** a piece of tissue and retrieve it if the doctors want to test it.

3) Endoscopy isn't generally painful but it can be **uncomfortable**. People are unlikely to volunteer for it for no reason, so it's **not** used for general screening but only for investigating patients who have **some symptoms**.

4) It's commonly used to detect **cancer of the colon** (large intestine) and is also used to investigate **bowel irritations** and **lung cancer**.

Genetic Screening is a Sensitive Issue

Screening for **genetic diseases** like **cystic fibrosis** has been covered on pages 130-1. It may be suitable for screening **embryos** before implantation in **IVF treatment**, and for parents with a **known risk** of passing on a serious genetic condition to their children. The extraction of cells for genetic analysis carries some **risk** for both the mother and the foetus in a normal pregnancy, though, so it's **not** done routinely.

Other types of **genetic screening** involve testing children or adults for '**dormant**' **genetic diseases**, or testing potential parents to assess the risk that they will have a child with a serious genetic defect. This testing is **safe**, but has **ethical** considerations:

- It must always be **voluntary** — it's unethical to force people to undergo genetic screening.
- People are likely to need some **counselling** after being told that they're at risk.
- The information must be **confidential**.
- If information on health problems is known, **insurance companies** may demand that people tell them before quoting for insurance.

People who think they might be at risk of developing a genetic disease or of having a child with one can be referred to a **genetic counsellor**. A genetic counsellor has to have a sound knowledge of human genetics, and must be able to explain all the genetic ratios and probabilities in a way that their patients can understand. They also need to be able to deal sensitively with worried or distressed patients, and be able to explain all the **legal** and **medical options** — e.g. if a scan has revealed that a pregnant woman's baby has a serious genetic condition, the genetic counsellor would have to explain all the options to the parents, while ensuring that they are allowed to make their own choice.

Practice Questions

Q1 What is a mammogram?

Q2 What name is given to a picture produced by ultrasound?

Q3 Give two reasons why ultrasound might be better than x-rays for screening women under 50 for breast cancer.

Q4 Name two diseases that are screened for using endoscopy.

Exam Questions

Q1 Explain the reasons why screening for breast cancer is restricted to women over 50
and is only carried out once every three years.   [5 marks]

Q2 Why is ultrasound preferred to x-rays when doctors want to see a picture of a developing foetus?   [3 marks]

Q3 Huntington's disease is a dominant inherited disease involving the degeneration of nervous system cells, including brain cells, usually beginning at age 40+. The disease is fatal. Genetic testing can detect the disease, but there is no known treatment. Suggest **two** advantages and **two** problems that could result if a genetic screening programme for Huntington's disease was introduced.   [4 marks]

The only screen I'm interested in is one with Coronation Street on it...

Just kidding. You can't fail to appreciate modern medical techniques that can save lives by detecting the very earliest stages of a serious illness. They could even save your life one day, so it shouldn't be too much of a hardship to learn a couple of pages on them. There's not that much to learn — the picture of the surprisingly trim pregnant woman takes up a lot of room.

Bacteria, Viruses and Fungi

Unless stated otherwise, pages in this section are for OCR, AQA B and Edexcel.
*A **microorganism** is basically any organism that can only be seen with a **microscope**.*
They're mostly the single-celled type of beastie — bacteria, fungi, protoctista and the like.

Bacteria are Prokaryotic Microorganisms

All bacteria are **single-celled**, although some kinds of bacteria can stick together to make **clusters** or **chains**.
Bacteria are really, really **small** — they range from about 0.1 to 2μm.

Here are the **features** shared by **most** bacterial cells:

1) They have a **cell wall** made of **peptidoglycan**.

2) They divide by **binary fission**.

3) They **don't** have **membrane-bound organelles** (e.g. a nucleus), so **can't** do endocytosis, mitosis or meiosis.

4) Their **DNA** consists of a circular molecule with **vital** genes (the bacterial **'chromosome'**), and smaller rings called **plasmids**, which contain other, non-vital genes, e.g. for antibiotic-resistance.

5) Since there's no nucleus, **transcription** occurs in the **cytoplasm**.
Translation occurs on **70S ribosomes** (smaller than the 80S ones in eukaryotes).

6) **Genetic recombination** can occur by **conjugation**, where two cells join up to exchange parts of their DNA. Genes can also be **recombined** by exchanging plasmids, by transfer in viruses called **bacteriophages**, or by taking up DNA from the **surroundings** (e.g. from dead bacteria).

And here are some **special** features found in **only some** kinds of bacteria:

1) **Bacterial flagella** may be present.

2) **Mesosomes** are **folds** of the cell membrane — their purpose isn't known for sure, but they may be associated with respiration.

3) **Bristle**-like projections called **pili** and **fimbriae** allow the bacteria to **stick** to surfaces. (Fimbriae are **shorter** than pili and there are **more** of them.)

4) There may be a layer of **polysaccharide** (called a **glycocalyx**) around the cell wall. It could be a loose **slime** layer or a more solid **capsule**. It could **defend** the bacteria from white blood cells or stop it **drying out**.

5) Bacteria can be **heterotrophic** (consume organic molecules), **photoautotrophic** (make food using light energy) or **chemoautotrophic** (make food using energy from oxidation reactions).

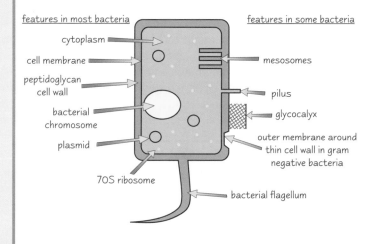

features in most bacteria — cytoplasm, cell membrane, peptidoglycan cell wall, bacterial chromosome, plasmid, 70S ribosome, bacterial flagellum

features in some bacteria — mesosomes, pilus, glycocalyx, outer membrane around thin cell wall in gram negative bacteria

Bacteria are Classified on the Basis of Cell Wall Structure

Bacteria can be classified as either **Gram-positive** or **Gram-negative**:

1) **Gram-positive** bacteria have a **thick** cell wall. **Gram-negative** bacteria have a **thinner** cell wall coated with an outer **membrane**.

2) To test whether bacteria are Gram-negative or positive, they can be stained purple with **crystal violet** solution, then washed with Gram's **iodine** solution, then **ethanol**, then counter-stained with the red stain **safranin**.

3) If, after this process, they end up **purple**, they're **Gram-positive**. If they end up **red**, they're **Gram-negative**.

Gram-positive wall

this stains purple with crystal violet (the red stain doesn't show over the purple)

thick cell wall
phospholipid membrane
cytoplasm

Gram-negative wall

the outer membrane of this (with the purple stain) is washed away by ethanol — the cell wall then stains red

thinner cell wall
cytoplasm

Bacteria, Viruses and Fungi

Fungi are Eukaryotic Organisms

Some types of fungi are big and **multicellular**, like mushrooms, and others are smaller and even **single-celled**. Here are their general characteristics:

1) Mature fungi are made up of many thread-like structures called **hyphae**. Together, these hyphae form large **webs** called **mycelia**.

2) Cell walls are made of **chitin**. Most fungi have partition walls called **septa** along the length of hyphae.

3) **Vacuoles** are usually present in the cells of hyphae.

4) Cells are **diploid** and divide by **mitosis**. Genetic recombination occurs by **meiosis** to form **spores**, and **conjugation** (where DNA is passed between cells).

A fungal hypha

cytoplasm · cell membrane · cell wall · vacuole · septum · nucleus

Yeasts are single-celled fungi. Unlike **moulds** and other types of fungi, they have cell walls made from complex **polymers of mannose and glucose**. Yeast cells divide by **budding**, involving a kind of asymmetrical mitosis.

A budding yeast cell

new cell budding

Viruses are Just Particles of Nucleic Acid and Protein

Viruses are microorganisms that are **not** made of cells, so can be described as **acellular**. All viruses are **pathogens** (infectious organisms that usually cause disease), whereas only **some** types of bacteria, fungi and protoctists are pathogens. Some viruses only infect bacteria. These are called **bacteriophages**, or sometimes just **phages** for short.

Viruses are **tiny**, even compared to bacteria. The biggest virus is only 0.2 μm across, and most are much smaller. This means that they can only be seen with an **electron microscope**.

Their basic **features** are:

1) A core of **nucleic acid**, either DNA or RNA, sometimes with some **protein**.

2) An outer coating of **protein** called a **capsid**, made up of protein units called **capsomeres**.

3) All viruses are **infectious** — they invade cells. Some viruses have an extra layer, called an **envelope**, made of **membrane** from the cell membrane of a previous host cell.

4) They have **no** organelles, **no** cytoplasm and **no** plasma membrane.

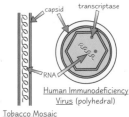

capsid · reverse transcriptase · RNA

Human Immunodeficiency Virus (polyhedral)

Tobacco Mosaic Virus (helical)

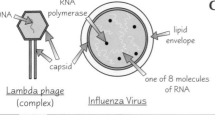

DNA · RNA polymerase · lipid envelope · capsid · one of 8 molecules of RNA

Lambda phage (complex) · Influenza Virus

Capsids and **envelopes** may:

1) **Protect** the virus from **chemicals** (e.g. enzymes) while outside a host.

2) Enable the virus to **bind** to host cell membranes, ready for **penetration**.

3) Assist in **penetration** of a host cell.

(See page 192 for the life cycle of a virus.)

Practice Questions

Q1 Give two differences and two similarities between bacteria and fungi.

Q2 Describe the procedure used to distinguish between Gram-positive and Gram-negative bacteria.

Q3 Explain why viruses can be described as 'acellular'.

Q4 List the basic features of a virus.

Exam Questions

Q1 Explain the functions of viral capsids and envelopes. [3 marks]

Q2 a) Describe two possible functions of the layer of polysaccharide around a bacterial cell wall. [2 marks]
b) Draw a labelled diagram to show the layers around the cell wall of a Gram-negative bacterium. [3 marks]

Bacteria — Unhygenix's wife...

A lovely couple of pages on bacteria, fungi and viruses. A lot of this stuff shouldn't be too new. You'll know a lot about bacteria and their microorganism friends from all your past Biology studies. It makes it a bit easier for you, but it doesn't mean you can get by with what you know already. So go on — learn those nasty, long words. You know you want to.

Cell Culture

Loads of kinds of microorganism can be grown in the lab in cultures. Which means you have to learn all about it.

Microorganisms have Special **Chemical Requirements**

Autotrophic organisms (plants, algae and some kinds of bacteria) need a supply of simple inorganic molecules with which to build organic molecules. So, if they're being grown in a lab, there should be a ready supply of these things. Here are the **elements** they need and where they get them from:

1) **Carbon** and **oxygen** (in all organic molecules) are supplied by **carbon dioxide**.

2) **Hydrogen** (also in all organic molecules) is supplied by **water**.

3) **Nitrogen** (in amino acids and the bases of nucleotides) is supplied by **nitrate**.

4) **Phosphorus** (in nucleotides) is supplied by **phosphate**.

5) **Sulphur** (in just two kinds of amino acids) is supplied by **sulphate**.

> **Heterotrophic organisms** (animals, fungi and most bacteria) take in organic molecules, and some inorganic materials. They can make a variety of other types of organic molecules through their metabolism, depending on which enzymes they have.
>
> **Chemicals** that microorganisms need in order to grow are called **essential nutrients**. For heterotrophs, these include **essential amino acids** and **essential fatty acids**. **Macronutrients** are needed in large amounts, but **micronutrients** are needed in tiny amounts. Many essential micronutrients needed by heterotrophs are **vitamins**.

Microorganisms can be **Grown** on **Different Media**

Microorganisms are cultured for a variety of **reasons**, including:

- **identifying** the species of microorganism in a medical sample — to **diagnose** the disease
- producing **useful** substances, such as **antibiotics**
- **food production**, such as making cheese and wine

> In the lab, microorganisms are cultured **in vitro** ('in glass', like on an agar plate). The material where microorganisms are grown is known as the **medium**. It can be liquid, in the form of a **nutrient broth**, or solid, as in **agar jelly**.

Agar jelly is made by heating a solution of agar, a polysaccharide, which solidifies to jelly when cooled to 42°C. Special nutrients are added before the agar sets.

> **Selective media** contain substances which selectively **prevent** the growth of certain microorganisms, whilst allowing others to grow. This means that specific types of microorganism can be **isolated** and **cultured**.
>
> **Indicator media** contain a chemical **indicator** (e.g. pH indicator) that changes colour due to chemicals made by certain types of microorganisms. For example, **EMB agar** contains lactose, sucrose, eosin and methylene blue. Methylene blue inhibits Gram negative gut bacteria, and only some types of Gram positive ones can use lactose or sucrose. Of these, *Escherichia coli* produce black colonies with eosin, whereas *Salmonella* produce pink ones.

Other **Conditions** are Needed for **Optimum Growth**

1) The optimum **temperature** for growth varies between microorganisms, with some having extremes. **Psychrophiles** grow best at low temperatures, **thermophiles** grow best at high temperatures.

2) Some microorganisms are **aerobic**, but others are **anaerobic**. **Obligate aerobes** can only respire aerobically and so will die if oxygen is not present. **Obligate anaerobes**, however, die in the **presence** of oxygen. If necessary, oxygen concentration in a culture can be increased by **aeration**.

3) Most species have an optimum **pH** of between 5 and 9, but **acidophilic** forms have a low optimum pH. Usually fungi are more tolerant of low pH than bacteria.

4) **Disinfectants** inhibit growth of microorganisms, and **antibiotics** kill them or inhibit their replication. (See the next page.)

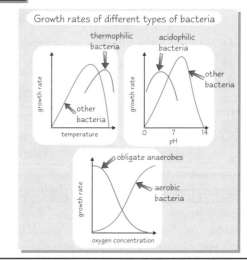

Growth rates of different types of bacteria

Cell Culture

Aseptic Technique is used to Culture Microorganisms

Aseptic technique is handling a microorganism culture in a way that minimises **contamination**.
Sterilisation is the removal or killing of **all** microorganisms on an object, so that after sterilisation,
there aren't any microorganisms to reproduce, even if the most favourable conditions occurred.
The equipment and media used are sterilised by heating to 121°C for 15 minutes in an **autoclave**.

Other precautions taken to minimise **contamination** by other microorganisms include:

1) Lab workers pay close attention to **personal hygiene** (e.g. clean hands, hair net).

2) Windows and doors are kept **closed**.

3) All containers holding microorganisms have **lids** to minimise
contamination by 'fall out' of microorganisms from the **air**.

4) Working close to a **Bunsen flame** ensures that many of the microorganisms
in 'fall out' are **killed** before they land on working surfaces.

5) Surfaces are swabbed with **disinfectant**.

6) Culture **containers** are made of smooth glass or stainless steel as **scratches** can
harbour stray microorganisms.

7) The microorganisms can be transferred from medium to medium on **wire inoculating
loops** that are first **sterilised** by heating them to red hot in a Bunsen flame. Alternatively,
a sterilised **Pasteur pipette** can be used for transferring a known volume of culture
(necessary for quantitative work), and the culture is then spread on agar using a sterilised
spreader made from a bent glass rod. This gives a fairly even **lawn** of microorganisms.

Disinfectants and Antibiotics Affect Growth of Bacterial Lawns

The **effectiveness** of antibiotics and disinfectants at preventing bacterial growth
can be discovered by adding them to bacterial lawns on sterile discs of filter paper.

The disinfectant or antibiotic **diffuses** out of the disc to create an **inhibition zone**, where
bacteria can't grow. The **bigger** this is, the more **effective** the antibiotic or disinfectant.

Practice Questions

Q1   Name five elements which microorganisms need to build organic molecules. Where do they get them from?

Q2   Explain the meaning of the terms 'selective media' and 'indicator media'.

Q3   Explain the difference between obligate aerobes and obligate anaerobes.

Q4   What is meant by sterilisation of laboratory apparatus?

Q5   How can you test how good a disinfectant or antibiotic is?

Exam Questions

Q1   A student was provided with a monoculture of the mould *Aspergillus nidulans* on an agar plate in a Petri dish.
She was asked to transfer a sample of the *A. nidulans* culture onto a separate sterile agar plate.

a) Describe the stages in the aseptic transfer of the sample from the monoculture to the sterile new plate.   [5 marks]
b) Suggest a method that was used to sterilise the agar plate.   [2 marks]

Q2   a) Explain why a culture of bacteria in a nutrient broth showed increased population growth
when oxygen was bubbled through the medium.   [3 marks]
b) Suggest how the results would be different for a bacterial culture of an obligate anaerobe.   [2 marks]

Glasgow, City of Culture 1998 — Great conditions for bacterial growth...

*This is the sort of stuff that's really boring... but really important. I know you want to get on with learning about nasty
diseases and other gruesome stuff, but you've got to take the time for cell cultures. If it wasn't for cell cultures, we
wouldn't be able to diagnose lots of diseases, or make antibiotics to cure them, or, erm, make cheese. So there.*

Measuring Bacterial Growth

It's a lot easier to measure the population growth rate of a bacterial culture, than of, say, lions. Or whales. Or people.

Bacterial Cultures Show Typical **Population Growth Curves**

You might recognise this stuff from page 24, but it's here again for you dirty skivers who've not covered it yet. Following the initial introduction of bacteria to the medium, the following **phases** of **population growth** are shown:

1) The **lag phase** occurs when there is a very **small** early **increase** in population density. **Reproduction rate** is **low** because it takes time for enzymes to be made for DNA replication and to use any new food.

2) The **exponential phase** occurs during the most **favourable** conditions. Here there is a **doubling** in population size per unit time (one cell divides to produce two cells, each of which can divide further in the same amount of time). There is sufficient **food** to support the growth, and **competition** for food is at a minimum.

3) The **stationary phase** is approached as **death rate** increases and becomes **equal** to the reproductive rate. This happens because **food** gets used up and poisonous **waste products** build up.

4) **Decline phase** occurs where death rate is **greater** than reproductive rate, due to the further depletion of food and accumulation of excreted waste.

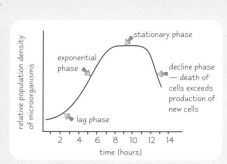

Depending on culture conditions, many bacteria produce substances called **secondary metabolites** towards the end of the exponential phase and into the stationary phase. These are not essential for growth and reproduction, but some, such as antibiotics, can help bacteria **survive** the stressful conditions at this point in the population growth curve.

Growth Rate can be Expressed by Calculating a **Growth Rate Constant**

Any **exponential increases** in a population size (or any other factor) can be **quantified** by working out the $\log_{10}$ **values** of the population sizes. If we plot the $\log_{10}$ values of the population size of an exponentially growing population on a graph, we end up with a **straight line**.

This kind of conversion then lets us calculate something called the **growth rate constant**. It's equal to the number of **generations** that the population goes through in a particular length of time. It's calculated by using the following **formula**.

$$\text{growth rate constant} = \frac{\log_{10} N_1 - \log_{10} N_0}{t \times \log_{10} 2}$$

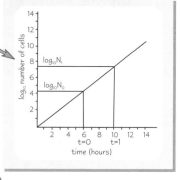

Look at the straight line graph shown above. Here's what the **formula** means:

- $\log_{10} N_1$ is the log of the number of cells at a point in time ($t = 1$)
- $\log_{10} N_0$ is the log of the number of cells at an earlier point in time ($t = 0$)
- t is the **time** taken between these points

$\log_{10} N_1 = 7.3$
$\log_{10} N_0 = 4.3$
$t = 10-6 = 4$ hours

therefore the growth constant rate is:

$$\frac{7.3-4.3}{4 \times 0.301} = \frac{3}{1.204} = 2.49$$

Because this is a log graph, you can simply **read them** off from the axes — you **don't** need to calculate them. Similarly, $\log_{10}$ of 2 = 0.301. (You won't even need to work this out; any question on this will give you the $\log_{10}$ value of 2.) The calculation value of the growth rate constant in this time interval is shown under the graph.

Bacteria with **Two Food Sources** often Show **Two Exponential Phases**

If bacteria are grown on media with **two** different **carbon sources**, they will often use one before starting on the other. This is called **diauxic growth**.

E. coli grown on a mixture of **glucose** and **lactose** will use up the glucose first. This is because the presence of glucose **represses** the gene that codes for the production of the enzyme **lactase**. The lactase is finally made during the second lag phase when all the glucose is used up (see p.44 for more about this).

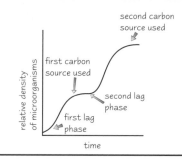

Not for AQA B or OCR

Measuring Bacterial Growth

There are Different Ways to Find out Population Sizes of Bacterial Cultures

Some techniques are designed to determine the actual **population density**.
This involves working out the number or mass of **cells** in a particular volume of medium.

1) Direct **cell counts** are possible by taking a tiny sample of the medium and examining it under a **microscope**. You can then count the number of cells you see in a fixed volume of the medium using a **haemocytometer** (a grid of microscopic chambers, each with a known area and depth). It's best to do a number of counts and find the **average**, for it to be **accurate**. This number can then be **multiplied** to give the number in the sample.

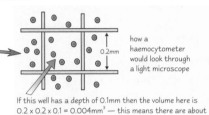

how a haemocytometer would look through a light microscope

If this well has a depth of 0.1mm then the volume here is $0.2 \times 0.2 \times 0.1 = 0.004mm^3$ — this means there are about 5 cells per $0.004mm^3$, which is about 1250 cells per mm^3.

2) The **density** of cells can be measured with a **turbidometer** (colorimeter) — the more **turbid** (cloudy) the medium is, the less **light** is transmitted through the turbidometer. The turbidity is then **compared** to the turbidity of a **known** number of bacteria.

3) It's possible to measure **fungal** cultures by simply **weighing** them, but this isn't usually sensitive enough for bacteria.

The Dilution Plating Method tells us the Number of Viable Cells

All the methods above tell us the total **quantity** of cells in a sample of culture. The **dilution plating method**, however, can tell us the number of **viable** (living) **cells** in a medium — those that can divide. Here's how it's done:

1) A known volume of the culture is **diluted** by a known amount. It's then **further diluted**, again by a known amount each time.

2) A known volume of the final dilution is spread over an **agar plate** and **incubated**. Each single viable cell found in this final dilution will **reproduce** to produce a visible **colony**.

3) The number of colonies is **counted**. By knowing the **dilution factors** used each time, it is possible to work out the **total** number of cells in the first sample.

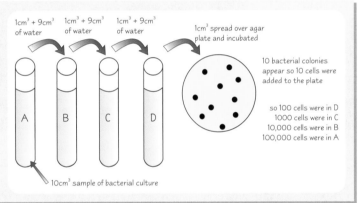

$1cm^3 + 9cm^3$ of water $1cm^3 + 9cm^3$ of water $1cm^3 + 9cm^3$ of water

$1cm^3$ spread over agar plate and incubated

10 bacterial colonies appear so 10 cells were added to the plate

so 100 cells were in D
1000 cells were in C
10,000 cells were in B
100,000 cells were in A

$10cm^3$ sample of bacterial culture

Many types of bacteria are potentially dangerous. This means that any labs growing bacteria must have built-in safety features that prevent contamination of workers and the environment. Regions of negative air pressure are created in microbiology cabinets, so that contaminated air from the lab flows into the cabinet through air flow hoods, instead of spreading around the environment.

Practice Questions

Q1 Which factors result in lag, exponential, stationary and decline phases in a bacterial population growth curve?

Q2 What is meant by the term 'secondary metabolite'? Give an example of a secondary metabolite.

Q3 Describe the basis for the turbidity technique in establishing the density of a bacterial culture.

Q4 Describe one safety feature of a microbiology lab.

Exam Questions

Q1 A bacterial culture was grown in a laboratory fermenter for 10 hours.
   a) Calculate the growth rate constant where $\log_{10} N_1 = 8.6$, $\log_{10} N_0 = 3.4$ and $\log_{10}$ of 2 = 0.301. [3 marks]
   b) Outline the dilution plating method that was used to establish the density of bacterial cells at the two points in time. [8 marks]

Q2 a) What is meant by diauxic growth? [2 marks]
   b) Draw a graph for a bacterial culture that illustrates diauxic growth, labelling the phases. [3 marks]

Log graphs? OK, that's it — I want out...

Now, don't be scared off by that nasty-looking formula. It's really not that bad — all you have to do is read your figures off the graph and then plug them into the formula. Or if you're lucky, you might be given the figures to just plug in. Anyway, you've dealt with much worse in your GCSEs, so you can definitely handle it. And then the rest of the page is dead easy.

Industrial Growth of Microorganisms

The large-scale culturing of microorganisms means we can produce lots of important substances... like yoghurt. Great.

Many *Industries* Grow Microorganisms on a *Large Scale*

All sorts of microorganisms can be grown to provide useful **products**. Many of these products are important in the **food industry**, such as alcohol in wine-making or lactic acid in yoghurt manufacture. Others are grown for **medical** applications, such as for antibiotic production.

These substances are produced in the metabolic reactions of the microorganisms.

Large cultures of microorganisms are grown in vessels called **bioreactors** (also see page 104). The culture methods used are designed to **increase** the growth rate of microorganisms through creating **optimum conditions** (see page 178).

There are two main **culture methods** that are used:

Batch culture occurs in a **fixed volume** of medium. **Oxygen** is usually added during the growth of the microorganism and **waste gases** removed. Growth occurs up to the **stationary** phase until food becomes depleted and excretory products accumulate. Eventually **new cultures** must be established to start the process over again in a different batch. It is easier to **control** conditions in a batch, and **isn't costly** to start again if contamination occurs.

In **continuous culture**, fresh, sterile medium is added to the culture at a constant rate. Used-up medium and dead cells are removed at a constant rate. Cells are kept in the **exponential** growth phase and the culture lasts for far **longer** than with the batch method. This method is more **productive** than batch culture, and **smaller** vessels can be used. However, if **contamination** occurs, the whole lot will be lost, which is **very costly**.

Conditions in a *Bioreactor* are Controlled to *Optimise Microorganism Growth*

A typical **bioreactor** consists of the following **components**:

Bioreactors are also called fermenters.

1) A motor drives the rotation of an **impeller**, which **stirs** the culture and continually brings fresh medium in contact with the microorganisms.

2) Cool water is circulated through a **cooling jacket**, which reduces the risk of **overheating**, especially in larger bioreactors. Fermentation releases a great deal of heat, which is transferred to the cool water in the jacket by conduction and is carried away as the water flows out.

3) A **pH controller** contains a **pH probe** to detect changes in pH (e.g. caused by the acids produced by fermentation). It automatically delivers a **neutralising** quantity of **base** when required. Constant pH is needed for optimum enzyme activity.

4) **Sterile air** is delivered through an **aerator** in order to provide **oxygen** to maximise aerobic respiration, and thus maximise growth.

Large-scale fermenters are difficult to keep **sterile** and heating and cooling can be slower, making it more difficult to **control** conditions.

The whole bioreactor is usually made of smooth stainless steel. This can be sterilised before use by pumping **steam** through the apparatus. Product harvest from a bioreactor occurs through a **tap**.

After this, the product is **purified** and sometimes chemically **modified**. This series of treatments after harvesting is called **downstream processing**.

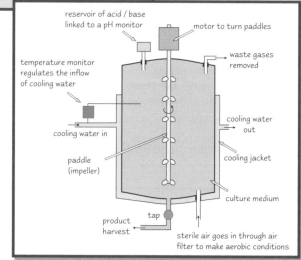

reservoir of acid / base linked to a pH monitor

motor to turn paddles

temperature monitor regulates the inflow of cooling water

waste gases removed

cooling water in

cooling water out

cooling jacket

paddle (impeller)

culture medium

product harvest

tap

sterile air goes in through air filter to make aerobic conditions

Industrial Growth of Microorganisms

Microorganisms are **Screened** for **Suitability**

Microorganisms are **screened** to maximise the production of certain substances. **Antibiotic production**, for example, can be detected in a Petri dish by placing samples of the antibiotic-producer with colonies of antibiotic-sensitive bacteria. Colonies disappear as they die in the vicinity of the antibiotic-producer.

Similarly, microorganisms can be screened for **enzyme production**. For example, protease production by a bacteria can be detected by its ability to clear a cloudy suspension of the protein casein.

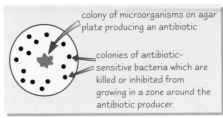
colony of microorganisms on agar plate producing an antibiotic

colonies of antibiotic-sensitive bacteria which are killed or inhibited from growing in a zone around the antibiotic producer

Penicillin is Produced by a Specific Type of **Mould**

Penicillin is an **antibiotic** made by a mould called *Penicillium chrysogenum*. It's grown in **bioreactors** by **batch culture**.

Here's how penicillin is made:

1) Organic **carbon** and **nitrogen** sources are added to the medium, as well as **oxygen** for aerobic respiration.
2) **Downstream processing** of the penicillin begins by **filtering** the mould mycelium from the medium.
3) The penicillin is **extracted** in an organic solvent.
4) It is **concentrated** and **crystallised**.
5) It may be chemically **modified** so that its effects can overcome penicillin-resistant strains of bacteria.

Mycoprotein is a **Foodstuff** Produced from Another Type of **Mould**

Mycoprotein is compacted mycelium (see p. 177) with a similar consistency to **meat**. You'll know it by the name of **Quorn** — a vegetarian source of **protein**. It's obtained from the growth of a **mould** called *Fusarium graminareum* by **continuous culture**:

1) A **continuous culture** of the mould is grown in a **bioreactor** at around 30°C and at a slightly acidic pH. **Carbon** is provided as glucose syrup, and **nitrogen** as gaseous ammonia. Vitamins, salts and oxygen are also given.
2) Large quantities of **protein** are made, but so are large amounts of **RNA**, which could be harmful. This RNA is **reduced** by adding **ribonuclease**, an enzyme that hydrolyses RNA to nucleotides.
3) The hyphae are **filtered** and **compacted** into a 'cake', which is sold as **Quorn**.

Practice Questions

Q1 Describe how each of the following components of an industrial bioreactor will optimise the growth rate of the cultured microorganism: water cooling jacket, pH controller, stirrer.

Q2 What is meant by the term 'downstream processing'? Describe the downstream processing of penicillin.

Q3 Name two kinds of fungi that are grown in industrial fermenters. What useful products do they synthesise?

Q4 Explain why ribonuclease must be added to cultures used in mycoprotein production.

Exam Question

Q1 a) Distinguish between the terms batch culture and continuous culture. [2 marks]
b) Give one advantage each for batch culture and continuous culture. [2 marks]
c) Explain why the antibiotic production of a *Penicillium* bioreactor would decrease if the flow of sterile air into the reaction vessel were interrupted. [3 marks]

Quorn — truly a gift from the gods for all the veggies of the world...

Sausages... mmm... bacon... mmm... chicken fillets... mmm... spare ribs... mmm... ham slices... Yep, there's absolutely nowt they can't create with Quorn these days. And it tastes good too. Anyway, all this stuff is fairly straightforward again. Just remember the bioreactor bits and that it's all about making sure as many microorganisms are grown as possible.

Biotechnology in Food Production

Mmmm... yoghurt, cheese, bread, beer. Mmmm... beer. Made by fermenting yeast? Mmmm... yeast.
These two pages are for OCR and Edexcel.

Bacterial Fermentations are used in Yoghurt and Cheese Production

When microorganisms are cultured, they often make a substance known as a **fermentation product**. This fermentation product is made under anaerobic conditions, when the respiratory substrate is **not** completely **oxidised**. Two common fermentation products are **lactic acid** and **ethanol**. These are important in **yoghurt** and **cheese** production.

Microorganisms can be genetically modified for food production.

Here's how **yoghurt** is made:

1) Milk is screened for **pathogenic bacteria** and **antibiotics**.
   (If antibiotics have been given to cows, they may be present in the milk.)

2) It is **heated** to around 90°C for 30 minutes to **kill** bacteria, **denature** milk proteins and **reduce** oxygen levels.

3) It is **cooled** to 45°C. Lower temperature and oxygen levels are needed for the **starter culture**.

4) **Starter culture** is added, including *Lactobacillus bulgaricus* and *Streptococcus thermophilus*. These are **incubated** at around 32°C to 40°C. *Lactobacillus* **hydrolyses** proteins to peptides, (which are used by the *Streptococcus*), and *Streptococcus* makes **methanoic acid** (for growth of *Lactobacillus*). *Lactobacillus* produces **lactic acid**, so the pH drops and a sour flavour develops. Both produce **ethanal**, which provides more **flavour**.

5) Sterilised **flavourings** (e.g. fruit) are added.

6) The yoghurt is **stored** at 2°C to reduce activity of the bacteria.

Cheese is also made by the **fermentation** of milk:

1) The bacteria *Streptococcus lactis* and *Leuconostoc lactis* are added. *S. lactis* produces **lactic acid** and *L. lactis* produces CO_2.

2) **Rennet** (see p. 104) is added, which **coagulates** the milk, separating solid **curd** from liquid **whey**.

3) The curd is **milled** (broken to tiny pieces) and **salted**.

4) Hard cheeses are **ripened** — the bacteria are allowed to continue reproducing, making **flavourings** such as aldehydes and ketones.

5) **Blue cheeses** are inoculated with the **mould** *Penicillium roqueforti*.

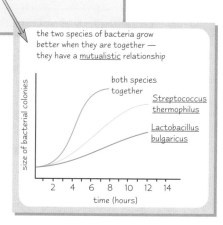

the two species of bacteria grow better when they are together — they have a <u>mutualistic</u> relationship

both species together

<u>Streptococcus thermophilus</u>

<u>Lactobacillus bulgaricus</u>

size of bacterial colonies

2 4 6 8 10 12 14
time (hours)

Yeast Fermentations are used in Brewing and Bread-Making

The **yeast** *Saccharomyces* is used in brewing **beer**. Its fermentation product is **ethanol**.

1) Germinating barley seeds (the **malt**) contain both **starch** and the **amylase** enzyme needed for the hydrolysis of starch into soluble sugars. When this malt is **mashed**, it produces a sweet **liquor**.

2) The liquor is **fermented** by yeasts, which switch to anaerobic respiration as oxygen is used up — so converting sugars to ethanol and carbon dioxide.

   • **Top fermentation** involves *Saccharomyces cerevisiae*, which produces a **froth** on the surface of the fermenting vessel. This is how **ales** are made.

   • **Bottom fermentation** uses *S. carlsbergensis*, producing **less froth** so yeast sinks to the bottom. This takes longer at lower temperatures and makes lighter **lagers**.

3) Yeast is **removed**, the surplus being used in the food industry.

4) The beer is **conditioned** before sale.

look no froth!

ale   lager

The yeast *S. cerevisiae* is also used in dough to produce light (leavened) **bread**. Enzymes in wheat flour **hydrolyse** starch, mainly to maltose and some glucose. The yeast converts these sugars to carbon dioxide and some ethanol. It is the **carbon dioxide** that makes the bread **rise**. As the carbon dioxide expands, it gets trapped in the dough, making it lighter.

Holes in the bread, which make it nice and light, are made by carbon dioxide bubbles in the dough.

Biotechnology in Food Production

Meats can be Tenderised by Adding Enzymes

Meat can be **tough** because of fibres of **collagen** and **elastin**, which are proteins. In some countries (although not in the UK), meat is **tenderised** by injecting it with a **protease** called **papain**. This hydrolyses the fibrous proteins to peptides and amino acids, so reducing the toughness.

Julio sometimes preferred his meat to be on the hard side.

There are Issues Surrounding Biotechnology in Food Production

Most issues in food production are to do with **genetically engineering** organisms for food:

1) Some people think it's **morally wrong** to genetically modify organisms in general, especially if it **harms** the organism involved.

2) Genetically engineered organisms could **escape** into the wild, where some scientists think they might **mutate** or **interbreed** with natural forms. There could be unforeseen **environmental effects** on natural communities. For example, herbicide-resistant genes in crop plants could transfer to weeds.

3) **Inserted** genes could also affect the expression of other genes inside the engineered cell, with unforeseen effects.

Remember that it's not all bad — the biotechnology industry provides more efficient ways of producing food, and provides skilled jobs, and so economically benefits countries. Sometimes more nutrient-dense foods, such as mycoprotein, can be made without the need to even grow crops or farm animals. Wow.

Practice Questions

Q1   Name a product of bacterial fermentation and one of yeast fermentation that are used in food production.
Q2   Name both bacteria used in cheese production.
Q3   Give one similarity between the biological basis of yoghurt production and that of cheese production.
Q4   Describe the role of yeast in brewing.
Q5   Describe a moral issue associated with using biotechnology in food production.

Exam Questions

Q1   Brewing begins with the mashing of malt (germinating barley seeds) to produce sweet liquor.
    a) Describe the processes that occur in order to produce the sweetness.   [3 marks]
    b) A yeast culture is used to act on the sweet liquor to produce ethanol. This requires anaerobic conditions. Explain how these conditions arise in the fermentation vessel.   [2 marks]
    c) Suggest how the quality of the brew would be different if the concentration of oxygen in the culture was higher. Explain your answer.   [2 marks]

Q2   Yoghurt production uses a starter culture of two kinds of bacteria:
Lactobacillus bulgaricus and *Streptococcus thermophilus*.
    a) Describe the role of each in affecting the flavour of the yoghurt.   [4 marks]
    b) Explain how these kinds of bacteria are mutually supportive.   [4 marks]
    c) Flavourings, such as fruit, are often added at the end of yoghurt production. Explain why these must be sterilised first.   [2 marks]

Mmm... cheese... a combination of bacteria and gone-off milk... great...

Here's a useful tip for you: If you ever decide to make bread, and it goes a bit wrong and isn't cooked properly, don't feed the nasty uncooked mess to the ducks. The reason why (and it's a good one) — ducks can't burp, so when the yeast carries on making carbon dioxide inside them, they'll eventually explode. Erm, don't try this out just to see if it's true either...

Biotechnology in Medicine

These two pages are for OCR.
This stuff's quite interesting. It's about the practical applications of all that boring biotechnology rubbish.
The type of stuff that could save your life, or at least tell you when you're pregnant. So read on...

Proteins can be used to Detect Molecules that are Substrates

Some **proteins**, such as enzymes and antibodies, have specific shapes that fit specific **substrates**. The **active sites** of such proteins bind to specific substrate molecules of **complementary shape**.

If this binding results in an observable **reaction**, it can be a good way of detecting the presence of the substrate in a sample mixture. This technique is used a lot in medicine for **diagnosis** of certain conditions. For example, **diabetes mellitus** is diagnosed by detecting **glucose** in urine (see page 187 for full details).

A **biosensor** is a device that uses this principle. It uses a protein (enzyme or antibody) to detect a specific kind of chemical and gives an indication of the **concentration** of the chemical. The more chemicals that bind to the biological molecules, the bigger the observable change (e.g. colour change).

In more complex biosensors, the chemical 'signal' is picked up by a **transducer**. The transducer then converts the chemical signal into a measurable **electrical value**.

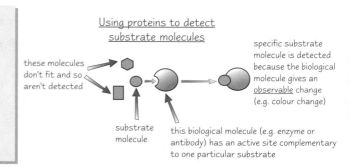

Using proteins to detect substrate molecules

these molecules don't fit and so aren't detected

substrate molecule

this biological molecule (e.g. enzyme or antibody) has an active site complementary to one particular substrate

specific substrate molecule is detected because the biological molecule gives an observable change (e.g. colour change)

Monoclonal Antibodies are Really Useful in Diagnosis...

Antibodies are proteins that are produced by white blood cells called **lymphocytes**. As part of the **immune response**, they bind to **antigen** molecules in the same way that enzymes bind to their substrates.

Monoclonal antibodies are antibodies produced by the descendants of a **single B-lymphocyte**. They bind tightly to really **specific** molecules. This means they're useful in **diagnosis** — for example, they can be used to **distinguish** between molecules on the surface of cancer cells and molecules on the surface of normal cells.

(See pages 194-197 for more on the immune response.)

... and can be Produced in the Lab

A **lab animal**, usually a mouse, is used to produce the **monoclonal antibodies** for biosensors:

1) The mouse is injected with the molecule that needs to be detected in the biosensor. This molecule acts as an **antigen**, so stimulates the mouse to produce the **B-lymphocytes** needed to make specific **antibodies** in response to the injected molecule.

2) The mouse is **killed** and the lymphocytes in its spleen are **washed out** and separated by **centrifugation**.

3) On their own, lymphocytes can't divide outside the body, so the lymphocyte cells are **fused** with cancerous cells called **myeloma cells** to form **hybridoma cells**. The fusion is encouraged by adding a substance called a **fusogen**, which makes cell membranes fuse. The hybridomas have the key characteristics of lymphocytes (i.e. produce **antibodies**) and myeloma cells (i.e. continue **dividing**).

4) The hybridoma cells are **separated** and **cultured** individually. Only some produce the desired **antibody**. These **clones** are cultured further to produce the antibody in quantity.

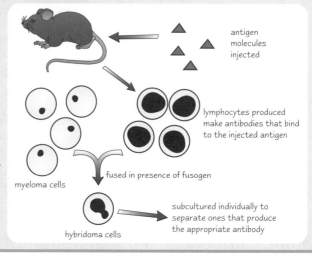

antigen molecules injected

lymphocytes produced make antibodies that bind to the injected antigen

myeloma cells

fused in presence of fusogen

hybridoma cells

subcultured individually to separate ones that produce the appropriate antibody

Other proteins can be mass produced from **genetically engineered microorganisms**, especially bacteria. The techniques are described on page 104.

Biotechnology in Medicine

Biosensors can Detect Glucose to Diagnose Diabetes Mellitus

Diabetes is diagnosed when **glucose** is detected in a person's **urine**. The glucose biosensor is a pad which gets dipped in urine. It's impregnated with an enzyme called **glucose oxidase**, which speeds up (catalyses) this reaction:

$$\boxed{\text{Glucose}} + \boxed{\text{Oxygen}} \longrightarrow \boxed{\text{Gluconic acid}} + \boxed{\text{Hydrogen peroxide}}$$

There are two types of glucose biosensor. They tell us there's **glucose** in the urine by detecting the products **gluconic acid** or **hydrogen peroxide**.

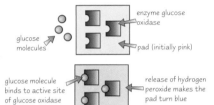

The simple glucose biosensor

- In the **simple** biosensor, the **hydrogen peroxide** reacts with a pink dye in the pad and makes the pad turn **blue**. This reaction is catalysed by another enzyme called **peroxidase**.

- In the more **complex** biosensor, there's a **transducer** which is sensitive to the levels of **gluconic acid**. The more glucose present, the more gluconic acid is produced, so the greater the electrical signal generated.

enzyme glucose oxidase

glucose molecules

pad (initially pink)

glucose molecule binds to active site of glucose oxidase

release of hydrogen peroxide makes the pad turn blue

(Other molecules have a different shape, so don't bind and don't change the colour of the pad.)

These techniques are much more **specific** than another type of sugar test, the **Benedict's test**, which will show a similar result with **any** type of **reducing sugar**. This is because the biosensors use an **enzyme**, but the Benedict's test doesn't.

Monoclonal Antibodies Form the Basis of Human Pregnancy Tests

Pregnancy tests work in a similar way to glucose biosensors. The molecule being detected in pregnancy tests is a hormone called **human chorionic gonadotropin (HCG)**. It's produced by cells of the placenta and embryo, and ends up in the mother's urine, so it's a good indication of a pregnancy. Here's how it works:

1) The sensitive pad is dipped in a **urine** sample. The pad contains a **monoclonal antibody** that binds specifically to any **HCG** present in the urine.

2) The pad is then washed and dipped into a solution which contains another **antibody** with an **enzyme** attached. The second type of antibody binds to the first if the first is bound to an **HCG** molecule.

3) The pad is then washed again and dipped into a chemical that is the **substrate** for the attached enzyme. When this substrate binds to the enzyme it makes the pad turn **blue**.

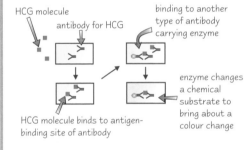

HCG molecule

antibody for HCG

binding to another type of antibody carrying enzyme

enzyme changes a chemical substrate to bring about a colour change

HCG molecule binds to antigen-binding site of antibody

Practice Questions

Q1 What makes monoclonal antibodies?

Q2 What's a hybridoma cell?

Q3 Explain why it is better to detect glucose with a biosensor than with Benedict's test.

Q4 Describe the mechanism by which the pregnancy test kit can diagnose a pregnancy.

Exam Question

Q1 Laboratory mice were used to produce the monoclonal antibodies used in the development of a pregnancy test kit. The mice were injected with an antigen to stimulate the production of antibodies. The B-lymphocytes extracted from the mouse were mixed with myeloma cells and a quantity of polyethylene glycol, a fusogen, was added.

a) What are monoclonal antibodies? [2 marks]

b) Name an appropriate molecule that could be used as an antigen. [1 mark]

c) Describe the immune response brought about by injection of the antigen. [3 marks]

d) Explain the purpose of the fusogen polyethylene glycol. [2 marks]

Weeing on sticks — don't you learn about fun things in Biology...

Pregnancy tests are surprisingly accurate. Apparently, after you wee on the stick and all that, if it tells you you're pregnant then you almost certainly are. A false positive is really unlikely. However, if it tells you you're not, then you still might be, especially if you're testing quite early on (the embryo doesn't make HCG at day one). So try again a week later.

Biotechnology in Industry and Public Health

These two pages are for OCR and AQA B.
Not a very exciting couple of pages I'm afraid... But you've got to learn it — I'll be asking questions later.

Enzymes and Microorganisms are used in Industry

1) Many industrial processes are **catalysed** by enzymes.

2) Some of these processes use **isolated enzymes**, as opposed to using whole **microorganisms** for their enzymes. These are often **extracellular enzymes** — the ones that are naturally **secreted** from cells and so work outside of cells.

3) As these enzymes are **naturally** secreted from cells, they don't need to be manually **extracted**, which is good as extracting enzymes can be complicated.

4) Isolated enzymes are used in a wide variety of **industries**, including the production of food, textiles, leathers and medicines.

Mmm... I'd like a lovely pint of mashed germinating barley seeds fermented by yeast, please.

> Some industrial processes (like **brewing**) use whole **microorganisms**. These microorganisms have **enzymes** that catalyse the reactions in the industrial processes. So it's really the enzymes, not the microorganisms themselves, that are useful.
>
> **Isolated enzymes** are **easier** to use than microorganisms — it's easier to **isolate** and purify their products, and as only a single reaction is catalysed, it's easy to control **conditions** to optimise this.

Isolated Enzymes can be Immobilised

When enzymes are used in industrial processes, they end up **dissolved** in **solution** with their substrates and products. The **product** needs to be **separated** from this mixture, which can often be quite complicated.

Many industrial processes therefore use **immobilised enzymes**, which don't need to be separated out afterwards. These are enzymes attached to an **insoluble material**, which could be **fibres** (e.g. of collagen or cellulose) or **silica gel**, or they could be encapsulated in **alginate beads**. (Alginate is a jelly-like substance.)

> The substrate solution can be run through a column of **immobilised enzymes**.
>
> The active sites of the enzymes are still available to catalyse the conversion of substrate into product, but only **product**, not product and enzyme, emerges from the column.

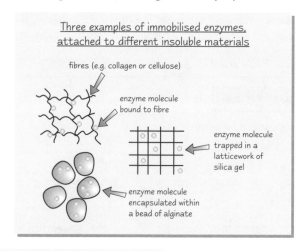

Three examples of immobilised enzymes, attached to different insoluble materials

fibres (e.g. collagen or cellulose)

enzyme molecule bound to fibre

enzyme molecule trapped in a latticework of silica gel

enzyme molecule encapsulated within a bead of alginate

This method has several **advantages**:

1) The insoluble material with attached enzymes can be washed and **re-used**. The enzyme molecules remain **active** and so don't need to be continually purified (which can itself be expensive).

2) The enzymes **don't contaminate** the product.

3) The immobilised enzymes are more **stable** and less likely to denature at high temperatures or extremes of pH.

> Immobilised lactase is an immobilised enzyme used industrially to break down the lactose in milk, so that dairy produce can be made safe for those who are lactose-intolerant.

Biotechnology in Industry and Public Health

Skip this page if you are doing Edexcel or AQA.

Microorganisms are used to Break Down Sewage

Sewage consists of **domestic waste** (human excretion and detergent) and **industrial waste** (loads of random stuff).
Microorganisms are needed to **break down** sewage in order to make it safe. There are three main stages:

1) **Primary treatment** involves **screening** the sewage through grids and then **sedimentation tanks**. Large particles sediment out into solid **sludge**. The remaining **liquid effluent** is passed to the second stage.

2) **Secondary treatment** involves the use of aerobic **microorganisms** to break down the organic matter in the effluent, converting it into harmless inorganic materials. There are two ways of doing this:
   - The **activated sludge method** occurs in large tanks kept aerobic by aeration or spinning paddles.
   - The **filter bed method** involves trickling the effluent over layers of **porous** material which is coated with the microorganisms.

   The **solid** material recovered from these processes is added to the **sludge** from primary treatment.

3) **Tertiary treatment** only occurs if the water from secondary treatment is discharged close to a source of **drinking water**. The water is **chemically treated** to remove harmful substances, such as phosphates.

The **sludge** from primary treatment is acted on by **anaerobic bacteria**, producing a solid material that's useful for conditioning **soil** for agriculture, and even **methane** production (see below).
(There's more on sewage treatment on page 116.)

Microorganisms can be used to Produce Biofuels

Biofuels are organic molecules, made by organisms, that can be burned to release **energy**.

1) **Biogas** consists of a mixture of methane and carbon dioxide. Methane is generated by anaerobic bacteria called **methanogens**, such as *Methanobacterium* or *Methanococcus*. It is a **cheap** source of fuel in many countries, where **animal waste** can be used as the main substrate.

2) **Gasohol** is 90% **petrol** and 10% **ethanol**. The ethanol is produced by the **fermentation** of material rich in carbohydrate (such as starch and cellulose from plant matter). **Yeasts** are used to ferment this material. Some countries, such as Brazil, have made a special effort to produce gasohol as a **substitute** for pure petrol. **Ethanol** is a **cleaner** fuel because it doesn't release acidic gases when burned. This means that **gasohol** is **cleaner** than petrol.

Practice Questions

Q1 Most immobilised enzymes used in industry are described as extracellular. What does this mean?

Q2 Explain what is meant by the immobilisation of enzymes.

Q3 Name one commercially used enzyme that is immobilised.

Q4 Distinguish between the terms sludge and sewage effluent.

Q5 Describe the role of aerobic bacteria in the treatment of sewage effluent.

Exam Questions

Q1 An experiment was undertaken to compare the hydrolysis of starch by immobilised amylase and a solution of amylase.

  a) Describe one way that the amylase could be immobilised.   [2 marks]
  b) Suggest why the rate of hydrolysis would be slower for the immobilised enzyme.   [2 marks]
  c) Give three advantages of immobilising an enzyme in this way from a commercial point of view. [3 marks]

Q2 a) Describe the process by which gasohol is produced.   [4 marks]
   b) Give two reasons why gasohol is considered to be more environmentally friendly than petrol. [2 marks]

You don't get extra marks for making your own biogas...

Oh goody, we're still on about microorganisms and enzymes and exciting stuff like sewage... Don't worry, it does get better — the next lot of pages are about diseases and other medical things. Well, it might get better, but it's still not excitement central, so I give you full permission to go and make a cup of tea now. And put the telly on for a while. You deserve it.

Bacterial Disease

These two pages are for AQA B and Edexcel.
Now you're going to learn all about nice bacterial diseases and how much they make you throw up, swell in funny places or, erm, excrete excessively. You're also going to learn how to avoid these lovely afflictions, should you so desire.

Some Types of *Bacteria* are *Pathogenic*

Some bacteria are **parasites** — they **invade** a living body and obtain **nourishment** from it. A **pathogen** is any microorganism that is a **parasite** of a plant or animal and causes **disease**. Whether bacteria are pathogenic depends largely on the structure of their **cell wall** and **capsule**. These have features that enable the bacteria to:

- invade the body of the **host**
- **attach** to the cells of the host
- some species can even invade the individual **cells** of the host

Bacteria that are better equipped to invade a host and cause a disease there are said to have high **pathogenicity**. The **number** of bacteria needed to cause a certain disease is an indication of the **infectivity** of the bacteria. The **invasiveness** is the ability of bacteria to **spread** within the host.

Here are examples of the **infectivity** of two types of bacteria that cause food poisoning:

- *Salmonella typhi*, with **high infectivity** (around 0.1 to 1 million cells needed) causes **typhoid fever**.
- *Salmonella enteritidis*, with **low infectivity** (around 1 to 10 million cells needed) causes **salmonellosis** (Salmonella food poisoning).

Many Pathogenic Bacteria Produce *Harmful Toxins*

Most kinds of pathogenic bacteria produce chemicals as **by-products** of their metabolism. Chemicals that can cause harm inside the body of the host are called **toxins**.

Exotoxins are soluble chemicals that are **released** from the cells. They are produced by bacteria such as *Staphylococcus* and *Vibrio cholerae*, which causes cholera.

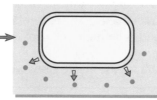

Exotoxins released from bacterium (e.g. <u>Staphylococcus aureus</u>) quickly cause food poisoning.

Endotoxins are **retained** inside the bacterial cells, but are released when the cell walls and membranes of these bacteria are damaged. *Salmonella* produces endotoxins.

Usually **higher** concentrations of endotoxins than exotoxins are needed to produce the symptoms of disease.

Endotoxins inside cell, released on death of bacterium (e.g. <u>Salmonella enteritidis</u>) cause food poisoning after a delay.

Bacterial Toxins have Different *Effects* on the Body

The **effects** of bacterial toxins on the body range from vomiting to fever to swellings and other nasty stuff.

Food poisoning is caused by toxins from bacteria in food or water. The exotoxins of *Staphylococcus*, for example, produce vomiting and diarrhoea within a few hours of eating contaminated food.

Salmonella enteritidis bacteria, on the other hand, occur in undercooked poultry. They stick to the gut epithelium where they are taken up by phagocytosis. They divide and produce endotoxins that cause inflammation and fever. Diarrhoea occurs a day or so after eating contaminated food.

Typhoid fever is also caused by a type of *Salmonella* bacterium.

Cholera is caused by *V. cholerae*, mainly found in water. The exotoxins cause diarrhoea.

Tuberculosis is caused by the bacterium *Mycobacterium tuberculosis*. It's contracted simply by breathing it in. The bacteria penetrate into **unactivated alveolar macrophages**, where they can stay **dormant** for years until the host's immune system becomes weakened, e.g. by old age. They can then replicate inside the macrophages and spread to **lymph nodes** and other tissues. Affected tissues are damaged, and symptoms include weight loss and coughing.

Bacterial Disease

Try to **Avoid** Getting a **Bacterial Infection**

Here are a couple of tips on reducing your chance of getting a **bacterial disease**:

1) Food (especially poultry) must be thoroughly **cooked** to make it safe to eat. This **kills** bacteria, such as *Salmonella*. It's also important to make sure that frozen meat is properly **thawed** before cooking, or the centre might not reach the right **temperature** for the bacteria to be killed.

2) Many disease-causing bacteria, such as those causing typhoid, can live in water, and so can be transmitted by **drinking** from infected water supplies. Infection might happen if the water is **contaminated** by raw sewage, for example — something that often happens in less economically developed countries. It's therefore important to develop proper **sewage systems** that discharge downstream of drinking water sources. Sewage can also be **treated** to filter out and kill microorganisms (see pages 116 and 189).

Some individuals are **carriers** of disease-causing organisms — they **don't** show the **symptoms** of the disease. This could be because they're in the very **early** or **late** stages of the disease. Some individuals simply never develop the disease, but they could be **spreading** it without knowing.

Transient carriers only carry the pathogen for a brief time, but **chronic carriers** have it for months or years. Carriers of typhoid may continue to pose a threat in a population, even after all the sufferers have been cured.

Treating **Diarrhoea** Means Replenishing Lost **Body Fluids**

Diarrhoea is one of the most obvious symptoms of food poisoning. It quickly leads to **dehydration** and loss of **salts**. Treatment therefore consists of a form of **rehydration therapy** — consuming fluids that replace both lost water and lost salts. **Antibiotics** rarely help in cases of food poisoning because the diarrhoea makes it difficult for the drug to be **absorbed** into the bloodstream from the gut cavity.

The diarrhoea sufferer's best friends.

Practice Questions

Q1 Distinguish between the terms pathogenicity, infectivity and invasiveness as they are applied to disease-causing bacteria.

Q2 Distinguish between exotoxins and endotoxins, naming examples of bacteria that produce each type.

Q3 Describe how *Mycobacterium* bacteria can invade the body to cause tuberculosis.

Q4 Give two ways of preventing bacterial infection.

Exam Questions

Q1 Describe how the bacteria *Salmonella enteritidis* can bring about food poisoning. [4 marks]

Q2 Describe how *Staphylococcus* bacteria can bring about food poisoning. [3 marks]

Weeing on sticks, then biogas, now diarrhoea? Help me please...

Not a happy couple of pages — nasty diseases, food poisoning, diarrhoea. Still, at least it's a bit more interesting than reading about the use of enzymes in industry (*yawn*). Here's a clever little tip for you — learn how to spell diarrhoea before your exam. You'll only waste time worrying and crossing out different spellings otherwise. So — D-I-A-R-R-H-O-E-A.

Viral Disease

These two pages are for AQA B. If you're doing Edexcel or OCR you need to know the first 3 sections on this page.

Viruses are acellular microorganisms, so they're not cells like bacteria. They're not even living things. So they don't work like bacteria. So don't ever, ever, say they do. Or I'll come round your house with a big stick and a vivid imagination.

All Viruses **Invade** Living Organisms

Here's a bit about the basic **structure** and **workings** of **viruses**. You might have seen this stuff before on page 177, but it's repeated here in case you missed out page 177. (You big skiver).

1) Viruses only become **metabolically active** when they've **invaded** a living cell.

2) Because they invade cells, viruses are **pathogenic** — they can cause **disease**.

3) Viral particles are called **virions**. Most are less than 0.2 μm in diameter, so they're only visible through an **electron microscope**.

4) All viruses contain **nucleic acid** (either DNA or RNA). These are the molecules that replicate in order for the viruses to **reproduce**.

5) Each particle is surrounded by a protein coat called a **capsid**, composed of protein units called **capsomeres**.

6) Some viruses, in addition, are surrounded by another layer of **lipid** called the **envelope**. This is made from the **cell membrane** of the host cell that the virus invades.

Viruses **Replicate** Inside their **Host Cells**

Viruses can only **reproduce** and make **protein** inside a **host** cell. They don't have the equipment, such as **enzymes** and **ribosomes**, for doing this on their own, so they use those of the **host**. The **life cycle** of a virus is an '**infection cycle**':

1) The capsid or envelope of the virus particle **attaches** to a **receptor** molecule on the host cell membrane.

2) The virus **penetrates** the cell and the capsid **uncoats** to release the **nucleic acid** into the host cell's cytoplasm.

3) The nucleic acid **replicates** and is **transcribed** to make **mRNA**. **Translation** occurs on the host's **ribosomes** to make more capsid **proteins**.

4) New viruses are **assembled**. Then they are **released** from the cell by **lysis** (splitting) of the cell or by **budding** from the cell.

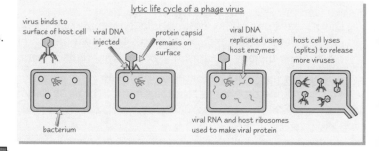

HIV is the Virus which leads to AIDS

Human immunodeficiency virus (HIV) affects the white blood cells known as **helper T-cells** in the immune system. It eventually leads to **acquired immune deficiency syndrome (AIDS)**. AIDS is a condition where the immune response worsens and eventually fails. This makes the sufferer more **vulnerable** to other infections, such as pneumonia.

HIV is a **retrovirus**. This means it carries an enzyme called **reverse transcriptase**, which does this:

1) Inside the white blood cell, reverse transcriptase is used to make a **complementary strand** of DNA from the **viral RNA template**.

2) From this, **double-stranded DNA** is made, which is **inserted** into the human DNA.

3) The DNA stays in the white blood cell for a long while (the **latency period**).

4) Eventually, after a long period of time, the virus particles **multiply**. The DNA **codes** for the synthesis of **viral proteins** to make new viruses that burst out, killing the cell, and then **infect** and **destroy** other white blood cells.

The **Flu** Virus also Contains **RNA**

The **influenza virus** infects **epithelial cells** of the nose, mouth, throat, trachea and bronchi. The virus fuses with the **endocytotic vesicle** so that it can penetrate the **cytoplasm**, and then the viral RNA enters the **nucleus**. Here it uses its own **viral RNA polymerase** to replicate the **viral RNA**. Then influenza proteins are made on host **ribosomes**. The assembled viruses are released from the dying cell by **exocytosis**. The accumulation of dead cells makes **mucus** build up — the most obvious symptom of flu. (Remember that, unlike HIV, the flu virus **doesn't** have reverse transcriptase.)

Viral Disease

It's **Difficult** to **Treat** Viral Diseases

Diseases caused by viruses are **difficult** to **treat** because the viral pathogens are actually **inside** the host cells. Some actually **incorporate** their nucleic acid into the host DNA too, which means that the host cell is able to **produce** even more viruses. Here are a few ways of dealing with viral diseases like flu:

1) **Drugs** used to treat viral diseases are designed to target infected cells in a similar way to special white blood cells called **killer cells** (a type of T-lymphocyte). These actively **seek out** and **destroy** infected host cells by **lysis**. They recognise the **protein coat molecules** of the virus that were left behind when the virus invaded the cell. Other types of drugs act as **inhibitors** of specific viral **enzymes**, such as reverse transcriptase.

2) Treatment of viral diseases is more often about **relieving** the **symptoms** than destroying the virus. Apart from mucus production, the symptoms of **flu** are fever, headache and muscular pain, which can be treated by **painkillers** (e.g. aspirin or paracetamol). It's important to drink a lot of **fluids** too to replace those that are lost.

3) Flu is **infectious** during the time you're showing symptoms, and for about a week afterwards. This means that its **spread** can be controlled by **isolation** of sufferers, although that's not much fun.

4) **Vaccination** against flu is possible, but the vaccine must be based on many **antigens** as the virus **mutates** a lot.

Viruses are **Transmitted** in Different Ways

Given the **difficulty** of treating viral diseases with **drugs**, the best way to control them is by reducing their **spread**. Viral particles can be **transmitted** in a variety of ways, depending upon the type of virus:

1) Some viruses, such as **flu**, are transmitted through **air** and so are contracted via the **respiratory system**.
2) Some, such as **HIV** or **herpes**, are transmitted by **sexual intercourse** or **blood transfusions**.
3) Others, such as **yellow fever**, are transmitted by **biting insects**.
4) A few, such as **hepatitis A**, are transmitted via contaminated **food** or **water**.

This means that different **control strategies** are needed depending on how the virus is **transmitted**.

Transmission of **HIV**, for example, can be reduced by using **barrier contraceptives** like condoms, heat-treating blood products and not sharing hypodermic needles. Transmission of viruses spread by **biting insects** might be reduced with **insecticide**.

The main way of reducing the spread of viruses, though, is simply to **isolate** the sufferer in **quarantine**, so they can't pass it on.

Practice Questions

Q1 Describe the infection cycle of a virus.
Q2 HIV is a retrovirus. What does this mean?
Q3 Explain why infection with HIV can lead to the development of AIDS.
Q4 Explain why most viral infections are more difficult to treat than bacterial infections.

Exam Questions

Q1 Explain what is meant by the term 'latency period' with reference to HIV infection. [2 marks]

Q2 AZT (azidothymidine) is an example of a drug that has been shown to prolong survival of AIDS sufferers. It is thought to be an inhibitor of reverse transcriptase. Explain how this effect reduces the number of viral particles that can be released from infected cells. [3 marks]

You can't avoid all viruses, so just learn to love 'em. Little darlings.

You see that question on AIDS drugs? Well, don't take that as evidence that there's a cure for AIDS, cos there's not. All the AIDS drugs do is prolong the life of sufferers. Although, obviously, that's a massive step in the race to find a cure, so it's not to be scoffed at. I'm just reminding you not to write that there's a cure for AIDS, cos, I've said it before — there's not.

Protection Against Disease

These two pages are for AQA B and Edexcel.
You'll know a lot about white blood cells and their disease-fighting friends from your earlier studies.
This stuff's just the same really, but with more long words to learn. Lucky you.

The **First Line of Defence** Stops Pathogens from **Infecting** the Body

1) The **skin** acts as a barrier, preventing the **entry** of microorganisms. The outermost layers make up the **epidermis**, where the cells closer to the surface are strengthened by the protein **keratin**. **Sebaceous glands** of the lower **dermis** layers produce water-proofing oils that also keep hair follicles relatively free of **bacteria**.

2) **Sweat**, **tears** and **saliva** contain an enzyme called **lysozyme**, which causes the **lysis** (splitting) of bacteria, and prevents them from infecting the living cells of the skin, respiratory and digestive tracts.

3) **Mucus** produced in the **respiratory tract** can trap bacteria and other microorganisms, preventing further invasion of the body. Mucus also contains **lysozyme**.

4) **Cilia** lining the trachea beat steadily, causing a net flow of **mucus** (with its entrapped microorganisms) towards the back of the throat, where they are **swallowed**. This prevents microorganisms from reaching the **alveoli**.

5) The concentrated **hydrochloric acid** of the **stomach** denatures the enzymes of microorganisms. **Pepsin** in the stomach also **hydrolyses** them (breaks them down by reacting them with water).

White Blood Cells of the Immune System Destroy Pathogens

There are loads of types of **leucocytes (white blood cells)** that defend the body against the microorganisms which manage to get through the first line of defence. These leucocytes are all produced in the **bone marrow**. Some kinds, though, only become functional once they've passed through special **lymphoid tissue**, such as the **thymus gland**.

Most leucocytes are **granulocytes**, so called because they have a granular cytoplasm due to the presence of **vesicles**, such as lysosomes. Here are three types of **granulocytes**:

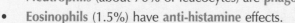
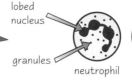

- **Neutrophils** (about 70% of leucocytes) are **phagocytic**.
- **Eosinophils** (1.5%) have **anti-histamine** effects.
- **Basophils** (0.5%) produce **histamine**. (**Histamine** is the substance that makes blood capillaries leaky during **inflammation**.)

lobed nucleus

eosinophil

granules

neutrophil

basophil

Other leucocytes are **agranulocytes**:

- **Monocytes** (4%) are **phagocytic**.
- **Lymphocytes** (24%) are involved in antibody production (**B-lymphocytes**) and cell-mediated immune responses (**T-lymphocytes**).

monocyte

B or T-lymphocyte

Phagocytosis is where Phagocytes Destroy Invading Pathogens

Some white blood cells, **neutrophils** and **monocytes**, are **phagocytic**. Here's what they do:

1) **Local** infections of bacteria cause **basophils** to produce **histamine**, which brings about **inflammation**.

2) Inflammation means that blood capillaries become dilated and **leaky**, allowing **phagocytes** to leave the blood and reach the **site** of infection. (Phagocytes squeezing through capillary wall pores is known as **diapedesis**.)

3) Phagocytes **engulf** microorganisms like bacteria, enclosing them within **food vacuoles**. **Lysosomes** inside the phagocytes **fuse** with the food vacuole. These contain **hydrolytic enzymes** to break down the microorganism.

Lymphocytes have Various Immune Responses

Each type of lymphocyte attacks a **specific** type of foreign particle (**antigen**). Lymphocytes belong to several classes:

1) **Killer T-lymphocytes (cytotoxic T-lymphocytes)** bind to cells (e.g. those infected by viruses) and kill them by **lysis**.

2) **Helper T-lymphocytes** stimulate **B-lymphocytes** to **divide** into antibody-producing plasma cells. B-lymphocytes can't work without them.

3) **Plasma B-lymphocytes** secrete **antibodies** into blood plasma.

4) **Memory B and T-lymphocytes** remain in the blood for many years and produce a more effective **secondary immune response** to an infection, which is the basis for **long-term immunity**. See pages 196-197 for more details.

Protection Against Disease

Invading Foreign Particles **Stimulate** the Body's **Defence Response**

If the body is invaded by foreign particles, the body's **defence system** is activated.
Phagocytes are the first line of defence, but a more **long-term** defence system is the
production of **antibodies** — which leads to **immunity** to certain diseases.

1) The body defends itself when invaded by pathogens by setting up an **immune response**.

2) An **immune response** is a cellular response to the presence of an **antigen**.
It's purpose is to remove the antigens or make them **harmless**.

3) An **antigen** is a particle that is detected by the immune system as being **foreign**, or '**non-self**'.

4) Cells and molecules that are part of the body are described as **self**, since the immune system
has become desensitised to them. Antigens are therefore described as **non-self** particles.

5) If non-self antigens are detected, the **helper T-cells** are activated and secrete **cytokines** to
stimulate the appropriate **B-cell** (with complementary shaped antibodies) to divide by mitosis
and form plasma and memory cells. The **antibodies** then seek out and **destroy** the foreign
cells, or coat them and make it easier for phagocytes to ingest them.

Antigens Trigger Immune Responses

1) Don't get confused about what antigens are — whole microorganisms **aren't** antigens.
Antigens are specific **molecules** — normally **proteins**. They are found on the **cell walls**
of a bacteria or the **capsids** of viruses, for example.

2) However, some **inorganic** substances, such as nickel, can act as antigens.
This is why some people develop **allergies** to metal earrings.

3) Foreign invaders are normally **microorganisms**, so it's definitely a **good** thing that the
immune system destroys them — it means you can fight off the diseases they cause.

4) Unfortunately, the immune system responds in exactly the same way when the body
receives a **transplanted organ**. **Rejection** of the organ is simply a natural immune
response to the **foreign antigens** on the cell surface membranes of the organ.

See the next couple of pages and page 132 for a bit more.

Practice Questions

Q1 Describe three ways in which the body prevents entry of microorganisms.
Q2 Describe the process of phagocytosis.
Q3 Distinguish between 'self' and 'non-self' particles.
Q4 What is an antigen?

Exam Questions

Q1	Explain the role of lysosomes in phagocytosis.	[3 marks]
Q2	Describe the process of diapedesis.	[2 marks]
Q3	Define the term 'antigen'.	[3 marks]

My Auntie Jen doesn't get on with my Auntie Biddy...

Phew — so many big words on this page. Never mind, it shouldn't be too hard to fit them all in your head. You know a lot about white blood cells and immunity already from AS level and even your GCSEs... so it's just a case of adding a few more words into that big bank of knowledge in your head. I'd learn how to spell those long words too, if I were you.

Cell and Antibody Mediated Immunity

These two pages are for AQA B and Edexcel.
This bit's all about the two types of immune response — the one with B-lymphocytes, and the one with T-lymphocytes.

There are **Two** Types of **Immune Response**

Remember that B-lymphocytes are to do with blood, and T-lymphocytes are to do with tissues. Easy eh?

The two types of immune response are the **humoral** and **cell-mediated** responses.
Humoral responses involve **B-lymphocytes**, and **cell-mediated** responses involve
T-lymphocytes. Here's how these different types of lymphocytes are **made** and start to **work**:

1) Both types of lymphocyte involved are made in the **bone marrow**.
2) However, **T-lymphocytes** need to pass to the **thymus gland** (in the chest cavity) during infancy to start working.
3) **B-lymphocytes** mature in the **bone marrow**, **lymph nodes**, and **foetal** (but not adult) **liver**.
4) When lymphocytes have become functional, they are said to be '**competent**'.

B-Lymphocytes *Release* Antibodies *into* Blood *Plasma*

1) The **humoral response** (also called the **antibody-mediated** response) occurs when the presence of an antigen stimulates **B-lymphocytes** to produce and secrete **antibodies** into the blood **plasma**.
2) The antibodies circulate in the bloodstream and **bind** to specific **antigens**.
3) When this happens, the antigen-carrying particle is made **harmless** in some way.
4) For example, this response might encourage **phagocytosis** of bacteria by other white blood cells, or might activate other proteins that **destroy** the antigen in other ways.

Antibodies are also called immunoglobulins.

Antibodies are **Y-shaped proteins** with a **quaternary structure**. This means that each molecule consists of more than one **polypeptide chain**. In fact, each has **four** chains joined by **disulphide bridges**.

Each type of **antibody** has a specific **antigen-binding site** with a specific **amino acid sequence**. This part of the antibody **binds** to the antigen, in the same way that the active site of an enzyme binds to a substrate.

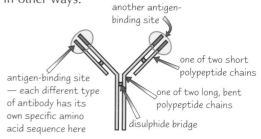

another antigen-binding site

one of two short polypeptide chains

antigen-binding site — each different type of antibody has its own specific amino acid sequence here

one of two long, bent polypeptide chains

disulphide bridge

T-Lymphocytes *have* Receptors *that Bind to Antigens*

The **cell-mediated response** involves **killer T-lymphocytes** (see p.194). T-lymphocytes have **receptors** on their surface which recognise and bind with particular **antigens** on the surface of other cells (in a similar way to how **antibodies** work). So instead of producing antibodies to attack pathogens, killer T-lymphocytes bind to them **directly**. Killer T-lymphocytes can then **destroy** the foreign cell. Some T-lymphocytes have receptors that bind to cells infected by **viruses**, others destroy **cancer cells**. They bind specifically to the viral molecules placed on the cell surface membrane of the cell. The **virus-infected cells** are then destroyed. T-lymphocytes are associated with immune responses in **tissues** rather than in blood.

Antibodies *make Antigens* Harmless

When an antibody **binds** to an antigen, it brings about a change in the body that makes the antigen **harmless**. This works in one of several ways:

1) **Agglutination** happens — the antibodies stick to the surface of the microorganisms, making them **clump** together and rendering them harmless. (This also happens in incompatible **blood transfusions** — antibodies in the recipient's blood respond to antigens in the donated blood, making donated blood agglutinate.)
2) The antibodies sticking to the surface of the microorganism cells can also sometimes bring about **lysis** (breaking up) of the cells.
3) **Phagocytosis** by phagocytic white blood cells (e.g. neutrophils) is stimulated or speeded up.
4) **Precipitation** happens to soluble toxins — the toxins, which were originally **dissolved** in solution, become **solid** and therefore can't do any harm.
5) **Bacteria** are **prevented** from attaching to host cell membranes.

Cell and Antibody Mediated Immunity

New Antigens Trigger the Primary Immune Response

Lymphocytes are **activated** by the presence of **antigens**. This is known as '**clonal selection**' because the antigen causes the **selection** of an appropriate type of lymphocyte for **cloning**. If the body **hasn't** been exposed to a particular kind of antigen before, then a **primary immune response** is set up by the lymphocytes:

1) **Exposure** to a specific antigen causes an antibody or a receptor on a T-cell to **bind** with it. At this early stage, both the antibodies and receptors are **attached** to lymphocyte cell membranes.

2) This binding stimulates the lymphocytes to undergo **mitosis**, producing **clones** of themselves.

3) **B-cell** clones produce the **specific** type of **antibody** to bind to the **specific** antigen present. **T-cell** clones have the **specific receptors** on their surface.

4) Many of the **B-lymphocytes** then become **plasma cells**. This means that they **secrete** their antibodies into the blood plasma. They only last for a few days. The **T-lymphocyte** clones just retain their **receptors**.

5) The antibodies bind to the rest of the antigens away from the cell, producing the **primary immune response**.

A Second Infection Produces a Secondary Immune Response

Not all the lymphocytes stimulated during the primary immune response become plasma cells. Some become **memory cells** instead. These **don't** secrete their antibodies during the first infection, but instead stay **dormant** in the blood. The plasma cells **die** once the infection is over, but the memory cells don't. If there is ever **another infection** by microorganisms with the same antigens, the memory cells can respond **immediately** by producing **antibodies** and **dividing** to create more plasma cells.

This is known as a **secondary immune response**. This time they secrete **more** antibodies, and do it more **quickly**. This secondary immune response is therefore **greater** and **faster**.

Genes occasionally **mutate** and, when they do, this could affect the **shape** of the antigens that are produced. This means that the **antibodies** produced by a host body will no longer **bind** to the antigens, and so an immune response won't happen.

Pathogens that mutate a lot can therefore do a lot of **damage** to their host. There are, for example, many mutated forms of the **flu virus**, which makes it really hard for the body to develop **immunity** to flu.

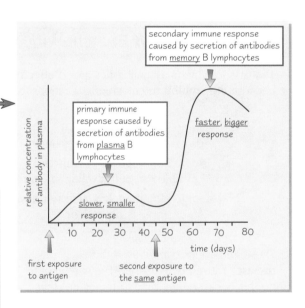

Practice Questions

Q1 Describe the structure of an antibody molecule.

Q2 Describe the difference in function between B-lymphocytes and T-lymphocytes.

Q3 Which is faster, the primary or secondary immune response?

Exam Question

Q1 a) With reference to the clonal selection theory, explain how an antibody-mediated immune response is set up when a body is invaded by a specific type of antigen. [6 marks]

b) Explain why antibodies don't bind to other molecules that are naturally found in the body. [2 marks]

Wish you could develop immunity to revision...?

Urgh, I bet you're ready for a nap now, aren't you? Lymphocyte this and lymphocyte that. Antibody this, antigen that. Zzzzzz. Well, I'm afraid there's more of it to come. Still two more pages on this topic before you get a new one. Sorry. But, anyway, here's an interesting fact — oh, erm, actually I can't think of one. Sorry again. Oops, I'm sorrying. Sorry.

Use of Antibiotics

These two pages are for AQA B and Edexcel.
These pages are all about antibiotics — what they do and where they come from. There's also a bit about antibiotic resistance, which is quite scary — the evolution of super-bacteria which aren't killed off by antibiotics.

Antibiotics Kill or Prevent the Growth of *Microorganisms*

Antibiotics are **chemicals** produced naturally by a range of different microorganisms (especially fungi and bacteria). They either **kill** or inhibit the **growth** of other types of microorganisms.

- **Microbicidal** antibiotics **kill** microorganisms.
- **Microbistatic** antibiotics inhibit **growth**.

Antibiotics are **secondary products** of the **metabolic reactions** of a microorganism, and are released from their cells. Although the microorganism uses up **energy** producing the antibiotic, it is an **advantage** to do so because antibiotics prevent the growth of other kinds of microorganisms. This reduces the numbers of **competitors** for food.

Antibiotics are often very **specific** in their mode of action. They usually work by acting as **inhibitors** of enzymes and other types of proteins. Some antibiotics are so **damaging** to the affected cells that the cells are **killed**.

Antibiotics **Inhibit Enzymes** Involved in Crucial **Metabolic Reactions**

The reason why many antibiotics are so **effective** at reducing growth or killing microorganisms is because they **inhibit** the metabolic reactions that are **crucial** to the growth and life of the cell.

1) Some inhibit enzymes that are needed to make the chemical **bonds** in bacterial **cell walls**. This stops the cells from **growing** properly. An example is **penicillin**.

2) Some inhibit **protein synthesis** by binding to bacterial **ribosomes**. Examples are **tetracycline** and **streptomycin**.

3) Some inhibit **nucleic acid synthesis** by binding to bacterial **DNA** or **RNA polymerase**. An example is **rifampicin**.

These antibiotics can be used to treat **diseases** caused by bacteria. They are able to do this because they bind to molecules that are **only** found in **bacterial** cells, rather than the host cells.

For example, bacteria have **70S ribosomes**, whereas the eukaryotic host has different shaped **80S ribosomes**. This means that the antibiotic will work to **kill** the **bacteria**, but will leave the host cells alone.

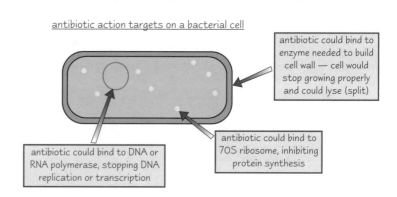

antibiotic action targets on a bacterial cell

antibiotic could bind to enzyme needed to build cell wall — cell would stop growing properly and could lyse (split)

antibiotic could bind to DNA or RNA polymerase, stopping DNA replication or transcription

antibiotic could bind to 70S ribosome, inhibiting protein synthesis

Antibiotics Vary in their **Range of Effects**

Some antibiotics have a **broader range** of effects than others:

- Antibiotics that target **specific** reactions in **specific** microorganisms are called **narrow spectrum antibiotics**.
- Other antibiotics inhibit enzymes that occur in a **wide range** of microorganisms, and so can be used to treat a **wide range** of different disease-causing pathogens. These are called **broad spectrum antibiotics**.

Use of Antibiotics

Antibiotics can be Made in Bioreactors

1) The bacteria and fungi that produce antibiotics are **cultured** in industrial **bioreactors** in order to produce **large quantities** of particular antibiotics for use in **medicine**.

2) The conditions in the bioreactor need to be **controlled** in order to optimise the growth rate of the microorganism.

3) Because antibiotics are secondary metabolic products, they tend to be produced in the **later** stages of the **exponential** growth phase.

(See page 182 or 104 for more on bioreactors.)

Mutated Genes Cause Antibiotic Resistance

Microorganisms that are resistant to antibiotics have **mutated genes** causing a change in their chemical makeup. The antibiotic is prevented from **binding** to the enzyme, but the reason why depends on the type of **mutation**.

For example, bacteria that are resistant to **penicillin** have a gene causing them to produce an enzyme called **penicillinase**. This **breaks down** the penicillin molecules before they can inhibit the enzyme that builds the bacterial cell wall.

Antibiotic Resistance Spreads Quickly Through Bacterial Populations

Bacteria reproduce very quickly, and every time the cells divide, the DNA replicates. **Natural selection** causes the incidence of antibiotic resistance to rise quickly. This is because exposure of a population to an antibiotic **kills** those bacteria **without** the antibiotic resistance gene. Then the **resistant** ones, which obviously survive, can reproduce without **competition**.

Genes for resistance can also spread because of the exchange of **plasmids**. Plasmids are small **rings of DNA** in bacterial cells, and many contain antibiotic-resistance genes. Plasmids are exchanged when two bacteria join together in a process called **conjugation**.

Overuse of antibiotics has resulted in **antibiotic resistance** being a big problem in medicine. There are currently certain strains of disease-causing bacteria (e.g. the *Mycobacterium* that causes tuberculosis) that have **evolved** resistance to most common antibiotics. This makes the treatment of these diseases increasingly difficult.

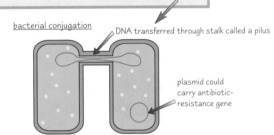

bacterial conjugation

DNA transferred through stalk called a pilus

plasmid could carry antibiotic-resistance gene

Practice Questions

Q1 Distinguish between the effects of the antibiotics penicillin, tetracycline and rifampicin.

Q2 Explain why antibiotics that target protein synthesis will only affect prokaryotic bacterial cells, rather than the eukaryotic host cells.

Q3 What is the difference between broad spectrum and narrow spectrum antibiotics?

Q4 Antibiotics are described as secondary metabolites. What does this mean?

Exam Questions

Q1 Explain how resistance for penicillin can spread through a population of bacteria. [4 marks]

Q2 What is the difference between a microbistatic antibiotic and a microbicidal one? [2 marks]

My Auntie Biotic's really well cultured... (oh shut up now with the stupid lines on stupid things... they're not at all funny)

Hooray! You've finished this section! So what was your favourite bit? I quite liked page 184 because now I feel I know how to brew beer. I wouldn't advise having a go at home though, partly because I shouldn't condone an underage interest in alcohol, but also because the dog always knocks over your big brewing vessels, and the resulting mess just stinks.

The Cardiovascular System

These pages are for Edexcel. AQA A need to know the bit about how heart rate is controlled by the nervous system.

The heart is a really important organ in exercise because it pumps blood to the respiring muscles.
Heart rate is controlled by a fairly complex mechanism — and I'm afraid it's part of your course…

The **Heart** is a **Pump** made of **Cardiac Muscle**

The walls of the heart are made of **cardiac muscle**. The heart has four chambers — two **atria** and two **ventricles**. These chambers **contract** and **relax** in a sequence (the **cardiac cycle**) to send blood around the body. When cardiac muscle contracts, a region of **high pressure** is created, which **forces** the blood out into a region of **lower pressure**.

The **wave of contraction** starts in a part of the **wall** of the **right atrium** called the **sinoatrial node** (SAN). Here, **sodium ions** continually diffuse into the muscle fibres, setting up **action potentials** (see p 72-73) in the fibres.

Like all muscle, cardiac muscle uses energy from ATP when it contracts. The contraction is brought about by **protein filaments** sliding over each other. But, unlike other muscle, cardiac muscle is **myogenic**, which means that it can contract **without stimulation** from the **nervous system**.

The structure of connections between the fibres in cardiac muscle means that it always contracts **smoothly** and **effectively**.

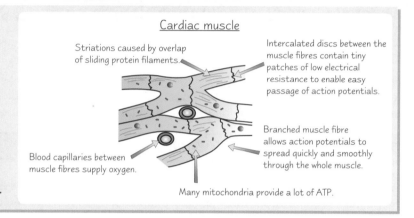

Cardiac muscle

Striations caused by overlap of sliding protein filaments.

Intercalated discs between the muscle fibres contain tiny patches of low electrical resistance to enable easy passage of action potentials.

Branched muscle fibre allows action potentials to spread quickly and smoothly through the whole muscle.

Blood capillaries between muscle fibres supply oxygen.

Many mitochondria provide a lot of ATP.

The **SAN** is Connected to the **Nervous System**

Although in theory cardiac muscle can contract on its own, its beat would be **irregular**. So in reality, it is **controlled** via **nerves** that connect to the SAN. The nerves carry **action potentials** from the **brain** which stimulate the SAN and cause the heart beat to **speed up** or **slow down** according to the body's needs.

The nerves that connect to the SAN are part of the **autonomic nervous system**. This system controls the **unconscious** activities of the body like heart rate, ventilation and digestion. The nerves are **motor nerves** carrying action potentials from the brain to the cardiac muscle. The part of the brain controlling autonomic activities is called the **medulla oblongata** (often shortened to medulla).

The activity of the **medulla** depends on action potentials being sent to it along **sensory nerves** from **receptors**. These are 'baroreceptors' in the aorta which are stimulated when the aorta wall stretches. (See the next page for more detail.)

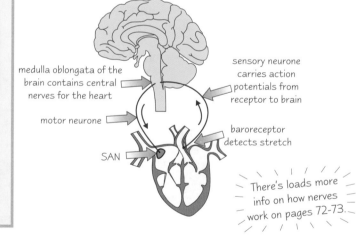

medulla oblongata of the brain contains central nerves for the heart

motor neurone

SAN

sensory neurone carries action potentials from receptor to brain

baroreceptor detects stretch

There's loads more info on how nerves work on pages 72-73.

Both **Sympathetic** and **Parasympathetic** Nerves Affect **Heart Rate**

There are two kinds of **autonomic** (unconsciously controlled) nerves — **sympathetic** and **parasympathetic**. They have different effects on heart rate —

1)   The brain's **medulla** contains control centres that **increase** activity in the body by sending action potentials along **sympathetic cranial nerves**. They also **decrease** it by sending them along **parasympathetic cranial nerves**.

2)   The **sympathetic nerves** secrete a neurotransmitter that **increases** the rate of cardiac muscle **contraction**.

3)   The **parasympathetic nerves** secrete a neurotransmitter that **relaxes** muscles and **slows down** the heart rate.

The Cardiovascular System

Autonomic Control of the Cardiac Cycle involves Reflex Responses

A **self-regulating mechanism** controls the cardiac cycle. **Baroreceptors** (stretch receptors) in the walls of certain blood vessels play an important role the cardiac cycle:

1) Blood entering the **aorta** stretches the **baroreceptors** in the aorta wall.

2) This causes an action potential to be sent through a **sensory nerve** to the **cardioinhibitory centre** in the **medulla**.

3) The medulla then sends an **impulse** along the motor **vagus nerve** (which is **parasympathetic**) to the **SAN**.

4) The neurotransmitter **acetylcholine** causes the SAN to slow the heart rate.

5) If the heart beats too **slowly**, blood **accumulates** in the **vena cava** putting pressure on the baroreceptors in the vena cava.

6) An **action potential** goes to the **cardioaccelerator centre** of the **medulla**, and then along the motor **accelerator nerve** (which is **sympathetic**) to the SAN. The neurotransmitter **noradrenaline** is released and the heart rate increases. This increase in heart rate is called the **Bainbridge Reflex**.

7) Swellings in the carotid arteries of the neck, the **carotid sinuses**, also have **baroreceptors**. They control the **pressure** of blood flowing to the head.

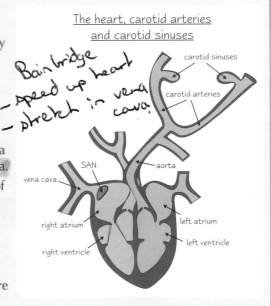

The heart, carotid arteries and carotid sinuses

Bainbridge
- speed up heart
- stretch in vena cava.

carotid sinuses
carotid arteries
SAN
aorta
vena cava
right atrium
left atrium
right ventricle
left ventricle

The **cardiovascular centres** of the medulla are also directly affected by substances in the blood that flows through them. During aerobic exercise the **oxygen concentration drops** and the carbon dioxide and lactate levels increase. These substances stimulate the **cardioaccelerator centre**, which brings about an increase in heart rate. Some **hormones**, such as **adrenaline** also act on the cardioaccelerator centre.

Artificial Pacemakers correct Slow Heart Beats

Sometimes **disease** of the heart can mean that the mechanisms that control heart beat **fail** to function properly. One **treatment** for this kind of disease is the use of an **artificial pacemaker**. Artificial pacemakers are **implants** near the collar bone which are connected to the vena cava. They use **electrical impulses** to stimulate the heart beat. Some pacemakers operate at a **fixed** rate whilst others detect when the heart **skips** a beat and then send a **correcting impulse**.

Practice Questions

Q1 What does the term myogenic mean?

Q2 What is a baroreceptor?

Q3 What is the difference between the sympathetic and parasympathetic nervous systems?

Q4 Explain the role of the sino-atrial node in controlling heartbeat.

Exam Questions

Q1 Describe the structure of cardiac muscle. [6 marks]

Q2 a) Dilation of the vena cava causes the heart rate to increase. Explain how this process happens, paying particular attention to the role of the autonomic nervous system. [6 marks]

   b) What would be the effect of severing the nerves from the medulla to the SAN? Explain your answer. [3 marks]

Once upon a time I was falling in love, now I'm only falling apart...

...nothing I can say, just learn a couple pages on the heart. Heartbeat... why do you revise when my baby misses me... Your exam is only a heartbeat away... baby... la la la la... when it comes your way... I hope this is cheering you up after another dull page of A2 Biology facts. I can do requests. Actually, it's probably just annoying you. I'd better stop.

The Pulmonary and Lymphatic Systems

These pages are only for Edexcel.
The pulmonary system's all about breathing, and the lymphatic system's all about draining your tissues of extra fluid. Easy.

Lungs are the Organs used for Gaseous Exchange

Lungs are composed of millions of microscopic air-filled sacs called **alveoli**. The structure of these is specialised for the function of gas exchange.

The alveoli walls are made of very thin **squamous epithelium** (that means it's made of a single layer of thin, flat cells). There are blood-filled **capillaries** (which are also lined with squamous cells) between the alveoli. The structure of the alveoli provides a **large surface area** for gaseous exchange. The short distance between the air and the blood also maximises the rate of gaseous exchange via **diffusion**.

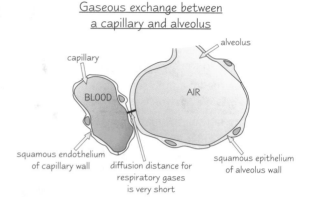

Gaseous exchange between a capillary and alveolus

Muscles Contract and Relax to allow Ventilation

1) **Inspiration** (breathing in) is due to two things — **contraction** of the **diaphragm** (which flattens it down, **increasing** the **volume** of the thoracic cavity), and **contraction** of the **external intercostal muscles** (which raises the rib cage and also **increases** the **volume** of the thoracic cavity).

2) **Expiration** (breathing out) involves **relaxation** of the **diaphragm** and **external** intercostal muscles.

3) Remember that air always flows from regions of **high pressure** to regions of **lower pressure** — just like the flow of blood in the heart.

The Medulla contains Control Centres for Ventilation

The muscles which bring about **ventilation** cannot work without the nervous system.

1) The **diaphragm** and **intercostal muscles** only contract if they receive **action potentials** from the **medulla**.

2) There are centres in the medulla that control the basic breathing **rhythm** and connect to the muscles by motor nerves of the **autonomic nervous system**.

3) The **inspiratory centre** stimulates **contraction** of the **diaphragm** and **external intercostal muscles** and **inhibits** the **expiratory centre**.

4) When inspiration is complete the inspiratory centre **stops inhibiting** the expiratory centre.

5) When the **expiratory centre** operates it **inhibits** the **inspiratory centre**, so the diaphragm and external intercostal muscles **relax**.

6) The inhibition means that both centres can't stimulate the lungs at the same time and so a **breathing rhythm** is established.

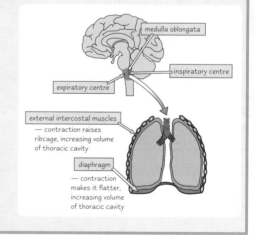

Just like the cardiac centre, the **respiratory centre** of the medulla is directly affected by levels of **respiratory gases** in the **blood**. When **carbon dioxide** levels are raised the **pH** of the blood falls. This is detected by **chemoreceptors** in the medulla, which makes breathing **rate** and **depth** increase. A drop in **oxygen** concentration of the blood has **little effect** on ventilation.

Autonomic Control of Depth of Breathing involves a Reflex Response

There are **stretch receptors** in the walls of the **bronchi** and **bronchioles** which are stimulated when the lungs inflate during inspiration. The receptors are connected to **sensory nerves** that send action potentials along the **vagus nerve** to **inhibit** the **inspiratory centre** in the medulla. This makes inspiration stop and so expiration starts. When the lungs **deflate** the **stretch receptors** stop sending **action potentials** to the inspiratory centre so inhibition stops and inspiration commences again.

The Pulmonary and Lymphatic Systems

The *Lymphatic System Drains* the *Tissues*

The lymphatic system
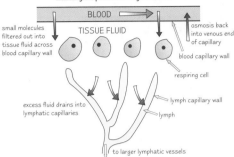

The lymphatic system acts as a kind of drainage system that removes **excess tissue fluid**. The colourless liquid that is drained away is called **lymph**.

The system is made up of special **lymphatic capillaries** that penetrate all the tissues of the body. The lymphatic capillaries join together to form larger lymphatic vessels which empty into two large **ducts** in the **chest** cavity. From here, the lymph is returned to the **blood** via two **subclavian veins**.

Lymph is Excess *Tissue Fluid* with added *Fatty Acids*

1) The **tissue fluid** between the cells is formed by **ultrafiltration** of the blood — cells and large molecules like proteins remain in the blood as it passes through blood capillaries, but **water** and other **small molecules** and **ions** squeeze out to form **tissue fluid**.

2) **Exchange** of substances occurs between tissue fluid and cells, and lots of the tissue fluid is **reabsorbed** into the **blood** at the **venous** end of the capillary due to **osmosis**.

3) **Excess tissue fluid drains** into the **lymphatic** capillaries because of the **pressure gradient**.

4) The lymph vessels in the **villi** of the **small intestine** are called **lacteals** — the lacteals absorb **fatty acids** from digested food and they enter the lymphatic system (see p. 96).

The *Lymphatic System* has a role in *Fighting Disease*

Some lymph vessels connect to clusters of tiny chambers that make up **lymph nodes**. The lymph nodes contain two kinds of **white blood cells** — **lymphocytes** and **neutrophils**.

As lymph flows through the lymph nodes these white blood cells can detect the presence of **antigens** and bring about an appropriate **immune response**.

Structure of a lymph node
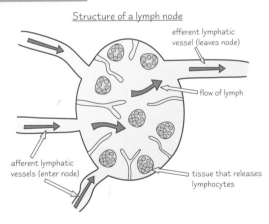

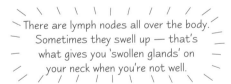
There are lymph nodes all over the body. Sometimes they swell up — that's what gives you 'swollen glands' on your neck when you're not well.

Practice Questions

Q1 Describe the roles of the intercostal muscles and the diaphragm in inspiration.

Q2 Describe the role of the medulla in maintaining the basic breathing rhythm.

Q3 What is the role of the lymphatic system?

Exam Questions

Q1 Explain how the stimulation of stretch receptors in the walls of bronchi and bronchioles causes expiration. [4 marks]

Q2 Explain how lungs are adapted to maximise gaseous exchange. [3 marks]

Q3 Explain the role of the lymphatic system in fighting disease. [3 marks]

Apparently, the 24ᵗʰ of January is the most depressing day of the year...

So, provided you're not revising this on the 24ᵗʰ of January, you've got no reason not to be joyous at the thought of learning pulmonary systems. Don't forget to learn about the lymphatic system too.

Exercise and the Cardiovascular System

These pages are for Edexcel.

When you do exercise your heart rate increases. This page goes into a bit more detail about how the increased heart rate actually helps the body to maintain increased levels of activity. Read on my friend…

Increased Muscular Activity Leads to Increased Oxygen Demand

Exercise involves lots of **muscular contraction** which leads to an **increased** demand for energy. The **extra** energy needs to be produced by the **oxidation** of **glucose** in **respiration** — and the oxygen needed for this has to be supplied by the **cardiovascular** system.

Oxygen is carried in the blood bound to **haemoglobin** inside **erythrocytes** (red blood cells). Haemoglobin **loads up** with oxygen in the **lungs** (where it has a **high** affinity for oxygen) and **offloads** it in the **tissues** (where it has a **lower** affinity for oxygen).

The **oxygen dissociation curve** shows how the willingness of haemoglobin to **combine** with **oxygen** varies depending on the partial pressure (concentration) of oxygen.

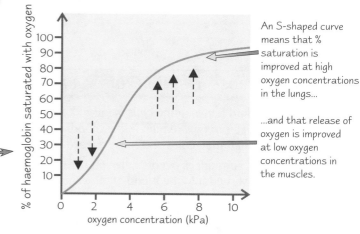

An S-shaped curve means that % saturation is improved at high oxygen concentrations in the lungs…

…and that release of oxygen is improved at low oxygen concentrations in the muscles.

Haemoglobin gives up its oxygen more readily at higher pressures of CO_2. This means that it's even more efficient at unloading oxygen in tissues when respiration has increased CO_2 levels.

Erythrocytes (red blood cells) have several features which make them suitable for **carrying oxygen**:

1) They're packed full of **haemoglobin**.
2) They **lack a nucleus, mitochondria, rough endoplasmic reticulum and Golgi apparatus** — this allows more **room** to carry oxygen.
3) They are **biconcave** in shape, which maximises the **surface area/volume ratio** for absorption of oxygen. The shape also allows them to be **flexible** enough to squeeze through capillaries.

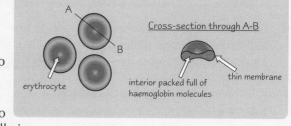

Myoglobin Stores Oxygen in Muscles

Myoglobin is a **blood pigment** that **stores oxygen** in **muscle tissues**. It has an **even greater affinity** for oxygen than haemoglobin. This means it acts as an oxygen **store** which only unloads when the oxygen concentration gets **really low**. When muscles are engaged in **vigorous exercise**, myoglobin releases its oxygen because the increased respiration rate quickly **uses up** the oxygen that is being provided by **haemoglobin** leaving an **oxygen deficit**.

Cardiac Output is a Measure of Oxygen Delivery to Cells

When **demand** for **oxygen increases** the **heart** has to deliver **more oxygenated blood** to the respiring cells. More blood (and therefore more oxygen) can be delivered if the heart beats **faster** and if each beat is **stronger**.

Stroke volume is the volume of blood pumped by each **ventricle** each time it **contracts**. The **product** of **heart beat rate** and **stroke volume** is the **cardiac output, Q**.

Q = stroke volume in cm³ x heart beat rate in beats per minute

So the cardiac output is the total **volume** of blood pumped by a **ventricle** every **minute** — cm³ per minute. Q gives a good indication of the **effectiveness** of delivery of blood (and therefore oxygen) to respiring cells.

Exercise and the Cardiovascular System

Blood Pumped **To** Respiring Cells has to be **Returned**

In order to maintain cardiac output, blood must be **returned** to the heart by the **veins** at the same **rate** that it is pumped out. This **venous return** needs assistance because there is **low blood pressure** in the **veins**. Also, blood needs to be returned to the heart from the lower limbs against the force of **gravity**.

Venous return is helped by:

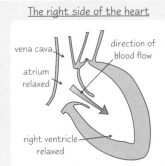

The right side of the heart

vena cava
atrium relaxed
direction of blood flow
right ventricle relaxed

1) **Relaxation** of **heart muscle** (diastole) makes the volume of the ventricles bigger. This **reduces** the **pressure** in the ventricles so that it drops below the pressure in the veins. Blood then flows from the veins into the heart because of the **pressure gradient.**

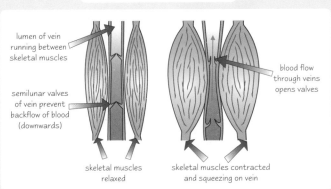

lumen of vein running between skeletal muscles
semilunar valves of vein prevent backflow of blood (downwards)
blood flow through veins opens valves
skeletal muscles relaxed
skeletal muscles contracted and squeezing on vein

2) **Skeletal muscle** helps pump blood **back** to the **heart**. When skeletal muscles **contract** they **squeeze** the veins that run between them **increasing** the **venous pressure**. The high pressure forces the blood to flow away **towards the heart**. When the muscles **relax** the veins **fill up** and the process can begin again.

There are many **valves** in the veins. They prevent **backflow** (e.g. due to gravity). Valves themselves **cannot contract** — i.e. they can't make blood flow like the muscles can, but they can stop it flowing backwards.

Practice Questions

Q1 Explain why contraction of muscles needs oxygen.
Q2 Describe how oxygen is transported round the body.
Q3 Define cardiac output.
Q4 Explain how skeletal muscles help pump blood back to the heart.
Q5 What is the purpose of valves in veins?

Exam Questions

Q1 a) The stroke volume of an athlete was 100 cm³. Her heart beat rate was 48 beats per minute.
    Calculate her cardiac output, showing your working.   [2 marks]
    b) Explain why cardiac output increases during exercise.   [6 marks]

Q2 What is the purpose of myoglobin in muscles?   [3 marks]

Surely the best exercise and the cardiovascular system page in the world...

...ever. Anyway, that oxygen dissociation stuff is a bit tricky, so if you're stuck have a look at pages 94-95. Make sure that you can use the calculation for stroke volume — you may need to use it in your exam. Also, don't forget about myoglobin — there's not too much to learn but it is easy to go think you know it when all you've really done is a quick skim-read.

Exercise and the Pulmonary System

These pages are only for Edexcel.

Exercise doesn't just affect the cardiovascular system. It causes changes in the pulmonary system too.
(If you were wondering, pulmonary means 'to do with the lungs'.)

Tidal Volume increases during Exercise

1) The volume of air **expired** during normal **ventilation** (breathing) is called the **tidal volume**. The tidal volume of an average person **at rest** is about **500 cm³**.

2) If someone breathes in as much as they can and then breathes out as much as they can, the volume **expired** is called the **vital capacity**. It can be as high as **5000 cm³**. The volume of this air which is **more** than the person **usually** breathes in and out, is called the **inspiratory** (and **expiratory**) **reserve** — it's only used when there is an **increased demand** for **oxygen**, e.g. during exercise.

3) Even with the greatest possible expiration, there is **always** some air left in the lungs (about **1200 cm³**) — this air is called the **residual volume**. This air ensures that the **alveoli** and **airways** always remain **open**.

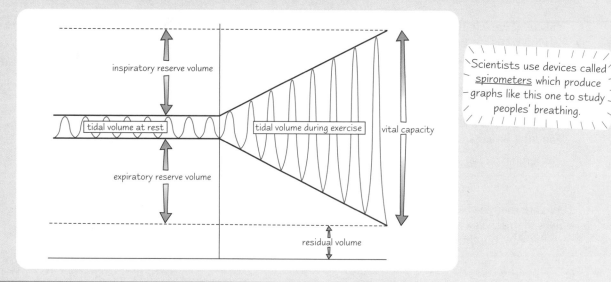

Scientists use devices called *spirometers* which produce graphs like this one to study peoples' breathing.

Minute Volume is a Measure of Air Replacement

More **oxygen** can **diffuse** into the blood if the volume of air that is taken in during ventilation is increased. If air stays in the lungs for some time, its oxygen concentration begins to decrease because it is continually diffusing into the blood. This means that the **concentration gradient** of oxygen between the air in the **alveoli** and the **blood** becomes less steep, so oxygen diffuses more **slowly**.

So, the **more often** air is breathed in and out, the **higher** the concentration gradient, and the **more** oxygen diffuses.

During **exercise**, the **tidal volume** increases, as does the **rate of ventilation**...

The **minute volume (VE)** is the product of the tidal volume and the rate of ventilation:

VE = ventilation rate (breaths per minute) × tidal volume (cm³)

The **minute volume** is basically the **volume** of air breathed in and out **per minute**. It **increases** during **exercise** — the **tidal volume** increases first and then the **ventilation rate** increases.

Exercise and the Pulmonary System

Training Increases the Effectiveness of Gaseous Exchange

Training (i.e. regular exercise) has some specific **effects** on the **pulmonary system**:

1) The **muscles used for ventilation get stronger** with **training.** This means that the **minute volume** (**tidal volume × ventilation rate**) increases. The increase in minute volume means that there is a **more effective** replacement of air in the alveoli during ventilation. This keeps the **concentration gradient** between **oxygen** in the **alveoli** and oxygen in the **blood** as **high** as possible.

2) More **capillaries** develop around the **alveoli** which increases the amount of **blood** available for **oxygen diffusion**.

3) In combination, this **improved ventilation** and **better blood flow** makes the concentration gradient of respiratory gases between the air in the alveoli and the blood in the capillaries **steeper**. This means they **diffuse faster.**

4) Remember that getting rid of **carbon dioxide** is just as important as getting in oxygen. The increases in blood flow to the alveoli and higher turn-over of air in the lungs brought about by **training** means that the **concentration** of carbon dioxide in the **blood** is much **higher** than in the alveoli so it can be efficiently expelled via **diffusion from** the **blood into** the **air** in the lungs.

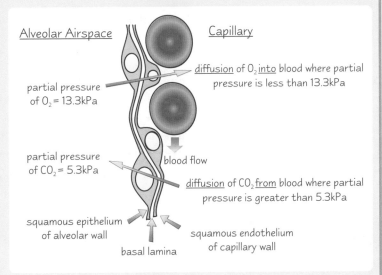

Alveolar Airspace | Capillary

partial pressure of O_2 = 13.3kPa

diffusion of O_2 into blood where partial pressure is less than 13.3kPa

partial pressure of CO_2 = 5.3kPa | blood flow

diffusion of CO_2 from blood where partial pressure is greater than 5.3kPa

squamous epithelium of alveolar wall

basal lamina

squamous endothelium of capillary wall

Ventilation requires Energy

Ventilation involves **contraction** of the **intercostal muscles** and the **diaphragm**, especially during inspiration. Inspiration is therefore an **energy-demanding** process — **respiration** is needed to provide energy for muscles to contract.

Practice Questions

Q1 What is the difference between tidal volume and vital capacity?
Q2 What effect does an increase in ventilation rate have on the concentration gradient between air in the alveoli and blood in capillaries of the lungs?
Q3 Give a definition of 'minute volume'.
Q4 Explain why ventilation needs energy.

Exam Questions

Q1 a) Describe the effects of exercise on: (i) ventilation rate (ii) minute volume.   [2 marks]
     b) What is the inspiratory reserve volume? What is the effect of exercise on this measurement?   [2 marks]

Q2 Describe the changes that occur in the alveolar tissues of the lungs as a result of training and explain how these changes affect gaseous exchange.   [6 marks]

Perhaps the finest exercise and the pulmonary system page in the universe...

Oh, but I think it is. Well, there's not too much to learn on these pages. Basically, it all comes down to concentration gradients — training increases the body's abilities to keep O_2 and CO_2 flowing in and out of the blood efficiently. And that means that your muscles can carry on contracting for longer.

Exercise and the Musculo-skeletal System

These pages are for people doing Edexcel.

Things are about to get complicated. Sorry. But bear with me, it's not too bad once you've got your head around it.

ATP *must be* Regenerated *in* Contracting Muscles

The **energy** for muscle contraction comes from the **breakdown** (hydrolysis) of **ATP**:

$$ATP \rightarrow ADP + P + energy$$

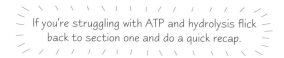
If you're struggling with ATP and hydrolysis flick back to section one and do a quick recap.

So much energy is required when muscles contract that the ATP that's available gets used up quickly. ATP has to be continually reformed so that exercise can continue — this happens in **three main ways**:

1) **ATP-PCr System** — a substance called **phosphocreatine** donates a **phosphate** group to ADP to reform ATP:

$$PCr \text{ (phosphocreatine)} + ADP \rightarrow ATP + Cr \text{ (creatine)}$$

This is good for regenerating ATP in **short bursts** of vigorous exercise like a 100m sprint. The system is **anaerobic** (it doesn't need any oxygen) and it's **alactic** (it doesn't form any lactate).

2) **Lactic acid system** — ATP is made by glycolysis (that's conversion of glucose into **pyruvate** in the **cytoplasm**). This is a good system for longer periods of vigorous exercise e.g. a 400m sprint. The **pyruvate** is then reduced to **lactate**. The **reduction reaction** involves the addition of **hydrogen** from the **reduced NAD** formed in **glycolysis**. The lactate (**lactic acid**) may **accumulate** in the muscles and cause **muscle fatigue**. After exercise has finished, oxygen is needed to **reoxidise** the lactic acid — that's called the **oxygen debt**. Some lactate is also carried to the **liver** where it is changed to **glucose** and then stored as glycogen.

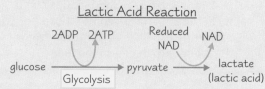
Lactic Acid Reaction

3) **Aerobic respiration** involves **glycolysis** as well as reactions inside the **mitochondria**. This generates a lot of ATP mostly via **oxidative phosphorylation** (see p7). Aerobic respiration only works in the presence of **oxygen** and during prolonged periods of exercise it is usually **dominant**.

Muscles *Send* Information *to the* Brain

The **sliding filament mechanism** is how muscle contraction occurs when **muscle filaments slide** over one another. Between these filaments are bundles of fibres of a different type called **spindles** — they're connected to **sensory receptors**. When spindles are **stretched** they trigger the firing of **action potentials** along **sensory neurones** to the **brain**.

This means that the brain constantly receives information on changes in muscle lengths all over the body and so it can respond and **coordinate** bodily **movement**.

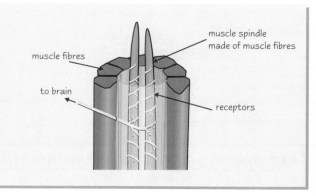
muscle spindle made of muscle fibres
muscle fibres
to brain
receptors

Exercise and the Musculo-skeletal System

A *Motor Neurone* with *its* *Connected Motor Fibres* makes up a *Motor Unit*

1) When an **action potential** arrives at the end of a motor neurone, a **neurotransmitter diffuses** across the **synapse** to create an action potential in the muscle **sarcolemma** (fibre membrane). This causes the muscle to **contract**.

2) In combination, the **motor neurone** and the **muscle fibres** are connected to make up a **motor unit**. Motor units with **few** muscle fibres per motor neurone can bring about **finely coordinated** movement, for example, movement of the **eye**.

3) When many **action potentials** arrive at muscle fibres in **quick succession** there is a **sustained contraction** called **tetanus** —it happens because the fibres do not have time to relax between contractions.

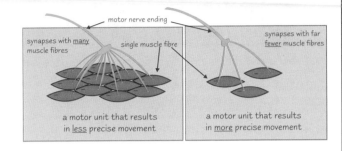

motor nerve ending

synapses with <u>many</u> muscle fibres | single muscle fibre | synapses with far <u>fewer</u> muscle fibres

a motor unit that results in <u>less</u> precise movement | a motor unit that results in <u>more</u> precise movement

Tetanus is also the name of a <u>disease</u> caused by <u>bacteria</u> found in soil. Tetanus stops the inhibition of nerve impulses to the muscles so they all <u>contract</u> (i.e. they go into tetanus). Tetanus is sometimes called 'lockjaw' because the muscles that close the jaw are much stronger than those that open it so sufferers end up not being able to open their mouths. Nice.

Muscles Don't Always Shorten when they Contract

There are two key types of muscle contraction —
1) **Isotonic** muscle contraction happens when muscles contract and shorten whilst maintaining a constant tension.
2) **Isometric** muscle contraction is when the tension of the muscle increases but the length doesn't change.
Isometric contraction happens when your arm holds a heavy weight. It's important in maintaining **posture** too.

Muscle Strength Depends on its Cross-Sectional Area

Muscles with a large cross-sectional area generate greater **shortening forces** so they are **stronger** than narrower muscles. Wider muscles have the same number of muscle fibres, but they have more **myofibrils**, the filaments which make up a muscle fibre. Although they generate greater forces, wider muscles fatigue quickly because of the **depletion** of ATP and glycogen and the build-up of **lactate**.

See pages 76-77 for more on muscles and the like.

Practice Questions

Q1 Describe two anaerobic sources of ATP.
Q2 What causes muscle fatigue?
Q3 Define the following terms: (i) motor unit (ii) tetanus.
Q4 What is the difference between isometric and isotonic muscle contraction?

Exam Questions

Q1 a) Explain why lactate can build up in skeletal muscle [3 marks]
     b) What is meant by the term 'oxygen debt'? [2 marks]

Q2 If creatine is taken as a dietary supplement, more short-term ATP can be made available to working muscles. Explain this effect. [3 marks]

Q3 Describe the principal sources of ATP for muscle contraction during a 10 minute period of running. [6 marks]

Just when you thought you were safe — ATP returns for seconds...

You probably thought you'd seen the last of ATP. If only. The trouble is that ATP is so central to Biology that it just keeps coming up again and again. Remember that the fundamentals of respiration are covered in section one, so you can always look at that to jog your memory. Other than that, there's only one thing that you can do — learn it.

Training

These pages are for Edexcel.

If you're into sport, or you're doing A level PE then you'll probably be familiar with some of the things on these pages. If you're not then perhaps this'll inspire you to take up marathon running.

Effective **Training** Requires a **Balanced Diet**

A **balanced diet** is one that includes all the necessary **food compounds** in the appropriate amounts to maintain **health**. The main food compounds are:

1) **Carbohydrates** and **lipids** are respired to release **energy**.

2) **Proteins** and the **amino acids** that they're made of are needed to make **enzymes**, **carrier molecules**, **muscle filaments** and **nucleic acids**.

3) **Vitamins** are **compounds** needed in very small quantities for specific functions (e.g. some are coenzymes which help enzymes work).

4) **Minerals** include **ions of salts** and are needed for various functions, for example, calcium is an important constituent of bones. Minerals are only needed in small quantities.

5) **Water** makes up about **80%** of cell contents. It's crucial because most **biological reactions** take place **in solution**. During exercise the body's demand for water rapidly increases.

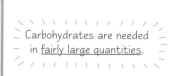
Carbohydrates are needed in fairly large quantities.

The quantity of these substances that the body needs depends on the type of activity — very strenuous exercise over a long period of time uses more fat than carbohydrate.

Anaerobic Conditioning is Important for **Short-Burst Exercise**

Training for sports that involve **brief bursts** of exercise tends to focus on **anaerobic conditioning**. Anaerobic conditioning increases the **efficiency** of ATP production via the anaerobic systems.

For example, **interval training** uses **short** (less than a minute) **bursts** of exercise separated by brief (30 second) periods of **rest**. The rest periods mean that there isn't much time for **lactate** to build up and so the production of **ATP** by the **ATP-PCr** system is **maximised**.

Aerobic Training is Important for **Prolonged Exercise**

The **aerobic system** is the main source of ATP for **long** periods of exercise. As you know, aerobic exercise increases the demand for oxygen at the contracting muscles. **Repeated aerobic training** has several **physiological consequences:**

vascular shunt

1) **Cardiac output** increases — due to an increase in **stroke volume**.

2) **Blood flow** to the **skeletal muscle** is increased. As a result blood flow to **less active organs**, e.g. the gut, is **reduced**.

3) **Venous return** to the **heart** is **improved**.

4) The **minute volume** of the lungs increases and the number of **alveolar capillaries** increases. In combination this means that there is a greater capacity to load the blood with oxygen and offload carbon dioxide.

5) The number of **mitochondria** per **muscle fibre** increases, so there is more **aerobic production** of **ATP**.

6) Aerobic training makes the muscles get increasing amounts of their **energy** from **lipids**, rather than **carbohydrates**.

7) Muscles take less time to become **fatigued**.

Aerobic training of muscles is sometimes called endurance training.

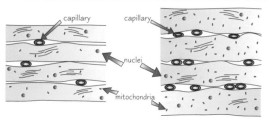

Untrained muscle has narrower muscle fibres, fewer mitochondria and fewer capillaries.

Trained muscle has wider muscle fibres (note that the number of fibres stays the same), more mitochondria and more capillaries.

Training

Training Improves *Performance*

1) **Training** involves **conditioning** the body for a **particular type** of exercise.

2) Different types of exercise use different methods of **reforming ATP**.
   This is because different types of exercise have different **energy demands**.

3) Short bursts of vigorous exercise use more of the **anaerobic** systems (ATP-PCr and lactic acid systems), whereas long periods of exercise, such as marathons, are more **aerobic**.

Dave believes the key to achieving peak performance is wearing the right pants.

Muscular Exhaustion is Called **Fatigue**

After strenuous activity, muscles can become **fatigued**.

> Fatigue has several different **causes**:
> 1) There is a **lack of available ATP**.
> 2) **Glycogen** stores have **run out**.
> 3) **High** levels of **lactate** have accumulated.

During exercise, **glycogen** stores are **hydrolysed** to provide the **glucose** that is needed for **respiration**.
Athletes can increase the amount of glycogen in their muscles before they exercise by **glycogen loading**.

To do this, athletes keep their glycogen levels **low** for a few days by reducing **carbohydrate** levels in their **diet**.
When they eat carbohydrates again, the body **overcompensates** and stores large amounts of **glycogen**.
The increased glycogen store means that muscles can be worked for **longer** without fatigue being a problem.

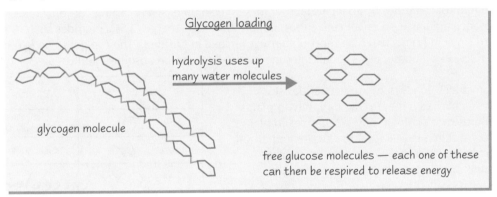

Glycogen loading

glycogen molecule

hydrolysis uses up many water molecules

free glucose molecules — each one of these can then be respired to release energy

Practice Questions

Q1 Explain the importance of carbohydrates, lipids and protein in the diet of an athlete.

Q2 What role do minerals play in diet?

Q3 Explain what is meant by anaerobic conditioning.

Q4 Explain what is meant by endurance training.

Exam Questions

Q1	a) Explain the training involved for a 100m sprint.	[5 marks]
	b) Explain what is meant by the term 'glycogen loading'.	[6 marks]
Q2	How does a trained muscle differ from an untrained muscle?	[3 marks]

Ooh, I like Dave's pants...

Well, there you go, the last page in the section. I'd say that revision is more like a marathon than a sprint so you'd better do some endurance training and increase the number of capillaries in your brain. Also, before the exam you can do some glycogen loading by eating lots of chocolate and pizza — it's the best way to ensure that you do well.

Balanced Diet

This section is for Edexcel students with some pages relevant to AQA B.

Food is made up of a mixture of carbohydrates, lipids, proteins, vitamins and minerals. They're all really important for being healthy. **These two pages are for Edexcel and AQA B.**

Food is a **Complex Mixture** of **Organic** and **Inorganic** Compounds

Organic food compounds belong to **four main groups**:

1)   **Carbohydrates** include **sugars** and **polysaccharides** (e.g. starch).

2)   **Lipids** include fats and oils, and the **fatty acids** they're made of.

3)   **Proteins** and the **amino acids** they're made up of.

4)   **Vitamins**.

And there are two types of inorganic compound:

5)   **Mineral salts** contain useful **mineral ions**.

6)   **Water**.

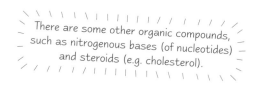
There are some other organic compounds, such as nitrogenous bases (of nucleotides) and steroids (e.g. cholesterol).

A **balanced diet** has all these different compounds in the right proportions for good health. Of course, these proportions are different for each person — they change because of things like age, gender and lifestyle. Some compounds can be made by the cells in the body, but you've got to get the others from your diet — these are called **essential nutrients** and they include vitamins, minerals and some types of amino acids and fatty acids.

Carbohydrates are our *Main Source* of *Energy*

Simple **sugars** are **monosaccharides** or **disaccharides** — these are soluble, sweet-tasting carbohydrates. **Glucose** (a monosaccharide) is used by all cells in respiration to release energy. Disaccharides such as **sucrose** (your bog standard sugar at home) make up most of the sugar intake in our diet. They've got to be hydrolysed into monosaccharides, like glucose, in the small intestine before they can be absorbed and used.

Long-chain carbohydrates (loads of monosaccharides stuck together) are **polysaccharides**. They're found mainly in plants — **starch** and **cellulose** are the main polysaccharides in plant matter. We hydrolyse starch in our gut into glucose and absorb that. But we can't break down cellulose. It goes straight through us, cleaning our insides a bit, keeping us regular ... nice. Sometimes it's called roughage or dietary fibre.

Sugar		Dietary Source
Glucose	⬡	grapes, other sweet fruits, onions
Fructose	⬠	sweet fruits, honey
Sucrose	⬡⬠	principal plant transport sugar, found in sap
Maltose	⬡⬡	germinating seeds

Lipids provide **Energy** too, but are also Needed for **Other Specific Tasks**

Lipids are **fats** and **oils** — they're compounds of glycerol and fatty acids. Weight-for-weight they contain more energy than carbohydrates. A **saturated** fatty acid has no carbon-carbon double bonds, **mono-unsaturated** has one, and polyunsaturated has two or more. **Polyunsaturated** fatty acids are the more healthy ones.

```
      H  H  H  H                    H  H  H  H                    H  H  H  H
      |  |  |  |                    |  |  |  |                    |  |  |  |
  H — C— C— C— C— COOH          H — C— C= C— C— COOH          H — C= C— C= C— COOH
      |  |  |  |                    |  |  |  |                    |  |  |  |
      H  H  H  H                    H  H  H  H                    H        H
  saturated fatty acid —        single C=C double bond in     more than one C=C double bond
  has no C=C double bonds        monounsaturated fatty acid    in polyunsaturated fatty acid
```

Fats are solid at room temperature, oils are liquid.

Phospholipids are the main molecules in **cell membranes** and they also help to carry lipids in the blood stream. (Lipids are absorbed first into the lymphatic system before moving into the blood — some are carried with proteins as lipoproteins.) **Steroids** are also fatty substances. The most famous is **cholesterol**, needed in cell membranes and to make certain hormones and bile salts — it's a real celebrity.

Balanced Diet

Many Different things Change your rate of Energy Release

The **metabolic rate** is the speed of all the reactions inside your cells. It varies a lot depending on what the body is doing. The **basal metabolic rate (BMR)** is a better comparative measurement — it's the rate of energy release when the body is at rest. It's worked out by finding the rate of **thermogenesis** (heat loss from the body, using a human calorimeter) or from the rate of **oxygen consumption**, and is usually expressed per kg of body mass. BMR varies with:

1) **Gender**: BMR is higher in males than females. Females have a higher proportion of **adipose** (fatty) **tissue**, which respires at a lower rate. If the adipose tissue effect is removed and BMR is just based on **lean body mass**, there is no difference in BMR between the sexes. Interesting...

2) **Age**: BMR falls as a person gets older

3) **Body mass**: BMR increases with increasing body mass (due to more respiring tissue)

4) **Disease**: fever makes BMR increase, as can an overactive thyroid gland

5) **Pregnancy**: BMR is greater during pregnancy and lactation

6) **Food intake**: digestion increases BMR

Type of activity	Amount of energy spent / MJ per day
Sedentary work	4.0
Moderate work	5.5
Heavy work	7.5

Exercise involves lots of muscle contraction, which uses up lots of energy. This means that you need to make more energy from your food. More intense exercise demands more energy, pretty obviously. So regular exercise increases your BMR.

Some Essential Nutrients need to come from your Diet

Proteins that you eat are hydrolysed in the gut into amino acids. These are absorbed into the blood and carried off to the cells to make proteins you need. Some kinds of amino acids can be made from other types in the cells; others — the **essential amino acids** — must be provided in the diet.

Vitamins are organic compounds that are needed in very small quantities to maintain health, often by acting as **cofactors** (molecules that help enzymes work properly). **Mineral ions** may act as cofactors too. The most important is **sodium** (from sodium chloride), needed to maintain the osmotic properties of blood and tissue fluid, and necessary to generate nerve impulses.

Essential Nutrient	Food Source
Vitamin A (retinol)	Fish liver oil, dairy products, liver, green vegetables
Vitamin C (ascorbic acid)	Citrus fruits, green vegetables, potatoes
Vitamin D (calciferol)	Fish liver oil, butter, eggs
Calcium	milk, cheese, bread
Iron	meat, liver, red wine
Iodine	Seafoods

Practice Questions

Q1 What is a balanced diet?

Q2 What are carbohydrates and lipids used for in the body?

Q3 What's the difference between metabolic rate and basal metabolic rate?

Q4 Why are some amino acids described as essential nutrients?

Exam Questions

Q1 a) What is the basal metabolic rate (BMR)? [1 mark]
    b) Explain why rate of oxygen consumption and heat loss can be used as indicators of BMR. [2 marks]
    c) Explain why BMR increases with increasing body mass. [2 marks]

Q2 Suggest one way in which the balanced diet of a two year old child would be different from that of a forty year old. Give a reason for the difference. [2 marks]

Q3 Explain why a diet rich in protein could lead to increased growth of working muscle, as well as increased urea production. [8 marks]

A balanced diet — a kilo of chocolate for every kilo of chips...

During the making of these pages, I've eaten a bar of Dairy Milk, two chocolate digestives, a slice of toast with jam and a banana. I've also had three cups of tea and a Diet Coke. Meanwhile, I've been sitting in one chair, and not moving much except to eat. So, anyway, I can't lecture you on healthy eating and exercise, but I can tell you that you need to know it.

Under-Nutrition

These two pages are for Edexcel and AQA B.

Vitamins, protein, calcium, iron and iodine are all essential to stay healthy. If you don't get enough of them you could be in trouble... bleeding gums and bow legs here we come...

A **Lack of Protein** can cause **Starvation**

A lack of protein in the diet is called **protein-calorie malnutrition**. It's common in the developing world. The diets of people in poorer countries may be particularly low in protein if protein-rich foods are expensive. Also, children need a greater proportion of protein than adults (so they can grow), so they may suffer more.

Conditions caused by lack of protein:

1) **Marasmus** is malnutrition caused by a diet that is low in protein and low in carbohydrate. Sufferers are very very underweight.

2) **Kwashiorkor** is malnutrition caused by low protein only. A common symptom is **oedema**. This is a swelling of tissues due to blood protein levels being very low. If blood protein levels are low tissue fluid will not re-enter the capillaries at the venous end (as it normally does) because the blood does not exert a big enough osmotic effect. The fluid accumulates, causing swelling. It can also reduce the production of digestive enzymes, so absorption of food from the gut is lowered.

Kwashiorkor sufferer has oedema of the stomach

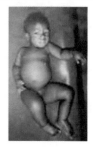

Malnutrition also occurs due to psychological disorders such as **anorexia nervosa** and **bulimia nervosa**. Anorexia nervosa results from an obsession with weight loss, leading to self-starvation. Bulimia nervosa involves bouts of binge eating, followed by self-induced vomiting. Both can be fatal.

Vitamin A is needed for healthy **Lining Tissue** and **Eyes**

Vitamin A (retinol) is a fat-soluble vitamin found in dairy products and fish liver oils. You can make vitamin A from a group of compounds called the **carotenoids**, found in orange and yellow coloured fruit and vegetables (e.g. carrots).

Vitamin A is needed to keep lining tissue (epithelium) healthy. It's also used to make the visual pigment called **rhodopsin**, found in the rods of the retina (see page 79). Vitamin A deficiency leads to a kind of blindness linked with an eye condition called **xerophthalmia**. The skin also becomes hard and flaky, making it prone to infection. Nasty.

Vitamin C is needed to maintain healthy **Connective Tissue**

Vitamin C (ascorbic acid) is a water-soluble vitamin especially found in citrus fruits, green vegetables and potatoes. Vitamin C is needed for the enzyme **hydroxylase** to work properly.

Hydroxylase is essential in maintaining the **collagen** fibres in **connective tissue**. It converts an amino acid called **proline** into another called **hydroxyproline**. **Hydrogen bonds** hold the polypeptide chains of the collagen fibres together. These bonds can only form from one hydroxyproline to another hydroxyproline on a different fibre.

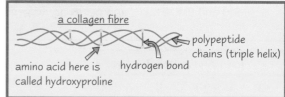

a collagen fibre

amino acid here is called hydroxyproline — hydrogen bond — polypeptide chains (triple helix)

© BIOPHOTO ASSOCIATES/ SCIENCE PHOTO LIBRARY

Deficiency of vitamin C means that hydroxylase won't work properly and hydroxyproline is not formed. This means the hydrogen bonds can't form and the **connective tissue weakens**. This can lead to a condition called **scurvy** which causes bleeding gums, bleeding under the skin, bleeding around the joints and poor wound healing.

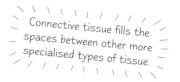

Connective tissue fills the spaces between other more specialised types of tissue.

AQA B ONLY

Vitamin D is needed for effective **Absorption** of **Calcium**

Vitamin D (calciferol) is a fat-soluble vitamin found in fish liver oil, eggs and butter. Also, when skin is exposed to ultra-violet light from the sun, it can make vitamin D from compounds related to cholesterol.

You need vitamin D to make a protein in the lining of the small intestine that binds to calcium ions in food. So vitamin D deficiency can lead to poor bone development — see opposite.

Under-Nutrition

Calcium is needed for healthy development of Bones and Teeth

Calcium is found in **dairy** products, **bread** and **flour**. It's absorbed from the small intestine through a **protein carrier**. This carrier can only be made if vitamin D is present in the diet (see last page). In fact, people can suffer from the effects of calcium deficiency because of lack of vitamin D, even though there might be plenty of calcium in the diet. Calcium deficiency in the diet leads to **rickets** in children and fracture-prone bones in adults (**osteomalacia**).

Calcium is also needed for **nerve** and **muscle** function — it plays an important role at synapses (see pages 74-75). This means that a deficiency in calcium (or vitamin D) can cause muscular spasms too.

Some chemicals (like **oxalate**) make calcium insoluble, which reduces its absorption. Spinach, for example, contains oxalate which reacts with calcium in this way. So while vitamin D is an **enhancer** of calcium absorption, oxalate is an **inhibitor**.

Bow legs are a common symptom of rickets.

© BIOPHOTO ASSOCIATES/
SCIENCE PHOTO LIBRARY

Iron is needed to make Haemoglobin

Iron is found in both meat and plants. It's needed for **haemoglobin** in red blood cells, so a deficiency of iron causes a form of **anaemia**, due to insufficient haemoglobin in the blood. Sufferers of anaemia are visibly paler and feel tired due to the fact that less oxygen is being carried to respiring cells.

Iron **absorption** is affected by various factors:

Iron from meat is found in an organic molecule called **haem**, which normally exists bound to a protein, e.g. haemoglobin. This haem-iron is more **effectively absorbed** than the non-haem iron of plants, so you tend to get less iron from a **vegetarian diet**. But, absorption of haem-iron is still affected by various **enhancers** or **inhibitors** (similar to what happens with calcium). Absorption also depends on the amount of iron **already stored** — if this amount suddenly goes down, e.g. during menstruation, the body absorbs a lot more.

Iodine is needed for healthy growth

AQA B ONLY

You get iodine in seafood. Iodine forms part of the hormone **thyroxine**.

Iodine deficiency in children causes poor growth and poor mental development — this is called **cretinism**. In adults, it causes **goitre**. One of the symptoms of goitre is a swelling in the neck region (see p162-163).

Practice Questions

Q1 What is the role of vitamin C in the manufacture of collagen?
Q2 Explain why deficiency of vitamin D may lead to problems with bone development.
Q3 Name two essential minerals that are affected by absorption enhancers and inhibitors.
Q4 What are the possible effects of iodine deficiency in children and in adults?

Exam Questions

Q1 a) Explain why people with vitamin C deficiency have poor wound healing. [4 marks]
b) Suggest two good sources of vitamin C. [2 marks]
Q2 Explain why a low protein diet could lead to oedema (fluid build up in the tissues). [4 marks]
Q3 Explain why a strict vegetarian diet could lead to anaemia. [6 marks]

So carrots make you see in the dark...

That's not totally true, but you can make vitamin A from a chemical in carrots, and that helps your eyes. There are loads of vitamins and minerals here to learn. You need to know about them, or you might accidentally get nasty problems from not eating the right things — like the student who got scurvy from eating nothing but porridge. Yuck.

Over-Nutrition

These two pages are for Edexcel and AQA B.
Loads of interesting stuff here on why it's bad to stuff your face with food.

Find your **Body Mass Index** to see if you're overweight or obese

The body mass index (BMI) is the official way of saying whether someone is overweight or obese.

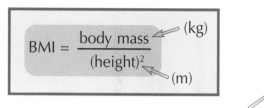

$$BMI = \frac{body\ mass\ (kg)}{(height)^2\ (m)}$$

The table shows how BMI is used to classify people's weight.

Body Mass Index	Weight Description
below 18.5	underweight
18.5 - 24.9	normal
25 - 29.9	overweight
30 - 40	moderately obese
above 40	severely obese

Obesity reduces your *Life Expectancy*

You've seen this stuff before (page 172), but here it is again for all you skivers who like to skip bits. The risk of early death is higher if you're obese. You're more likely to get **coronary heart disease**, **high blood pressure** and you're at an increased risk of **stroke** and some **cancers** (e.g. breast, cervical and colon). There may also be a higher risk of certain other conditions, such as **osteoarthritis**, **gallstones** and **diabetes type 2**. Not good at all.

Coronary heart disease is often caused by **atherosclerosis** (get ready for lots of silly words) ...

> **Atherosclerosis**
>
> 1) **Fatty deposits** called **atheromas** (also called atheromatous plaques) form beneath the artery endothelium lining, **narrowing** the vessel. This reduces the blood flow as there is increased resistance.
>
> 2) The fatty material mainly comes from cholesterol in the plasma, carried by **low-density lipoproteins (LDL)**.
>
> 3) If atherosclerosis occurs in the coronary arteries it can cause **angina pectoris** (chest pains).
>
> 4) If the membrane over the plaque ruptures (breaks open) then blood cells stick to the exposed material. This forms a blood clot. If this clot blocks the arteries supplying the heart it causes **myocardial infarction** (heart attack) and if it blocks arteries in the brain it causes **stroke**. Uh-oh.

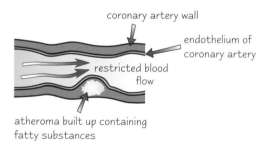

coronary artery wall

endothelium of coronary artery

restricted blood flow

atheroma built up containing fatty substances

Saturated fat in the diet can **increase** the risk of atherosclerosis (which is bad), whereas polyunsaturates, e.g. vegetable oil (and soluble fibre, e.g. from porridge) have been shown to **reduce** plasma LDL and cholesterol levels (which is good).

The effects of excess salt on health are less clear, but it may cause high blood pressure.

Lack of **Dietary Fibre** can cause disease of your **Colon**

Dietary fibre mainly comes from **non-starch polysaccharides (NSPs)** which are found in the cell walls of plant matter (e.g. in apples and carrots). **Insoluble fibre** is largely cellulose, **soluble fibre** includes pectins and gums (carbohydrates found in cell walls of fruits). Fibre reduces the amount of time that food is inside the intestine (transit time) so there's less time for **harmful substances** in the food to affect the gut wall. If you don't have enough NSPs in your diet, you're more likely to get **constipation**, **irritable bowel syndrome** or even **cancer** of the large intestine. So who's for a big bowl of bran then?

Over-Nutrition

Over-nutrition can cause Diabetes Mellitus in adulthood

Diabetes mellitus is a condition associated with abnormally high levels of glucose in the blood.
There are two types:

> **Type 1** is caused by a **lack of insulin being secreted**. This is often diagnosed in childhood and must be treated with **insulin injections**.
>
> **Type 2** is caused by a **lack of insulin receptors** on liver cells, so they can't absorb glucose properly. This develops in middle-age and is often associated with **obesity**. Type 2 must be controlled by **diet**, rather than with insulin injections — sufferers produce insulin properly, but their liver cells can't respond to it.

Obesity is treated by Exercising More and Eating Less — pretty much

Actually, it's not quite as simple as this.

Exercise

> The type of exercise is important — fairly **vigorous exercise** is needed to reduce body fat.

Dudley fitted some
vigorous exercise into
his lunch hour

Diet

> Weight should be lost **gradually** by reducing **energy-dense** foods (like carbohydrates and lipids).
>
> It's not a good idea to reduce **nutrient-dense** foods (such as those containing protein and vitamins) as it could lead to **deficiencies**.

Severe obesity may be treated using surgery, e.g. stapling of the stomach so less food can be eaten.

Practice Questions

Q1   Describe how body mass index is calculated, and how this value might give an indication of obesity.

Q2   What do these words mean?
a) atheroma   b) angina pectoris   c) myocardial infarction

Q3   What's the difference between type 1 and type 2 diabetes mellitus?

Q4   What are the main ways of tackling obesity?

Exam Questions

Q1   a) Explain how a diet high in saturated fat and cholesterol can lead to an increase in heart attack risk.   [8 marks]

b) Name one factor that will lower blood cholesterol levels.   [1 mark]

Q2   Fibre in the diet is said to decrease the 'transit time' through the intestine.
Explain what this means and why it is good for the health of the intestine.   [3 marks]

Put your hands up and slowly step away from the cheeseburger...

Even if you're not overweight it doesn't mean you're free to eat whatever you want — if your diet consists solely of chips and pizza, you won't be doing your heart much good. There's some pretty weird names on these two pages... atherosclerosis, diabetes mellitus, myocardial infarction — make sure you know how to spell them and, ideally, what they mean too.

Food Additives

These pages are for Edexcel only.

Two pages all about stuff we put in our food to try to make it tastier — but if you're having sprouts or liver for your dinner they aren't going to help you... nothing can.

Food Additives are in **Loads** of **Foods**

Food additives are chemicals that are added to processed foods to preserve them or to make them more palatable (i.e. improve flavour, texture or appearance). Food additives that are controlled by the European Union are given **E numbers**:

Additives		E Numbers
tartrazine (yellow) sunset yellow } colourings		E102 E110
sulphites sodium nitrite } preservatives		E221 to E227 E250
l-ascorbic acid tocopherols } anti-oxidants		E300 E306
emulsifiers and stabilisers		E400 to E483

Some food manufacturers don't add additives to their products. This is because the **long-term effect** of eating some of them isn't known. Also, some people are sensitive to them and get rashes or upset stomachs if they eat them.

There are **Different** kinds of **Sweetners**

The most obvious types of substances that taste sweet are **sugars** — monosaccharides like glucose and disaccharides like sucrose. But other things are sweet, and some of these are used as food additives.

Relative Sweetness

Sweetness is graded relative to that of **sucrose**, which is given a **relative sweetness** value of 1.0. Glucose is slightly less sweet (0.74), whereas fructose (sugar found in fruit) is much sweeter (1.73). The sweetness value is worked out by finding the lowest concentration at which the sweet taste can still be detected. This concentration will be greater for glucose than for fructose. Simple.

Jenny tested the relative sweetness of a baseball

Artificial sweetners can have very high relative sweetness values. An example is **aspartame**, with a relative sweetness value of 200. They're often used as low-calorie alternatives to sugars. Tiny amounts of aspartame can make food taste really sweet — it's often in those little sweetner things people put in their tea and coffee.
You need much more sugar to make something sweet, and that means more calories.

Many people have **Lactose-Intolerance**

If you don't have enough of the enzyme **lactase** in your intestine, you won't be able to hydrolyse lactose in milk properly — that's **lactose-intolerance**. Undigested lactose causes intestinal complaints (e.g. wind) because bacteria ferment it. Milk can be artificially treated with purified **lactase** in order to make it suitable for lactose-intolerant people. Lactose-intolerance is found in 10-15% of Northern Europeans, 50% of Mediterraneans, 95% of Asians and 80-95% of people of African descent.

Food Additives

Enzymes are used in Food Processing

During food processing, special enzymes can be used to change the carbohydrates (sugars), allowing manufacturers to **change the sweetness** of the food. The enzymes are produced by **microorganisms** grown in fermenters and are often **immobilised** (see p.188). Immobilising them means that can be **reused** and are more **stable**, which makes the whole process cheaper.

Examples of such enzymes are **glucose isomerase** and **amyloglucosidase**...

Glucose isomerase converts **glucose** into sweeter **fructose**. Fructose is an **isomer** of glucose — it has the same molecular formula but a different structural formula (shape). This means that less sugar (and so fewer calories) needs to be used to give the same degree of relative sweetness.

glucose molecule   fructose molecule

Amyloglucosidase hydrolyses **starch** into **glucose** for use in sweet drinks. Sometimes this glucose is then treated with isomerase to make fructose. Great.

Practice Questions

Q1   Explain how relative sweetness of a substance is defined.

Q2   Compare the relative sweetness of sucrose, glucose, fructose and aspartame.

Q3   What does amyloglucosidase do?

Q4   Describe how fructose is made from glucose syrup using immobilised enzyme.

Exam Questions

Q1   a) Explain why lactose in the diet causes indigestion in people with lactose intolerance.   [3 marks]

     b) How can dairy foods be made safer for lactose-intolerant people?   [2 marks]

Q2   a) The relative sweetness values for glucose, fructose and sucrose are 0.74, 1.73 and 1.0 respectively. Use this information to explain why fructose is considered to be a healthier alternative to sucrose.   [3 marks]

     b) High fructose corn syrup (HFCS) is used to sweeten many kinds of processed food. It is produced by immobilised glucose isomerase, which has an optimum activity at around 60 °C. Outline the advantages of using immobilised glucose isomerase of this temperature optimum.   [4 marks]

Eee by gum, look at all the E numbers in that...

So there you go — some sugars are sweeter than others. You learn something new every page, don't you? Mmmm... sugar. But it's not all good — some people can't eat things with the sugar lactose in cos it makes them fart. Then there's those E numbers — they might taste yummy, but they make some people ill. And those blue smarties, well, that's another story.

Other Additives

These pages are for Edexcel only.

Here's some more pages on stuff that's put into food. There are some scary chemical names here but it's really all about making food look nice, taste nice and last for longer — you still need to learn the names though.

Processed Food *often contains additional* Colourings

Colourings are added to processed food to change the colour and make it look more appealing. Many colourings are substances that occur naturally, such as **carotene**, but others are man-made dyes, such as **sunset yellow** and **tartrazine**.

Recent medical findings have suggested that there is a link between eating tartrazine and incidence of hyperactivity in children. Most manufacturers don't use this additive anymore.

Antioxidants *are added to* Preserve *food*

Antioxidants stop food going off. Oxidation processes cause food to brown and spoiling of drinks. Food goes 'off' because it reacts with oxygen in the air. This causes lipids to become oxidised and long-chain fatty acids to turn into shorter chain fatty acids. These compounds make food taste bad. Many antioxidants are naturally occurring vitamins. Examples of antioxidants added to processed food include **L-ascorbic acid** (related to vitamin C) and **tocopherols** (vitamin E).

browning of apples due to oxidation

Flavourings *and* Flavour Enhancers *are also added during Food Processing*

Flavourings are added to processed food to improve taste. Many of these are naturally occurring substances too, like **spices**, **herbs**, **oils** and **vanilla**. Others are artificial — for example, many **esters** give artificial fruit flavourings.

Flavour enhancers don't have much of a taste themselves, but they enhance the taste of other substances. The main ones are:

1) **Sodium chloride** ('normal' salt)

2) **Monosodium glutamate**

Think about it — you've been feeding yourself all sorts of chemicals for years and years. Great, isn't it?

This table gives some examples of where different flavourings and flavour enhancers are used:

Additive	Type	Uses
herbs / spices	flavouring	wide range of savoury and sweet foods
esters	flavouring	artificial fruit flavours in sweet foods and drinks
sodium chloride	flavour enhancer	wide range of uses: soy sauce, crisps, packet noodles, etc
monosodium glutamate	flavour enhancer	

Other Additives

Preservatives extend the Shelf-life of food

Preservatives are added to food to reduce the growth of microorganisms, such as bacteria and fungi (which make food go off). Vinegar, salt and alcohol are all used for this:

1) **Vinegar** has a low pH, which **reduces the enzyme activity** in microorganisms. This means they can't metabolise properly, which stops them breaking down and spoiling the food. Vinegar is used in **pickling**. →

The nicest pickled onions you have ever seen

2) **Salt** is used because the high solute concentration it creates makes the **bacterial cells lose water** by **osmosis**, so they can't grow and multiply effectively. Salt is usually used in solution as brine to preserve loads of things like tuna, olives and anchovies.

Some poisonous looking alcohol

3) **Alcohol** is **poisonous** to many microorganisms, because it is a waste product of anaerobic respiration (it's like lactate for us). Alcohol is often used to preserve fruit, e.g. pears in brandy.

Sulphites also act as preservatives. These are used to stop foods going brown (e.g. apples), but are also used in wines, beers and ciders. They act as **antioxidants** — they **reduce** the concentration of available **oxygen** in the food. But they're not all great... they can taste pretty nasty and have also been shown to reduce the concentration of the vitamin **thiamine** in foods.

Practice Questions

Q1 Name one example each of a flavour enhancer, a preservative and a colouring agent that might be used as additives in processed food.

Q2 Explain how antioxidants can stop food from going rancid.

Q3 What are the possible hazards associated with the food additive tartrazine?

Exam Question

Q1 a) Distinguish between a flavouring and a flavour enhancer, giving an example of each. [4 marks]

b) Explain how preserving works using (i) salt; (ii) vinegar; (iii) sulphites. [6 marks]

Préservatif — French for condom.......

Yes, it doesn't mean jam. Now there's a fact that could save you some embarrassment at the breakfast table on your French exchange. It doesn't have much to do with biology though. Actually, I suppose it does. But it doesn't have much to do with flavourings. Actually... I think I'll shut up now. (By the way, it's only 4 pages till the end of the book now.)

Food Storage

These pages are for Edexcel only.

This page is all about how fruit ripens and how to store your food — oh the excitement, it's almost too much for me.

Metabolic Processes continue in plants even After Harvesting

Fruits and vegetables contain living cells and the metabolism (chemical reactions) in these cells continues even after harvesting. This includes **respiration** (see section 1) — stored fruits take up oxygen and respire, releasing carbon dioxide. Also, fruits and vegetables lose water by **transpiration**. These processes affect how long the fruits and vegetables stay 'fresh' — tastes, texture, colour and mass are all affected.

Fruit softens and sweetens as it ripens

Ripening is a complex process but the basic effect is the fruit getting sweeter and softer. The whole point of a fruit is to attract animals to disperse the seeds within, so they've got to taste good.

1) During ripening, **ethene** is produced — this promotes the conversion of starch to sugar.

2) Genes are switched on to make enzymes that can **hydrolyse pectins** in the plant cells walls.

3) Pectins are sticky substances that make up the **middle lamellae** between cell walls.

4) As the pectins are hydrolysed cells become leaky and the whole fruit becomes **juicy**.

5) **Acids** in the fruit are also converted into **sugars**. (Sometimes **aromatic** compounds may be produced).

<u>cells in fruit</u>

cell sap in vacuole is the 'fruit juice'

cellulose cell wall

the middle lamella (contains lots of pectin)

cell membrane

When apples ripen they become sweeter and softer. The sweetness is caused by malic acid changing into sugar. Red apples become redder - the redness of the skin is because chlorophyll breaks down and the orange-red pigments (such as carotenoids) show through. So **colour** can be a **visible indicator of ripeness**. Yum yum.

Many Processes Reduce Food Quality

The quality of food varies, depending on the levels of:

1) **Respiration** — this changes the taste of food because it **uses up sugar**, changing the balance between sugars (sweet) and acids (sour). Other substances, like vitamins, can also be affected. Respiration can make the food **smell different** too (by formation of aromatic substances).

2) **Evaporation** — water is lost and the **food becomes dry**.

3) **Softening** — this mainly affects fruits (see ripening). Softer fruits will bruise more easily.

4) **Senescence** — the **deterioration of cells**. Senescence happens when catabolic (breakdown) processes are faster than anabolic (build up) ones.

The **growth of microorganisms** in food is often the reason for reduced quality. They will rapidly multiply in the food if the conditions are right. They need:

1) **A food source** — this could be any of the nutrients (carbohydrates, proteins and so on) in the food.

2) **Suitable temperatures** — to grow and multiply.

3) **Oxygen** — (if they are aerobic).

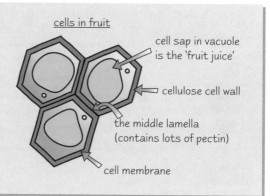

Ripening and senescence both encourage the growth of microorganisms, speeding up decay. Moulds, for example, grow really well on the surface of ripening fruits because of the sugars they release.

Food Storage

Temperature affects Food Spoilage...

A common method to stop things going off is to **refrigerate** the food at low temperatures. This slows down the multiplication of microorganisms. Alternatively, when the food is processed it can be **heated** to kill the microorganisms present. This is normally done in one of two ways:

1) **Pasteurisation** of milk involves heating it to around 65°C for 30 minutes. This kills harmful (pathogenic) bacteria, but the milk still needs to be refrigerated to slow down the multiplication of others.

2) **Sterilisation** of milk is heating it to 100°C. This kills off more bacteria than pasteurisation (but not their spores), so it doesn't need refrigerating. This means it can be **stored for longer** than pasteurised milk before it goes off. Sterilised milk tastes funny though, because the higher temperature affects the flavour.

...as does the Storage Atmosphere

Fruits and vegetables need fresh air to respire (aerobic respiration needs oxygen, remember). So if you put an apple in an **airtight box**, the oxygen will be used up and carbon dioxide produced. This means that respiration will slow down and your apple will last for longer. You can achieve this with food packaging...

The packaging material is a **barrier** between the food and its surroundings. The best material is **partly permeable** — this minimises water loss, but stops the humidity rising to such an extent that it would encourage the growth of moulds and other microorganisms. This is called **modified atmosphere packing (MA)**. There are 2 types:

1) **Passive MA** packing lets respiratory gases through as respiration inside the container continues to take place.

2) With **active MA**, however, the air inside is **removed** and **replaced** with gas of a different composition (e.g. nitrogen) that reduces the deterioration of the food (due to respiration) as much as possible. Sometimes **sachets of chemicals** are added that absorb the carbon dioxide or ethene.

Vacuum packaging has low permeability and keeps the oxygen concentration in the container very low.
Shrink packs are similar, except they are warmed to shrink the packaging around the food. This reduces the growth of aerobic microorganisms, but anaerobic ones might still survive.
The **plastic films** themselves are chosen depending upon their permeability, strength, stability (under storage conditions) and transparency.

Practice Questions

Q1 Outline the changes that occur in a fruit as it ripens.

Q2 Why is there an increased risk of mould growing on ripening fruits?

Q3 What is a modified atmosphere storage system?

Exam Questions

Q1 a) Pectinase is an enzyme that hydrolyses pectins in fruits.
Explain why its use increases the yield of juice that can be extracted to make drinks. [3 marks]

b) The taste of an apple depends upon the balance between sugar (sweet) and malic acid (sour).
Explain how the processes of respiration and ripening will affect the taste of the apple. [4 marks]

Q2 Give an account of the processes of pasteurisation and sterilisation and describe how the different techniques have implications for storage of food. [8 marks]

Lunchboxes — what can I say... Linford Christie...

OK, that was a strange diversion my brain just had. Erm, yeah, lunchboxes... they're a really clever invention. They stop the bread in your sandwiches from going stale, and your apple from going brown and mushy. Just don't pick up small animals and try to take them home in your lunchbox, cos after sitting in that airtight box all afternoon, they won't be so pretty...

Microorganisms in Food

These pages are for Edexcel only.

Some microorganisms are like magicians and can turn one food into another more tasty one — amazing.

Microorganisms are used to Modify foods

There are many types of foods that we produce by actually making use of the metabolic processes of microorganisms (e.g. beer, wine, bread, cheese). Some bacteria carry out **fermentation**:

> In the study of metabolism, fermentation means the same as anaerobic respiration. However, in biotechnology it is used for any metabolic process (aerobic or anaerobic) carried out by the microorganisms. Remember that, or you'll get confused.

Fermentation often **produces acids** that **flavour** the food and **stop it going off** as quickly — by inhibiting the growth of other microorganisms.

Bacteria are used to produce Sauerkraut and Yoghurt

Making Sauerkraut

Sauerkraut is a sour-tasting food made from chopped raw **cabbage** which is allowed to ferment under anaerobic conditions in a salty environment. The cabbage loses water by osmosis and specific bacteria grow:

1) Firstly the bacteria *Leuconostoc* converts sugar to lactic acid until the right pH is reached.

2) Then *Lactobacillus* uses some of the by-products from *Leuconostoc* to produce substances that give sauerkraut its distinctive flavour.

Making Yoghurt

The conversion of milk to yoghurt also depends upon bacteria:

Lactic acid bacteria are added to either skimmed or pasteurised milk. This is heated and the bacteria clot the milk into yoghurt. This is then cooled to 5°C.

1) Some of the bacteria convert **milk protein** into **amino acids**.

2) The amino acids are then used by other bacteria to convert **lactose** in the milk into **lactic acid**.

Soy Sauce is made using Bacteria, Yeasts and Moulds

Soya beans are used to produce **tofu** (bean curd) and **soy sauce**. Tofu is made by grinding soya beans in water, filtering, and coagulating (clotting) the protein by adding vinegar.

The distinctive flavour of soy sauce is made by a combination of fermentation reactions:

1) **Starch** (as flour) is added to ground boiled soya beans and the mixture **fermented** by microorganisms at about 30°C for up to a year.

2) The **mould** *Aspergillus oryzae* in the mixture produces amylase and protease enzymes, which produce sugars and amino acids.

3) The **bacteria** *Bacillus* and *Lactobacillus* make lactic acid.

4) The **yeast** *Saccharomyces rouxii* makes alcohol.

5) The final sauce is filtered and pasteurised.

Harriet wished she had some soy sauce to make her plate more tasty.

I know it seems like I've gone off on one a bit, but you do need to know these details about making sauerkraut, yoghurt and soy sauce.

Microorganisms in Food

Yeast is important in Breadmaking...

Yeasts ferment glucose into carbon dioxide and ethanol. In **breadmaking**, starch is hydrolysed into glucose by the enzyme amylase (found in wheat). The glucose is then fermented by the yeast, releasing **carbon dioxide**, which provides the pockets of gas to make the dough rise.

Ascorbic acid is added to speed up the processing time:

1) The ascorbic acid oxidises the S-H (sulphur and hydrogen) groups in a wheat protein called gluten. (Oxidising removes the H atoms).

2) This makes **disulphide bridges** form (bonds between sulphur atoms).

3) This **strengthens** the dough structure so that it **traps the gas** more easily.

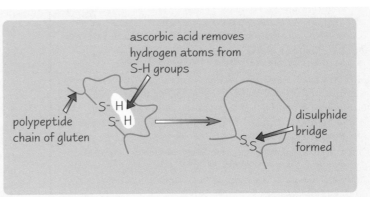

ascorbic acid removes hydrogen atoms from S-H groups

polypeptide chain of gluten

disulphide bridge formed

..... as well as in Winemaking

Yeast ferments glucose in winemaking, producing **ethanol** (that's alcohol to you and me). The anaerobic respiration happens after a period of aerobic respiration. The taste of the wine depends upon the balance of **sugars** and **acids**. Black grape skins are added to make red wines and sulphites are used to prevent the growth of wild yeasts and bacteria. Different flavours develop as the wine matures. It's a pretty complicated thing and all the subtle differences mean people can prattle on about wine for ages.

Gratuitous picture of a bottle of 1968 Claret

Practice Questions

Q1 Explain how the biotechnological definition of fermentation differs from the usual metabolic version.

Q2 What is sauerkraut?

Q3 What kind of bacteria are used in yoghurt production? What do they do?

Q4 Why is ascorbic acid added to bread dough?

Exam Questions

Q1 a) Sauerkraut production requires the addition of salt and the creation of an anaerobic environment. Explain the purpose of these two factors. [2 marks]

b) Explain how the anaerobic process differs between sauerkraut production and wine making. [2 marks]

Q2 Give an account of the role of microorganisms used in the production of soy sauce. [6 marks]

I can't really believe it, but.... THE END

Hello Edexcel student. Yes, it's just you and me now. All the other exam board students finished pages ago. Well I know it's been a real hard slog (if you started at page 1, anyway), but it's all over now. So, my young apprentice, go forth into the real world, sit your A2 Biology exam, maybe go on to university, and you can <u>really start</u> to learn about <u>life</u>. (Ooh, what a cheesy ending.)

Answers

Section 1 — Energy Processes

Page 3 — Energy and the Role of ATP

1   Maximum of 2 marks available.
ATP is a molecule made from adenosine diphosphate (ADP)
and phosphate, using energy from reactions like those of
respiration *[1 mark]*. The energy stored in the chemical bond
between the ADP and the phosphate can be released when it
is needed by a cell, by breaking the ATP back down into ADP
and phosphate *[1 mark]*.

2   Maximum of 3 marks available.
Because ATP is small and water-soluble, it can be easily
transported around a cell or between cells to the places
where there is a demand for energy *[1 mark]*. There it can be
rapidly converted back into ADP to release the energy stored
in the bonds *[1 mark]*. Because an enzyme is required for
this reaction, there is little risk of the ATP breaking down into
ADP and releasing its energy in the wrong place, wasting the
energy *[1 mark]*.

Page 5 — Glycolysis

1   Maximum of 6 marks available, for any 6 of the following:
The 6-carbon glucose molecule is first hydrolysed / split using
water *[1 mark]*, and then phosphorylated using phosphate
from 2 molecules of ATP *[1 mark]* to give 2 molecules of the
3-carbon molecule triose phosphate *[1 mark]*. This is then
oxidised by removing hydrogen ions *[1 mark]* to give 2
molecules of 3-carbon pyruvate *[1 mark]*. The hydrogen is
accepted by 2 molecules of the coenzyme NAD giving
reduced NAD *[1 mark]*, and 4 molecules of ATP are also
produced *[1 mark]*.

2   Maximum of 4 marks available.
The 3-carbon pyruvate is combined with coenzyme A
[1 mark] to form a 2-carbon molecule, acetyl coenzyme A
[1 mark]. The extra carbon is released as carbon dioxide
[1 mark]. The coenzyme NAD is converted into reduced
NAD in this reaction by accepting hydrogen ions *[1 mark]*.

*You need to know how many carbons there are in each
molecule. That decides the molecule's basic structure, and
it's the most important change from molecule to molecule in
these big chains you have to know about. Learn the names of
the molecules by all means, but if you forget one in the exam
and just put '3-carbon compound', I bet you get the marks.*

Page 7 — Krebs Cycle and Electron Transport Chain

1   Maximum of 14 marks available.
1 mark can be awarded for any of the following points, even if
the final answer is incorrect:
2 ATP are produced in glycolysis *[1 mark]*. 1 ATP is produced
per turn of the Krebs cycle *[1 mark]*, which happens twice
per molecule of glucose *[1 mark]* giving 2 ATP from the Krebs
cycle per molecule of glucose *[1 mark]*.
In the electron transport chain, 2.5 ATP are produced for
every molecule of reduced NAD coenzyme made in the
earlier stages of respiration *[1 mark]*, and 1.5 ATP for every
molecule of reduced FAD produced *[1 mark]*.
2 reduced NAD are produced in glycolysis *[1 mark]*,
1 reduced NAD is produced in the link reaction *[1 mark]*
and 3 in the Krebs cycle *[1 mark]*, but for every molecule of
glucose, 2 molecules of pyruvate are made by glycolysis
[1 mark], so the link reaction and Krebs cycle happen twice
per molecule of glucose *[1 mark]*.
So in total, 8 molecules of reduced NAD are produced by the
link reaction and the Krebs cycle *[1 mark]*. Adding the 2
reduced NAD produced in glycolysis gives 10 molecules of
reduced NAD *[1 mark]*. $10 \times 2.5 = 25$ ATP *[1 mark]*.
1 molecule of reduced FAD is also produced per turn of the
Krebs cycle *[1 mark]*, giving 2 reduced FAD per glucose
molecule *[1 mark]*. $2 \times 1.5 = 3ATP$ *[1 mark]*.
So in total, the electron transport chain produces $25 + 3 = 28$
ATP *[1 mark]*.
Adding the ATP produced in glycolysis and in the Krebs cycle
gives $28 + 2 + 2 = 32$ molecules of ATP in total *[1 mark]*.

Page 9 — Anaerobic Respiration

1 a) Maximum of 10 marks available from the following points:
The two forms of anaerobic respiration are alcohol
fermentation and the lactate fermentation *[1 mark]*.
Both are ways of releasing energy without using oxygen
[1 mark], and both take place in the cytoplasm rather than in
mitochondria *[1 mark]*. Both produce 2 ATP per molecule of
glucose *[1 mark]*, and both begin by using the process of
glycolysis *[1 mark]* to convert glucose into 2 molecules of
pyruvate *[1 mark]*.
They differ in what happens next — in alcohol fermentation,
carbon dioxide is removed from the pyruvate to give 2-carbon
acetaldehyde *[1 mark]*. A molecule of reduced NAD from
glycolysis is then oxidised back to NAD *[1 mark]*, and the
hydrogen ions it gives up are transferred to the acetaldehyde,
making ethanol (alcohol) *[1 mark]*. In lactate fermentation,
no carbon dioxide is given off *[1 mark]* — reduced NAD is
used to supply the hydrogen ions needed to reduce the
pyruvate to lactic acid/lactate *[1 mark]*. Alcohol fermentation
happens in plants and some micro-organisms, and lactate
fermentation happens in animals *[1 mark]*.
b) Maximum of 3 marks available.
Aerobic respiration produces 32 molecules of ATP per
molecule of glucose *[1 mark]*, and anaerobic respiration
produces only 2 molecules of ATP per molecule of glucose
[1 mark]. So in terms of ATP production, aerobic respiration
is 16 times more efficient than anaerobic respiration *[1 mark]*.
c) Maximum of 2 marks available.
The muscles of an Olympic sprinter running a 100 m race are
working so hard that oxygen is used up very rapidly *[1 mark]*.
If the muscles did not respire anaerobically, they wouldn't be
able to release any energy at all because there is no oxygen
available *[1 mark]*.

Answers

Page 11 — Measuring Respiration

1   Maximum of 2 marks available.

$$RQ = \frac{CO_2}{O_2} = \frac{102}{145} = 0.7$$

[1 mark for the showing the working, 1 mark for correct answer]

Try to find an easy way to remember that CO_2 goes on top of O_2 in the RQ equation — alphabetical order, perhaps? Two easy marks if you can remember it.

2   Maximum of 3 marks available.
The graph for sunflower seeds dips to a RQ of 0.7, so they must be respiring fats as well as carbohydrates **[1 mark]**. The graph for maize seeds never drops to an RQ of 0.7, so maize seeds don't use fat as a respiratory substrate **[1 mark]**. Sunflower seeds must therefore contain carbohydrates and fats as food stores for the developing seeds, while maize seeds only contain carbohydrates **[1 mark]**.

Page 13 — Photosynthesis

1   Maximum of 5 marks available, for any 5 of the following points:
Leaves are broad and flat to absorb as much light as possible **[1 mark]**. They're thin so CO_2 and water can reach inner cells easily **[1 mark]**. Leaves have veins that contain xylem (to bring water from the roots) and phloem (to carry away the sugars made in photosynthesis) **[1 mark]**. The cuticle on the upper epidermis protects against UV rays in light and resists dehydration **[1 mark]**. Leaves contain a lot of chloroplasts, which are the photosynthetic factories and contain light absorbing pigments **[1 mark]**. Air spaces between the cells allow easy diffusion of CO_2 and O_2 gases **[1 mark]**. Stomata allow exchange of gases between the leaf and the atmosphere **[1 mark]**.

2   Maximum of 5 marks available, for any 5 of the following points:
The stroma contains the enzymes needed for the Calvin cycle (the light-independent stage of photosynthesis) **[1 mark]**. The photosynthetic pigments needed for the light-dependent stage of photosynthesis are found on the chloroplast's thylakoid membranes **[1 mark]**. Thylakoid membranes are covered in stalked particles to make ATP **[1 mark]**. Starch grains found in the stroma act as a storage site for any carbohydrate made by photosynthesis and not used straight away **[1 mark]**. The chloroplast has a double membrane / envelope to allow reactants to be kept close to the reaction site **[1 mark]**. The thylakoids are surrounded by the stroma, so that products of the light-dependent reaction can be quickly used in the light-independent reaction **[1 mark]**.

Page 15 — Limiting Factors in Photosynthesis

1   Maximum of 2 marks available.
Plants need a constant supply of nutrients for growth, but this doesn't directly affect the rate of photosynthesis **[1 mark]**. The mineral magnesium, however, is important in chlorophyll and so it's essential to light absorption **[1 mark]**.

2   Maximum of 2 marks available.
CO_2 is a waste product of respiration, which happens in the mitochondria of plant cells **[1 mark]**. The plant will be using the CO_2 it produces itself for photosynthesis, as well as what it takes in from outside, so just measuring CO_2 uptake by a plant won't give an accurate measurement of the rate of photosynthesis **[1 mark]**.

I've not tested you on them here, but for goodness sake make sure you understand the graphs on p 14. Examiners love them — it's pretty likely that limiting factors will come up somewhere in your exams.

Page 17 — Photosynthesis — The Light Dependent Stage

1 a) In the thylakoid membranes of the chloroplasts **[1 mark]**.
  b) Maximum of 2 marks available.
     ATP **[1 mark]** and NADPH + H+ / reduced NADP / NADPH **[1 mark]**.
  c) Maximum of 3 marks available.
     ATP is produced by both cyclic **[1 mark]** and non-cyclic photophosphorylation **[1 mark]**. NADPH is produced by non-cyclic photophosphorylation **[1 mark]**.

Page 19 — Photosynthesis — The Light Independent Stage

1 Maximum of 5 marks available.
  a) Ribulose bisphosphate (RuBP) **[1 mark]**.
  b) NADPH/reduced NADP **[1 mark]**.
  c) Ribulose bisphosphate (RuBP) **[1 mark]**.
  d) the enzyme ribulose bisphosphate carboxylase **[1 mark]**.
  e) Ribulose bisphosphate (RuBP) **[1 mark]**.

2 Maximum of 3 marks available.
  a) Between points a and b, light was available and both stages of photosynthesis (light-dependent and light-independent) were happening. ATP and NADPH / reduced NADP were being supplied for the Calvin cycle **[1 mark]**.
  b) At point b the light faded and the light-dependent stage of photosynthesis stopped, but the light-independent stage / Calvin cycle continued until point c **[1 mark]**.
  c) Photosynthesis stopped at c as supplies of ATP and NADPH were exhausted and no more could be produced **[1 mark]**.

Answers

Section 2 — Populations and Interactions

Page 21 — Ecosystems and Energy Transfers

1   Maximum of 4 marks available.
    A habitat is the place where an organism or group of
    organisms live [1 mark]. An ecosystem is not just the place,
    it also includes other abiotic factors [1 mark] such as
    temperature, oxygen level, soil pH, exposure to wind, etc.
    [1 mark for a relevant example] and biotic factors / living
    things / communities that live there [1 mark].

2   Maximum of 4 marks available.
    Pyramids of number can go out of shape when a single large
    organism can feed many smaller organisms [1 mark].
    Pyramids of biomass can go out of shape if you have a
    population of short-lived and rapidly reproducing organisms
    [1 mark]. Pyramids of energy can never be out of shape
    because, due to energy wastage between trophic levels
    [1 mark], there always has to be more energy in the
    population being fed on than the population feeding on it
    [1 mark].

Page 23 — Nutrient Cycles

1   Maximum of 10 marks available.
    Carbon dioxide is removed from the atmosphere by
    photosynthesis [1 mark]. Carbon dioxide is returned to the
    atmosphere by respiration [1 mark]. Carbon dioxide not
    released by respiration is released when the organism dies
    [1 mark] by the respiration of decomposers [1 mark].
    If organisms do not decay, their carbon is not released
    [1 mark]. This carbon can eventually form fossil fuels
    [1 mark]. Humans burn large quantities of fossil fuels
    [1 mark] which releases a lot of carbon dioxide [1 mark].
    Deforestation can also raise carbon dioxide levels [1 mark]
    by killing trees that would absorb it [1 mark].

2   Maximum of 6 marks available.
    Nitrogen-fixing bacteria convert nitrogen into nitrates
    [1 mark]. This is important because plants can then absorb
    the nitrates [1 mark]. Microbial decomposers break down the
    bodies of dead plants and animals and release the nitrogen
    [1 mark] as ammonium compounds [1 mark]. The
    ammonium compounds are converted into nitrites and then
    to nitrates by nitrifying bacteria [1 mark]. Denitrifying
    bacteria convert nitrates into nitrogen [1 mark].

Page 25 — Population sizes

1   Maximum of 8 marks available.
    The lag phase occurs because organisms are adapting to a
    new situation [1 mark]. In nature, it is rare for populations to
    move into a completely new and unfamiliar set of
    surroundings [1 mark]. The exponential growth phase occurs
    because there is ample food supply and no predation
    [1 mark]. Neither situation usually occurs in nature [1 mark].
    The stationary phase occurs because the carrying capacity is
    reached / the food supply is limiting [1 mark]. In nature there
    are more factors involved / food supply is more variable
    [1 mark]. The death phase occurs because of build up of
    wastes [1 mark]. This can only happen in a closed system like
    a culture vessel OR in nature the waste is recycled OR in
    nature there is a much greater capacity for waste [1 mark].

2   Maximum of 4 marks available.
    A density dependent factor has more effect as population
    density increases [1 mark]. A relevant example (e.g. food,
    oxygen, minerals etc) [1 mark]. A density independent
    factor's intensity is unaffected by the density of a population
    [1 mark]. A relevant example (e.g. fire, flood, drought etc.)
    [1 mark].

Page 27 — Diversity

1   Maximum of 5 marks available from any of the following:
    Stability means that an ecosystem is more resistant to change
    and/or more likely to recover from damage [1 mark]. High
    diversity leads to high stability [1 mark]. High diversity
    means a large number of species [1 mark]. Food webs are
    more complex where diversity is high [1 mark]. There are
    more potential sources of food for animals where there is high
    diversity [1 mark]. If a species is severely reduced or wiped
    out, its predators are less likely to be severely affected
    [1 mark] and the effect on the ecosystem will be less
    [1 mark].
    Most of the above points refer to high diversity ecosystems.
    The marks can also be gained by making the opposite points
    about low diversity ecosystems.

2   Maximum of 8 marks available.
    Map the field [1 mark]. Divide it into numbered squares
    [1 mark]. Use a random sampling technique to select the
    squares to sample [1 mark]. Place quadrats in the selected
    squares [1 mark]. Quadrats should be divided into 100
    smaller squares [1 mark]. Count the number of squares of
    clover [1 mark]. Count only squares which are at least half
    occupied by clover [1 mark]. Average the count for all the
    quadrats used OR add the total number of squares covered in
    all the quadrats and convert to a percentage of the total
    number of squares sampled [1 mark].

Page 29 — Succession

1   Maximum of 8 marks available.
    Change of an ecosystem and the species in it [1 mark] over a
    period of time [1 mark]. Change goes through stages called
    seral stages [1 mark]. The process stops when a climax
    community is reached [1 mark]. Changes are brought about
    by the interactions of species [1 mark]. Climax may be
    climatic [1 mark] resulting from the climate [1 mark] or a
    plagioclimax [1 mark] resulting from human activity [1 mark].

2 a) Maximum of 3 marks available from the points below:
    Successful features — rapid growth [1 mark], rapid
    reproduction [1 mark], asexual reproduction [1 mark],
    efficient seed dispersal [1 mark] tolerant of harsh
    environmental conditions e.g. high salt levels and strong
    winds [1 mark].
  b) Maximum of 2 marks available.
    Reasons for disappearance — shaded by larger plants
    [1 mark], eaten by herbivores [1 mark], unable to compete
    for water or minerals with newly arrived species [1 mark].

Answers

Page 31 — Agriculture and Ecosystems

1 Max of 6 marks available from the following points: (at least 2 must be for benefits).
Benefits — Efficient food production *[1 mark]*, increased food production *[1 mark]*, cheaper food *[1 mark]*.
Problems — Pollution from waste *[1 mark]*, increased use of pesticides *[1 mark]*, increased use of inorganic fertilisers *[1 mark]*, problem with increased animal waste *[1 mark]*, destruction of hedgerows *[1 mark]*, unemployment of farm workers *[1 mark]*.

2 Maximum of 5 marks available from the points below:
Advantages — Improves soil structure *[1 mark]*, less likely to pollute ponds and streams *[1 mark]*, uses animal waste *[1 mark]*. Disadvantages — more difficult to store *[1 mark]*, more difficult to apply to land *[1 mark]*, can't be measured out so easily *[1 mark]*. Manure can have an unpleasant smell *[1 mark]*.

3 Maximum of 8 marks available.
Eutrophication results from pollution of fresh water with fertilisers *[1 mark]*. Fertilisers leach through the soil from farmland and get into the water *[1 mark]*. Nitrates *[1 mark]* in the fertilisers increase the growth of algae *[1 mark]*. This blocks out the light and plants below then die *[1 mark]*. This causes bacterial growth as bacteria feed on the dead plants *[1 mark]*. Bacteria use up oxygen *[1 mark]*. Lack of oxygen kills organisms in the pond or stream *[1 mark]*.

Page 33 — Controlling Pests

1 Maximum of 3 marks available from any of the following:
Non-toxic to humans *[1 mark]*. Specific to weed species being targeted *[1 mark]*. Biodegradable / not persistent *[1 mark]*. Easy to wash off / remove from crops before consumption *[1 mark]*.

2 Maximum of 5 marks available from any of the following:
Biological control is more difficult to apply *[1 mark]*. Biological control can be unpredictable — it's hard to predict all of the knock-on effects *[1 mark]*. Biological control is slower to work than chemical pesticides *[1 mark]*. Biological control does not control sudden outbreaks as well as pesticides *[1 mark]*. Biological control does not completely eliminate the pest *[1 mark]*. Switching from chemical pesticides to biological control will be costly / the farmer may have to purchase new supplies and equipment *[1 mark]*.

Page 35 — Managing Ecosystems

1 Maximum of 4 marks available for any of the following.
Growing fast-growing trees *[1 mark]* which will replace themselves quickly *[1 mark]*. Coppicing or pollarding *[1 mark]* which allows timber to be removed repeatedly without killing the tree *[1 mark]*, increases species diversity *[1 mark]* and allows light to reach the floor level *[1 mark]*.

2 Maximum of 3 marks available.
a) Mowing *[1 mark]*.
b) Using the grass / silage cut between May and September *[1 mark]*.
c) It increases biodiversity / allows more species to live in the field *[1 mark]*.

Page 37 — Investigating Ecosystems

1 Maximum of 3 marks available.
The light meter would take a measurement for one moment in time *[1 mark]* but light intensity changes over time *[1 mark]*. A single reading would not take these variations into account and would therefore not give an accurate representation of the genuine amount of light in the area *[1 mark]*.

2 Maximum of 2 marks available.
A transect would be used to discover a trend across an ecosystem *[1 mark]*. Any suitable example (e.g. distribution of organisms up a rocky shore, distribution of plants with increasing shade) *[1 mark]*.

3 Maximum of 2 marks available.
Population size = n_1 x n_2 / n_m, 80 x 100/10 = 800
[2 marks for correct answer or 1 mark for correct working].

Page 39 — Data Analysis

1 Maximum of 3 marks available.
Set up two greenhouses / two test areas within the greenhouse, one with the original watering system and one with the new automatic irrigation system *[1 mark]*. Measure the size / mass of a sample of the cucumbers from each greenhouse *[1 mark]*. Use the t-test to see if there is a significant difference between the means of the two samples *[1 mark]*.

2 Maximum of 3 marks available.
The new diet makes no significant difference *[1 mark]*. The probability value is more than 0.05 *[1 mark]*. The result would occur by chance more than 1 time in 20 / 5% of time *[1 mark]*.

Section 3 — Meiosis, Genetics and Gene Control

Page 41 — Meiosis

1 Maximum of 2 marks available.
Ovaries *[1 mark]* and testes *[1 mark]*.

2 Maximum of 4 marks available.
a) A gene is a section of DNA that controls one characteristic / controls the synthesis of one or more polypeptide(s) *[1 mark]*. An allele is one of the alternative forms of a gene *[1 mark]*.
b) Haploid cells have half the number of chromosomes as a normal cell *[1 mark]*, diploid cells have the full number of chromosomes — in pairs of homologous chromosomes *[1 mark]*.

3 Maximum of 3 marks available.
Meiosis halves the chromosome number *[1 mark]*, so it compensates for the doubling of chromosome number at fertilisation *[1 mark]*. Meiosis also increases variety by producing new combinations of alleles *[1 mark]*.

Answers

Answers

Page 43 — Variation

1   Maximum of 2 marks available.
    2^3 or $2 \times 2 \times 2$ *[1 mark]*, so 8 possibilities *[1 mark]*.

In a question like this, always show your working.

2 a) Characteristics that show continuous variation are more likely
     to be affected by the environment *[1 mark]*.
  b) Characteristics that show continuous variation usually involve
     more genes *[1 mark]*.

Page 45 — Environment and Phenotype

1 a) There is discontinuous variation / there are two distinct
     categories *[1 mark]*.
  b) There is continuous variation within each category *[1 mark]*.

2   Maximum of 2 marks available.
    Continuous variation *[1 mark]*, because there are no distinct
    categories *[1 mark]*.

3   Both generations have tall and dwarf plants. This suggests a
    genetic component as dwarf plants [similar to the parental
    plants] have appeared in the F_2 generation *[1 mark]*.

Page 47 — Inheritance

1   Maximum of 3 marks available — 1 mark for every two
    correct answers from the following.
    $I^A I^A$, $I^A I^O$, $I^B I^B$, $I^B I^O$, $I^A I^B$, $I^O I^O$

2   Maximum of 3 marks available.
    Carry out a test cross *[1 mark]*, by crossing with a white-
    flowered plant *[1 mark]*. If some of the offspring are white-
    flowered then the plant was heterozygous / If all purple-
    flowered then the plant was homozygous. *[1 mark]*

*This is a really common question in exams, so make sure you
answered it correctly.*

Page 49 — Inheritance

1   Maximum of 4 marks available.

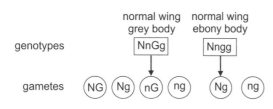

Ratio of 3 normal wings and grey bodies : 3 normal wings
and ebony bodies : 1 normal wings and ebony body :
1 vestigial wings and ebony body.
4 marks for all genotypes and phenotypes correct.
Deduct one mark for each mistake.

2   Maximum of 5 marks available for any of the points below.
    The recessive allele for haemophilia is carried on the X
    chromosome *[1 mark]*, so males will only have one copy of
    the allele and females will have two copies *[1 mark]*. This
    means that a female who inherits one copy of the
    haemophilia allele will also probably have a copy of the
    'normal' allele for factor 8, which is dominant *[1 mark]*. She
    will be healthy, but a carrier *[1 mark]*. Males who inherit one
    copy of the haemophilia allele will have the disease *[1 mark]*,
    so they can get it if they have a healthy father and a carrier
    mother *[1 mark]*. Females only stand a chance of having
    haemophilia if they're the child of a haemophiliac male and a
    carrier female *[1 mark]*, so the probability is much lower
    [1 mark].

Page 51 — Mutation and Phenotype

1   Maximum of 3 marks available.
    Three nucleotides form a single codon *[1 mark]*. All following
    codons are unchanged if the whole codon is deleted
    [1 mark]. If only one nucleotide is deleted there is a frame
    shift / all following codons are altered *[1 mark]*.

2 a) Maximum of 2 marks available.
     To reduce the radiographer's exposure to X-rays *[1 mark]*.
     X-rays are mutagenic / carcinogenic *[1 mark]*.
  b) Maximum of 2 marks available.
     Sunlight contains UV rays *[1 mark]*, which are mutagenic /
     carcinogenic *[1 mark]*.
  c) Maximum of 2 marks available.
     The X-rays cause mutation in the parent flies' sex cells /
     gametes *[1 mark]*. This leads to the production of genetic
     abnormalities in the next generation *[1 mark]*.

Page 53 — Frequency of Alleles

1   Maximum of 4 marks available.
    p = frequency of allele M,   q = frequency of allele N.
    The frequency of genotype MM = 0.36
    Therefore, $p^2 = 0.36$, and p = 0.60 *[1 mark]*
    So, q = 1-0.60 = 0.40 *[1 mark]*
    Frequency of blood group MN = 2pq = 2 × 0.6 × 0.4 = 0.48
    [1 mark]
    Frequency of blood group N = $q^2 = 0.4^2 = 0.16$ *[1 mark]*.

2   Maximum of 4 marks available.
    Frequency of the sickle-cell allele would decrease *[1 mark]*,
    because it now gives a disadvantage / no longer has any
    advantage *[1 mark]*. Frequency of the normal allele would
    increase *[1 mark]*. These changes are caused by directional
    selection *[1 mark]*.

Section 4 — Classification, Selection and Evolution

Page 55 — Classification

1   Maximum of 2 marks available.
    Traditional classification deals with features that are easy to
    observe *[1 mark]* whilst phylogeny deals with the genetic
    relationships between organisms *[1 mark]*.

2   Maximum of 5 marks available.
    a) phylum *[1 mark]*,   b) class *[1 mark]*,   c) family *[1 mark]*,
    d) Aptenodytes *[1 mark]*,   e) patagonicus *[1 mark]*.

*You'll be given all the info you need for this type of question —
you don't need to remember scientific names of organisms
but you do need to know the groups in the taxonomic system.*

Answers

Page 57 — Speciation

1 a) *Maximum of 3 marks available.*
   Darwin noticed that there were 14 different species of finch on the Galápagos islands [1 mark]. Each of these finches occupies a different ecological niche [1 mark]. The finches have different beaks which are adapted to eating different foods [1 mark].

   b) *Maximum of 4 marks available.*
   Darwin thought that all the finches on one island competed for resources [1 mark]. Some finches flew to other islands and established separate populations [1 mark]. Adaptation to different habitat / food sources gradually changed the beak shapes of the finches [1 mark]. This eventually led to the formation of new species [1 mark].

Page 59 — Natural Selection and Evolution

1 *Maximum of 6 marks available from the following points:*
   Natural selection has occurred [1 mark]. Before the drought individual birds showed variation in beak depth and length [1 mark]. When the drought happened birds with longer and deeper beaks had an advantage because they were able to obtain food [1 mark]. The birds with the longest and deepest beaks had the greatest chance of survival [1 mark]. The birds who survived produced offspring with larger beaks [1 mark]. The alleles for larger beaks would therefore be more prominent in future generations of Geospiza fortis on the island [1 mark]. This is an example of directional selection [1 mark].

   This question seems complicated because it is really long and it contains a table. Actually, it's not that complicated — it's just asking you to explain how natural selection can change the genetic make-up of a population.

Page 61— Adaptations to the Environment

1 a) *Maximum of 3 marks available.*
   In selective breeding individuals with desirable characteristics [1 mark] are interbred to produce offspring which meet the demands of consumers [1 mark]. Selective breeding over many generations makes farming more profitable [1 mark].

   b) *Maximum of 3 marks available.*
   Farmers should try to use heterozygous animals [1 mark] to avoid in-breeding [1 mark] which reduces the size of the gene pool [1 mark].

   You should be able to use all areas of knowledge and link ideas together.

2 *Maximum of 4 marks available from the following:*
   Polar bears have a small surface area : volume ratio [1 mark]. They have a layer of insulating fat [1 mark]. They have thick fur [1 mark] and a large body size [1 mark]. They have small ears [1 mark].

Section 5 — Control, Coordination and Homeostasis

Page 63 — Homeostasis and Temperature Control

1 a) *Maximum of 2 marks available.*
   A change in a factor brings about a response that counteracts the change / makes it opposite [1 mark] so that the factor returns to a norm [1 mark].

   b) *Maximum of 2 marks available from any of the following.*
   Body temperature [1 mark], blood glucose concentration [1 mark], water potential [1 mark]. Or other sensible answer.

2 *Maximum of 6 marks available from any of the following.*
   More food is consumed by the mouse because the mouse is an endotherm and the lizard is an ectotherm [1 mark]. The mouse has a higher metabolic / respiratory rate [1 mark] in order to generate more body heat [1 mark], so more food / glucose / carbohydrate is needed as a source of energy [1 mark]. The rate is temperature dependent in the lizard because the increase in external temperature raises the body temperature of the lizard, but not of the mouse [1 mark]. The lizard's metabolic / respiratory rate increases [1 mark] so its energy demand increases [1 mark].

Page 65 — The Liver and Excretion

1 a) *Maximum of 6 marks available from the following:*
   Protein in food is digested / hydrolysed in the gut / stomach / (small) intestine [1 mark] into amino acids [1 mark]. If a lot of protein is digested, excess amino acids will be deaminated / have amino groups removed [1 mark] in the liver [1 mark] to form ammonia [1 mark]. The ammonia reacts with carbon dioxide [1 mark] to form urea [1 mark], and more urea is excreted in the urine by the kidneys [1 mark].

   It's important to say that only the excess amino acids are deaminated.

   b) *Maximum of 3 marks available.*
   When amino acids are deaminated an organic acid is made [1 mark] which is changed into carbohydrate / sugar / glucose [1 mark], and the glucose is then built up / condensed into glycogen [1 mark].

2 *Maximum of 4 marks available.*
   Ammonia is very soluble / more soluble than urea [1 mark], so it can diffuse into the surrounding water [1 mark]. This means that conversion to urea isn't needed [1 mark], and so less energy is needed / used [1 mark].

Page 67 — The Kidneys and Excretion

1 a) *Maximum of 5 marks available.*
   Microvilli provide a large surface area of membrane [1 mark] so there are more carrier proteins (for glucose) [1 mark]. There are lots of mitochondria [1 mark] which provide energy / ATP for reabsorption [1 mark] by active transport [1 mark].

   b) *Maximum of 4 marks available.*
   More glucose passes into the blood, because it is actively reabsorbed [1 mark], but there is no carrier protein for urea [1 mark]. Some urea passes into the blood by diffusion [1 mark] because it's a small molecule [1 mark].

2 *Maximum of 2 marks available.*
   Either blood pressure is high(er) [1 mark]; so proteins get filtered out of blood [1 mark]. Or the wall of the capillary / basement membrane / wall of renal capsule is damaged / more perforated [1 mark] so that larger molecules are being filtered out [1 mark].

Answers

Page 69 — Water and Metabolic Waste Control

1   *Maximum of 10 marks available.*
*Strenuous exercise causes more sweating [1 mark] so more water is lost from the body [1 mark]. This increases the blood solute concentration / decreases blood water potential / makes blood water potential more negative [1 mark]. It also stimulates osmoreceptors [1 mark] in the hypothalamus [1 mark], which stimulates the pituitary gland [1 mark] to release **more** ADH [1 mark].*

The answer up to this point has explained the cause of the increase in level of ADH in the blood. After this, the answer explains the effect on the kidney.

ADH increases the permeability of the collecting ducts [1 mark] so more water is reabsorbed into the blood by osmosis [1 mark]. This means that less water is lost in the urine which prevents further dehydration [1 mark].

2   *Maximum of 5 marks available.*
*More sodium chloride is removed from the ascending limb [1 mark] of a longer loop by active transport [1 mark] so solute concentration rises / solute potential falls / water potential falls / becomes more negative in the medulla [1 mark]; so **more** water is reabsorbed from the collecting duct [1 mark] by osmosis [1 mark].*

Page 71 — The Nervous System and Receptors

1 a) *Maximum of 3 marks available for any of the following points:*
Receptors of the nervous system can communicate with effectors by releasing a chemical / hormone that binds to them [1 mark]. The chemical / hormone moves from the receptor to the effector by diffusion if the two cells are close together [1 mark], or if they are far apart the chemical / hormone is released into the blood and transported via mass flow [1 mark]. The other main way that receptors can send a message to effectors is to trigger an electrical impulse in a nerve, which is then passed through the nervous system until it reaches the effector and triggers a response [1 mark].
  b) *Maximum of 2 marks available.*
Plants don't have nervous systems, and have to release a chemical in order to respond to changes in their environment [1 mark]. This is much slower as it has to rely on diffusion [1 mark].

2   *Maximum of 4 marks available for any of the following points:*
Chemoreceptor cells have membrane-bound receptor molecules / proteins [1 mark]. The stimulus molecule must have a complementary shape [1 mark] to bind to the receptor molecule [1 mark] and create a generator potential [1 mark] to transmit a nerve impulse [1 mark].

Page 73 — The Nervous System — Neurones

1 a) *Stimulus [1 mark].*
  b) *Maximum of 3 marks available.*
*A stimulus causes sodium channels in the neurone cell membrane to open [1 mark] Sodium **diffuses** into the cell [1 mark], so the membrane becomes depolarised / more positive charge moves in (than moves out) [1 mark].*
  c) *Maximum of 2 marks available, from any of the following.*
The membrane was in the refractory period [1 mark] and so the sodium channels were inactive / recovering / couldn't be opened [1 mark]. Alternatively, the stimulus could have been lower than threshold level [1 mark].

2   *Maximum of 5 marks available, for any 5 of the following:*
Transmission of action potentials will be slower [1 mark]. Myelin insulates the axon / has high electrical resistance [1 mark] and there are gaps / nodes of Ranvier between sheaths [1 mark] where depolarisation happens /sodium channels are concentrated [1 mark]. So in an intact myelinated axon, saltatory transmission occurs / action potentials jump from node to node [1 mark]. This can't happen if the myelin sheath is damaged / more membrane is exposed [1 mark].

Don't panic if a question mentions something you haven't learned about. You might not know anything about multiple sclerosis, and that's fine because you're not supposed to. All you need to know about to get full marks here is the structure of neurones.

Page 75 — Synapses and the Reflex Arc

1   *Maximum of 8 marks available, for any 8 of the following:*
Arrival of the action potential causes calcium channels to open (in the presynaptic membrane) [1 mark] which makes calcium diffuse into the bouton / synaptic knob / cell [1 mark]. This stimulates the vesicles in the bouton to fuse with the presynaptic membrane [1 mark] and release the neurotransmitter [1 mark] by exocytosis [1 mark]. The neurotransmitter diffuses across the gap / cleft [1 mark] and binds to receptors on the postsynaptic membrane [1 mark] This stimulates opening of sodium channels (on the postsynaptic membrane) [1 mark] so sodium diffuses into the cell [1 mark]. The membrane then becomes depolarised [1 mark].

2   *Maximum of 4 marks available.*
Vesicles (containing neurotransmitter) are only found in the presynaptic neurone [1 mark], so exocytosis / release / secretion (of neurotransmitter) can only happen from here [1 mark]. The receptors (for the neurotransmitter) are only found on the postsynaptic membrane [1 mark] so only this membrane can be stimulated by the neurotransmitter / neurotransmitter can only bind here [1 mark].

Answers

Page 77 — The Nervous System — Effectors

1   Maximum of 10 marks available, for any 10 of the following:
The sarcoplasmic reticulum membranes become much more permeable *[1 mark]* and calcium ions diffuse out *[1 mark]*. They reach the actin filaments and bind to a protein called troponin *[1 mark]*, which causes another protein called tropomyosin to change position *[1 mark]* and unblock the binding sites on the actin filaments *[1 mark]*. The myosin heads attach to the binding sites *[1 mark]* to form actomyosin cross bridges between the two filaments *[1 mark]*. The myosin head then changes angle *[1 mark]*, pulling the actin over the myosin towards the centre of the sarcomere *[1 mark]*. The cross bridges then detach and reattach further along the actin filament *[1 mark]*. ATP provides the energy for this *[1 mark]*.

2   Maximum of 5 marks available.
More sarcoplasmic reticulum enables action potentials to be carried to more myofilaments *[1 mark]* so more calcium can be taken up by the cytoplasm *[1 mark]*. This means that there is more stimulation of the actin-myosin interaction *[1 mark]*, so more of the muscle / more myofibrils are stimulated to contract *[1 mark]*. This gives more efficient / faster / stronger muscle contraction *[1 mark]*.

Page 79 — The Mammalian Eye

1   Maximum of 8 marks available.
Cones are responsible for the high acuity of the human eye *[1 mark]*. They're densely packed at the fovea, where most of the light that enters the eye tends to focus *[1 mark]*. Each synapses with just one bipolar neurone *[1 mark]*, so it can send very detailed information to the brain *[1 mark]*. Rods are found in the more peripheral parts of the retina *[1 mark]*. They're more sensitive than cones, because it takes less light to activate the pigment inside them *[1 mark]*. Lots of rods converge onto the same bipolar neurone too *[1 mark]*, so even small responses from each of the rod cells is enough when combined for a message to be sent to the brain by that bipolar neurone *[1 mark]*.

Page 81 — The Autonomic Nervous System

1 a)   Maximum of 2 marks available.
It's involuntary *[1 mark]* and stereotypic / always the same kind of response *[1 mark]*.
b)   Maximum of 2 marks available.
Different muscles have different receptors *[1 mark]* which trigger different chemical effects inside the muscles *[1 mark]*.

Page 83 — The Central Nervous System

1 a)   Maximum of 8 marks available from any of the following.
Action potentials / nerve impulses arrive at sensory areas *[1 mark]*. There are different sensory areas for different parts of the body *[1 mark]* — the left hemisphere receives impulses from the right side of the body, and vice-versa *[1 mark]*. The more sensory cells (in a part of the body), the bigger the sensory area of the cerebrum *[1 mark]*. Association areas integrate information   — they pass action potentials / nerve impulses through motor neurones *[1 mark]*, so they control the kinds of responses *[1 mark]*. Examples of association areas are Broca's area (involved in speech) *[1 mark]* and Wernicke's area (language) *[1 mark]*.

b)   Maximum of 4 marks available.
Protein / amyloid deposits / plaques occur in the cerebral cortex *[1 mark]*. Protein tangles occur inside neurones *[1 mark]*. It causes loss of memory *[1 mark]* and changes in personality *[1 mark]*.

2   Maximum of 4 marks available.
The hypothalamus sends action potentials / nerve impulses to the pituitary gland *[1 mark]*. The signal / chemicals from the hypothalamus pass to the pituitary gland *[1 mark]*. The hypothalamus affects the activity of the pituitary gland *[1 mark]* so it allows the nervous system to control the endocrine system *[1 mark]*.

Page 85 — Hormonal Control

1   Maximum of 2 marks available, from any of the following.
It takes time for hormones to be synthesised *[1 mark]* and for the bloodstream to carry them (from gland to target) *[1 mark]*. The nervous system depends on faster action potentials / nerve impulses *[1 mark]*.

2   Maximum of 6 marks available.
The hormone diffuses into the blood capillaries *[1 mark]*. The hormone is circulated around the body by mass flow *[1 mark]*. The hormone diffuses out of the blood and binds to cell-surface receptors *[1 mark]* of the target cells *[1 mark]* with complementary shaped binding sites *[1 mark]*. This brings about a response in the target cells.

Page 87 — Blood Glucose Control

1 a)   Maximum of 10 marks available from the following.
Glucose is absorbed into the blood (from the gut) which makes the blood glucose concentration increase *[1 mark]*. This stimulates the beta cells *[1 mark]* of the islets of Langerhans *[1 mark]* to secrete insulin *[1 mark]*. Insulin is released into the bloodstream *[1 mark]* and binds to receptors *[1 mark]* on liver cells *[1 mark]*. This increases the permeability of the liver cells to glucose *[1 mark]*, converting glucose into glycogen / stimulating glycogenesis *[1 mark]*. This reduces blood glucose concentration *[1 mark]*.
b)   Maximum of 2 marks available.
There are no / insufficient receptors for insulin *[1 mark]* so insulin can't bind to / stimulate the liver cells *[1 mark]*.

2   Maximum of 5 marks available.
Exercise uses up glucose (by respiration) *[1 mark]* which lowers blood glucose concentration *[1 mark]* and stimulates the alpha cells *[1 mark]* of the islets of Langerhans *[1 mark]* to secrete glucagon *[1 mark]*.

Page 89 — Communication Systems in Plants

1   Maximum of 5 marks available.
Gibberellins stimulate the synthesis of amylase *[1 mark]* which hydrolyses (stored) starch to maltose *[1 mark]*. Glucose is respired *[1 mark]* to release energy / make ATP *[1 mark]* used in growth (during germination) *[1 mark]*.

Answers

2   Maximum of 5 marks available.
    Sunlight contains more red light than far red light *[1 mark]*,
    so exposure to sunlight converts PR into PFR *[1 mark]*. The
    nights are too short for PFR to be converted back into PR
    [1 mark]. More PFR form builds up in the plant during
    summer *[1 mark]*. PFR stimulates flowering in long day plants
    [1 mark].

    *It's really easy to get mixed up with PR and PFR, or just to
    write down the wrong one, even if you do know the difference.*

Section 6 — Physiology and the Environment

Page 91 — Water and Mineral Uptake in Plants

1   Maximum of 4 marks available from the following:
    Phosphate is needed in the photosynthesis and respiration
    reactions *[1 mark]*. Phosphate is needed for ATP, NADP and
    the phospholipids in membranes *[1 mark]*. Phosphate is
    needed for DNA and RNA synthesis *[1 mark]*. Nitrate is
    needed for the nucleic acids used in protein synthesis
    [1 mark]. Nitrate is needed in ATP and NADP *[1 mark]*.
    Magnesium is needed to make chlorophyll *[1 mark]*. These
    mineral ions are taken up via active transport *[1 mark]*.

Page 93 — Gas Exchange and Water Loss

1   Maximum of 3 marks available.
    a) High *[1 mark]*.
    b) High *[1 mark]*.
    c) Low *[1 mark]*.

    *Examiners like to ask questions where you can apply your
    knowledge. This question is actually fairly simple. It's just
    getting you to think about the adaptations that organisms
    have to make gaseous exchange as efficient as possible.*

Page 95 — Transport and Respiratory Gases

1   Maximum of 2 marks available.

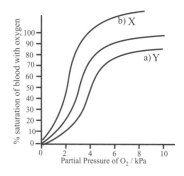

Page 97 — The Human Gut

1   Maximum of 4 marks available.
    Carnivores and ruminants have different teeth — carnivores
    have long sharp canines to tear meat whilst ruminants have
    more prominent incisors and a horny pad to enable them to
    eat vegetation *[1 mark]*. Carnivores have only one section in
    their stomachs whilst ruminants have four *[1 mark]*.
    Carnivores do not have the enzyme cellulase but ruminants
    host bacteria which do *[1 mark]*. Ruminants have a distema,
    but carnivores don't *[1 mark]*. Ruminants regurgitate food
    and chew the cud, carnivores do not regurgitate food
    [1 mark].

2   Maximum of 6 marks available.
    Amylase is used to hydrolyse starch to maltose *[1 mark]*.
    Maltase is used to hydrolyse maltose to glucose *[1 mark]*.
    Pepsin is used to hydrolyse protein to polypeptides
    [1 mark]. Trypsin is also used to hydrolyse protein into
    polypeptides *[1 mark]*. Chymotrypsin hydrolyses
    polypeptides into amino acids *[1 mark]*. Lipase hydrolyses
    fats into fatty acids and glycerol *[1 mark]*.

    *Bile salts are involved in digestion — they emulsify fats to
    fatty droplets — but they aren't enzymes.*

Page 99 — Control of Digestive Secretions

1   Maximum of 5 marks available.
    Reflex reactions produce saliva *[1 mark]* and gastric juices
    [1 mark]. Acetylcholine increases the muscle action in the
    stomach and intestines *[1 mark]*. Adrenaline inhibits the
    digestive system muscles *[1 mark]*. Nervous impulses from
    the autonomic nervous system control peristalsis *[1 mark]*.

2   Maximum of 3 marks available.
    a) Nervous or hormonal *[1 mark]*.
    b) Nervous reflex *[1 mark]*.
    c) Hormonal *[1 mark]*.

Page 101 — Metamorphosis and Insect Diet

1   Maximum of 6 marks available.
    Egg — peptidases, lipase *[2 marks]*, Caterpillar — peptidases,
    cellulase, and amylase *[3 marks]*. (Pupa — none) Butterfly —
    sucrase *[1 mark]*.

    *Make sure you know the enzymes relating to each stage of
    the insect's life — it comes up a lot in exams.*

2   Maximum of 3 marks available.
    a) Mandibles — the strong jaws that caterpillars use to cut
       through leaves *[1 mark]*.
    b) Imaginal discs — the clusters of cells which trigger the
       change into specific parts of the adult butterfly during the
       pupal stage *[1 mark]*.
    c) Maxillae — structures on jaws of the caterpillar which help it
       to recognise food, and push food into the mouth *[1 mark]*.

Section 7 — Gene Technology

Page 103 — Genetic Engineering

1 a) Maximum of 9 marks available.
    The DNA coding for the florescent protein is removed from a
    jellyfish cell *[1 mark]*. The gene is cut out using restriction
    endonucleases *[1 mark]*. A plasmid is cut open using the
    same restriction enzymes *[1 mark]*. The gene is then spliced
    into the plasmid *[1 mark]* using DNA ligase *[1 mark]*. The
    recombinant plasmid is added to a bacterial suspension
    [1 mark] in ice-cold calcium chloride solution *[1 mark]* and
    heated to 42 °C for 50 seconds *[1 mark]*, then cooled on ice
    [1 mark].
    Also accept alternative explanation:
    Isolate the correct mRNA in a jellyfish cell and use reverse
    transcriptase to make the complementary DNA *[1 mark]*.
    Use the enzyme DNA ploymerase to turn this single-stranded
    DNA into double stranded DNA *[1 mark]*. Insert this DNA
    into E. coli as described above.

Answers

b) Maximum of 2 marks available.
Could be used as a marker gene *[1 mark]*, as cells containing the recombinant DNA would glow *[1 mark]*.

If you have trouble remembering the sequence of events in transformation, try and make up a clever mnemonic to help.

2 Maximum of 6 marks available.
Reverse transcriptase catalyses the production of complementary DNA (cDNA) *[1 mark]* from messenger RNA (mRNA) *[1 mark]*. In genetic engineering, it is necessary to find the useful gene to remove, and there are probably only two copies in each cell *[1 mark]*, but there will be many mRNA molecules complementary to it *[1 mark]*. Finding the mRNA molecules will therefore be easier *[1 mark]*. The required gene can then be made if nucleotides and reverse transcriptase are put with the mRNA *[1 mark]*.

Page 105 — Commercial Use of Genetic Engineering

1 Maximum of 5 marks available.
An example of a GM human hormone is insulin / human growth hormone / growth hormone releasing factor (GRF) *[1 mark]*. The amino acid sequence of the hormone is converted into triplet codons and the gene is synthesised *[1 mark]*. A bacterial plasmid is cut with restriction endonuclease *[1 mark]*. The gene is spliced into the plasmid using the same restriction enzyme *[1 mark]*. The recombinant plasmid is introduced into the bacterium *[1 mark]*.

2 Maximum of 6 marks available.
Benefits of herbicide-resistant crops — **3** of: Weeds killed, but not crops *[1 mark]*. More potent herbicide can be used, so fewer applications necessary *[1 mark]*. Lower cost as fewer applications and less use of machinery, fuel and manpower *[1 mark]*. Less compaction of soil and damage of crops as less use of machinery *[1 mark]*. Improved yields due to the increased ability to kill weeds *[1 mark]*.
Drawbacks — **3** of: Recombinant resistance gene could spread to weed species *[1 mark]*. Gene could produce unforeseen side-effects in crops *[1 mark]*. Powerful herbicide doses could reduce biodiversity *[1 mark]*. Less incentive to research into selective weed killers *[1 mark]*.

You need to do a bit of lateral thinking in questions like this. Don't just put the obvious answers, and remember to give reasons — think A Level, not GCSE.

Page 107 — The PCR and Genetic Fingerprinting

1 a) Maximum of 1 mark available.
Genetic / DNA fingerprinting / profiling *[1 mark]*.
b) Maximum of 5 marks available.
A DNA sample would be taken from the suspect *[1 mark]*. The DNA from both samples would be cut into fragments by the same specific restriction endonucleases *[1 mark]*. The DNA fragments from both samples would be separated out by size using electrophoresis *[1 mark]*. Radioactive gene probes would be used to make the DNA fragment patterns of both samples visible *[1 mark]*. The bands on the photographic film would be compared to see if the pattern produced by the suspect's DNA matched that produced by the DNA found at the scene of the crime. *[1 mark]*.

'Explain' in an exam question tests your ability to apply your knowledge. You must give your answer in the context of the question. Simply describing the technique of genetic fingerprinting wouldn't get you full marks.

Section 8 — Ecology

Page 109 — Productivity

1 Maximum of 3 marks available.
In the tropical rainforest there are more plants (meaning a higher productivity) because there are more mineral nutrients available *[1 mark]* and there is more water available *[1 mark]*. The plants in the tropical rainforest also have a higher leaf area index, which allows them to photosynthesise more efficiently — desert plants have small leaves or spines to conserve water *[1 mark]*.

2 a) Maximum of 2 marks available.
Not all the available light is absorbed *[1 mark]*, because there aren't enough leaves. Some light that could have been absorbed will pass straight through the leaves or miss them altogether *[1 mark]*.
b) Maximum of 2 marks available.
Leaves in the lower layers don't get enough light because they're blocked by the leaves above *[1 mark]*. They use more food in respiration than they produce in photosynthesis, so energy is wasted *[1 mark]*.

Page 111 — Impact of Agriculture on the Environment

1 a) The BOD of a sample of water is its biological oxygen demand — the amount of oxygen removed from the sample in a given time *[1 mark]*.
b) Maximum of 3 marks available.
You would measure the oxygen concentration in the sample of water *[1 mark]*. You would then incubate the sample in the dark at 20°C for five days *[1 mark]*. Then you'd find the new oxygen concentration and use these figures to calculate how quickly oxygen was being used in the sample *[1 mark]*.

Answers

c) *Maximum of 6 marks available, for any 3 of the following reasons (2 marks each).*
The BOD test might give a misleading measurement if:
There are toxins present that kill bacteria [1 mark], because this will prevent the bacteria from decaying the organic matter and using up the oxygen [1 mark].
If the organic pollution levels are very high [1 mark], because this means that the bacteria don't have enough time (or oxygen) to decay all the pollutant [1 mark].
If the oxygen concentration is already very low [1 mark], because all the oxygen will be used up before decay has been completed [1 mark].
If there are chemicals present that use up oxygen [1 mark], because this will make the BOD seem artificially high [1 mark].
If the organic matter (e.g. cellulose) is difficult to decay [1 mark], because the bacteria will only be able to break it down slowly and will therefore use less oxygen, and this will make the BOD seem less [1 mark].

Page 113 — Harvesting Ecosystems

1 *Maximum of 4 marks available.*
Advantages of fishing — any 2 of the following 3 points:
An extra food resource, over and above farming [1 mark].
Allows a natural ecosystem to continue alongside food production [1 mark].
Encourages governments etc. to manage conservation of fish stocks [1 mark].
Disadvantages of fishing — any 2 of the following 3 points:
Overfishing can destroy the fish population [1 mark].
Fishing can disrupt the marine food web [1 mark].
The natural ecosystem can't be controlled like a farm can [1 mark].

If a question asks for advantages and disadvantages, try to give about equal numbers of both in your answer.

2 *Maximum of 2 marks available.*
Antibiotics could kill beneficial bacteria [1 mark], and overuse of the antibiotics might lead to antibiotic resistance in the bacteria [1 mark].

At first these problems might only affect the area around the fish farm, but the resistant bacteria could then travel through the water and affect a much wider area. If you mentioned this in your answer they might well give you a mark.

Page 115 — Pollution Caused by Industry

1 a) *Maximum of 2 marks available.*
More carbon dioxide production by people / cars / fossil fuels [1 mark], and fewer plants [1 mark].
 b) *Plants remove carbon dioxide from the atmosphere during photosynthesis [1 mark].*
 c) *Maximum of 2 marks available.*
Fossil fuels have been burnt in ever increasing amounts over the last 200 years, releasing extra carbon dioxide into the air [1 mark]. Widespread deforestation has destroyed a large percentage of the vegetation that used to remove carbon dioxide from the atmosphere [1 mark].

Page 117 — Conservation of Resources

1 *Maximum of 2 marks available.*
Sewage treatment depends on bacteria (and other micro-organisms) to break down organic waste [1 mark], and they may be killed by the toxic waste [1 mark].

2 *Maximum of 2 marks available.*
1 mark for an ethical reason, for example:
We have a moral duty to protect our planet / We owe it to future generations to preserve resources for their use.
1 mark for an economic reason, for example:
If we don't use energy resources responsibly now, they may run out and we won't be able to produce or run anything / We need to conserve minerals and raw materials or there won't be any left in the future to make new products / Some habitats, such as tropical rainforest, may be a source of valuable new resources, so they should not be destroyed.

Page 119 — Conservation of Species

1 *Maximum of 2 marks available.*
The human population has increased [1 mark], and industrialisation, agriculture, deforestation and other practices damaging to the environment have increased to support it [1 mark].

2 *Maximum of 4 marks available the following points:*
Visitors may damage habitats — perhaps just by repeatedly walking through them [1 mark].
Visitors may disturb animals / pick plants [1 mark].
Visitors may pollute / leave litter [1 mark].
Humans aren't part of the natural ecosystem [1 mark].
It may be necessary to limit the number of visitors or their access to certain areas in order to reduce this damage [1 mark].

Page 121 — Adaptations to the Environment

1 a) *Maximum of 3 marks available.*
Smaller shrews have a larger surface area:volume ratio [1 mark], so they lose heat more rapidly [1 mark]. They need to burn / respire more food to replace this heat energy [1 mark].
 b) *Maximum of 2 marks available.*
Shrews / mammals are warm-blooded [1 mark], so they lose more heat to their surroundings (than cold-blooded insects) [1 mark].

2 *Maximum of 2 marks available.*
Young maggots burrow into the tip and so get more food / are hidden from predators [1 mark].
Older maggots come to the surface so that the adult flies can escape / fly away when they hatch out [1 mark].

Answers

Section 9 — Applications of Genetics

Page 123 — Selective Breeding

1 a) *Maximum of 2 marks available.*
There was no genetic variation in this population *[1 mark]*.
So the observed variation was equal to the variation due to the environment (VE) *[1 mark]*.

b) *Maximum of 3 marks available.*
Total variation = genetic variation + environmental variation
= VG + VE *[1 mark]*.
So VG = 0.8 − 0.3 = 0.5 *[1 mark]*.
Heritability = VG/VP = 0.5/0.8 = 0.625 *[1 mark]*.

Page 125 — Artificial Insemination and Transplantation

1 *Maximum of 3 marks available.*
Different males would be bred with <u>same</u> female *[1 mark]*.
Egg mass of the (female) progeny would be determined *[1 mark]*. The males to be chosen for breeding should be the ones that produce (female) progeny that lay eggs of the largest mass *[1 mark]*.

2 *Maximum of 2 marks available.*
2 of: The eland undergoes the (traumatic) pregnancy so as not to put the rare bongo at risk *[1 mark]*. The embryos can be screened for genetic abnormalities before a pregnancy is allowed to continue *[1 mark]*. The number of young produced can be increased because more elands are available to act as surrogate mothers *[1 mark]*.

Page 127 — Genetic Diversity

1 *Maximum of 4 marks available.*
A sensible explanation including **4** of: Pedigree breeds are inbred so have lower genetic diversity *[1 mark]*. This means that a lot of their genes tend to be homozygous *[1 mark]*. Some of these genes are responsible for the disorders *[1 mark]*. These disorders are recessive *[1 mark]*; so are more likely to be expressed in homozygotes *[1 mark]*.

2 *Maximum of 2 marks available.*
2 of: They are small / take up little room *[1 mark]*.
They remain viable / alive at low temperatures *[1 mark]*.
They can survive for a long time in storage *[1 mark]*.

Page 129 — Resistance

1 a) *Maximum of 3 marks available.*
If dominant, both homozygotes and heterozygotes survive / carriers survive *[1 mark]*. If recessive, only homozygotes survive *[1 mark]*. This means that <u>more</u> rats in the population can pass on the allele if it is dominant *[1 mark]*.

b) *Maximum of 3 marks available.*
Individuals with the resistant allele are more likely to survive to breed (where warfarin is used) *[1 mark]*. These pass on the allele to their offspring *[1 mark]*; so offspring are resistant *[1 mark]*.

2 *Maximum of 4 marks available from any of the following.*
Bacteria reproduce quickly / have short generation times *[1 mark]*. Populations contain large numbers of individuals *[1 mark]*; so have a large gene pool / more likely to contain antibiotic-resistance genes *[1 mark]*. Antibiotic-resistance genes can spread via plasmids / horizontal transmission can occur *[1 mark]*. Bacteria can reproduce asexually, which means that if they have the resistance gene, all the clones they produce will too *[1 mark]*.

Page 131 — Human Genetics

1 a) *Maximum of 4 marks available.*
(let B = allele for sufferer, b = allele for non-sufferer)
Bb × Bb *[1 mark]*; show gametes from Bb: (B) and (b) *[1 mark]*; show genotypes of offspring: ½ Bb, ¼ BB and ¼ bb *[1 mark]*; ¾ chance of sufferer (Bb and BB) and ¼ chance of non-sufferer (bb) *[1 mark]*.

b) *Maximum of 2 marks available.*
It develops in later life, after those affected may have had children *[1 mark]*. So children may inherit the disease before it is known that the parents are sufferers (as compared to other diseases, where sufferers may decide not to have children so as not to pass the disease on) *[1 mark]*.

2 *Maximum of 2 marks available.*
Pedigree analysis determines the <u>probability</u> of inheritance of a genetic disorder *[1 mark]*; whereas genetic screening determines the actual presence of a genetic disorder or the allele for the disorder in an individual *[1 mark]*.

Page 133 — Human Genetics

1 *Maximum of 5 marks available.*
Mother has genotype I^OI^O, father has genotype I^AI^B *[1 mark]*. Son has 50% chance of being blood group A (I^AI^O) and 50% chance of being B (I^BI^O) *[1 mark]*. Father has <u>both</u> A and B antigens *[1 mark]*. Son produces either anti-A <u>or</u> anti-B antibodies *[1 mark]*, so would reject transfusion in both cases *[1 mark]*.

In questions about the genetics of blood groups, make sure you don't confuse genotypes (e.g. I^OI^O) with phenotypes (e.g. group O).

2 *Maximum of 5 marks available.*
Close family members share similar alleles in the MHC *[1 mark]*, so produce similar proteins / antigens *[1 mark]*, which are more likely to be recognised as self *[1 mark]*. So a transplanted organ is less likely to stimulate an immune response *[1 mark]* and so is less likely to result in rejection *[1 mark]*.

Answers

Section 10 — Mammalian Physiology and Behaviour

Page 135 — The Liver

1 a) Maximum of 8 marks available from any of the following.
When the blood glucose level rises, insulin is secreted from
the pancreas *[1 mark]*. This causes the conversion of glucose
to glycogen / glycogenesis *[1 mark]*, resulting in the increased
uptake of glucose in liver cells *[1 mark]*. When the blood
glucose level falls, glucagon is secreted from the pancreas
[1 mark]. This causes the conversion of glycogen to glucose /
glucogenesis *[1 mark]*. Glucose is made from a non-
carbohydrate source / gluconeogenesis *[1 mark]*. Adrenalin
secretion causes this conversion of glycogen to glucose
[1 mark]. Also one mark for reference to islets of Langerhans
[1 mark], alpha, beta cells / pancreas, in secreting hormones /
detecting blood glucose *[1 mark]*, homeostasis *[1 mark]*,
negative feedback *[1 mark]*.

*Whenever you have to write a long answer like this, write down
an outline of the points in rough first, to make sure you're
well organised in your answer.*

2a) Maximum of 2 marks available for any of the following.
Haemoglobin is broken down / converted to bile pigments
[1 mark], bilirubin and biliverdin *[1 mark]*. Bile pigments are
not excreted *[1 mark]* because liver cells are not able to
remove these from the blood *[1 mark]*.

b) Maximum of 2 marks available for any of the following.
The liver process detoxifies / eliminates drugs *[1 mark]*.
The drug may build up in the body / isn't broken down as
quickly *[1 mark]*. There will be increased side effects if a full
dose is given *[1 mark]*. The therapeutic / beneficial effects of
the medication will last longer *[1 mark]*.

Page 137 — The Skeleton and Joints

1 a) Maximum of 2 marks available.
After the menopause, the deficiency of oestrogen increases
the risk of osteoporosis-related bone fractures *[1 mark]*, as
parathormone, which is normally inhibited by oestrogen,
removes calcium *[1 mark]*.
b) Maximum of 2 marks available from any of the following.
In osteoarthritis, bone deformities *[1 mark]* rub against
ligaments / soft tissues / other bones *[1 mark]*. Receptors /
nerves are stimulated *[1 mark]* and send impulses to the brain
[1 mark].

Page 139 — The Mammalian Ear

1 a) Organ of Corti / hair cells *[1 mark]*.
b) Maximum of 3 marks available from any of the following:
An increased volume in the median canal *[1 mark]* displaces
/ moves the basilar membrane *[1 mark]*. This leads to loss of
contact with nerve endings / variable pressure on nerve
endings *[1 mark]*, which causes varying levels of stimulation /
over / under-stimulation *[1 mark]*.
c) Maximum of 2 marks available.
Treatment with diuretics causes reduction in the volume of
endolymph *[1 mark]* and therefore reduction in pressure
[1 mark].

d) Pressure of endolymph on cupola / hair cells pushed over /
displaced *[1 mark]*.

*When you are given a piece of text, highlight the key terms
and write some short notes or draw a quick sketch. You must
refer to the text given — it often gives you clues to the
answers.*

Page 141 — Simple Behaviour Patterns

1 Maximum of 3 marks available from the following.
Human babies are born with the innate (genetic) ability to
speak *[1 mark]*. However, they have to learn language
[1 mark]. Which language they learn depends on their
environment *[1 mark]* — Japanese babies learn Japanese, but
a Japanese baby brought up by German-speakers would learn
German *[1 mark]*.

2 Maximum of 3 marks available.
Operant conditioning could be used in dog training to reward
[1 mark] or punish specific behaviours *[1 mark]*. E.g. the dog
could be given a biscuit each time it offered a paw to its
owner, and it would learn to perform this behaviour to receive
a biscuit *[1 mark]*.

Section 11 —
Growth, Development and Reproduction

Page 143 — Growth and Development

1 Maximum of 5 marks available, for any 5 of the following
points:
Puberty begins when the hypothalamus begins to release
GnRH *[1 mark]*.
GnRH stimulates the pituitary gland *[1 mark]*.
The pituitary gland begins releasing FSH and LH *[1 mark]*.
FSH and LH stimulate the testes *[1 mark]*.
The testes start releasing testosterone *[1 mark]*.
Testosterone stimulates the testes to produce sperm *[1 mark]*.
Testosterone leads to male secondary sexual characteristics
(e.g. deeper voice, pubic hair) developing *[1 mark]*.

2 a) 0 – 1 years *[1 mark]*.
b) 6 cm/year *[1 mark]*.
c) 17 years *[1 mark]*.

3 Maximum of 8 marks available for any 8 of the following:
Older people tend to be less mobile because their tissues
begin to degenerate *[1 mark]* and the structural proteins
become harder and less elastic *[1 mark]*. This makes them
stiffer and less supple *[1 mark]*. They also experience more
health problems as their organs begin to function less
effectively *[1 mark]*, and their immune system becomes less
efficient *[1 mark]*. Accuracy of DNA replication also declines
with age *[1 mark]*, which can lead to the production of
abnormal cells which don't work properly *[1 mark]*, or die
[1 mark], or become cancerous *[1 mark]*.

Answers

Page 145 — Asexual Reproduction

1 a) Maximum of 5 marks available.
Hydra – Budding *[1 mark]*.
Potato – Tubers *[1 mark]*.
Aphid – Parthenogenesis *[1 mark]*.
Cod – Sexual *[1 mark]*.
Amoeba – Binary fission *[1 mark]*.

b) Maximum of 5 marks available.
The parent plant produces tubers, which are swollen underground stems *[1 mark]*. The tubers act as a food / starch store over the winter *[1 mark]*. Buds on the tubers sprout in spring *[1 mark]*, and each sprout can produce a new plant *[1 mark]*. New plants are genetically identical to / are clones of the parent *[1 mark]*.

Page 147— Artificial Propagation and Cloning

1 Maximum of 5 marks available.
Cytokinin – hormone which promotes cell division *[1 mark]*.
Grafting – procedure to join two parts of different plants *[1 mark]*.
Auxin – hormone used to promote root growth *[1 mark]*.
Micropropagation – tissue culture used to produce plant clones *[1 mark]*.
Layering – technique of staking horizontal runners *[1 mark]*.

Page 149— Plant Pollination

1 a) Maximum of 4 marks available.
i) diploid / 2n *[1 mark]*.
ii) haploid / n *[1 mark]*.
iii) haploid / n *[1 mark]*.
iv) haploid / n *[1 mark]*.

b) Maximum of 2 marks, for any of the following points:
Maize is wind-pollinated *[1 mark]*.
The pollen will be carried in all directions by the wind *[1 mark]*.
There's more chance of pollen from one of the plants reaching another if each plant is closely surrounded by others in all directions *[1 mark]*.

2 a) Maximum of 2 marks available.
Cross-pollination is where the pollen from one plant is transferred to a flower on a different plant *[1 mark]*.
Self-pollination is where pollen is transferred to a stigma of a flower on the same plant *[1 mark]*.

b) Maximum of 6 marks available.
Methods evolved by plants that avoid self-pollination include protandry *[1 mark]*, where the anthers mature before the stigmas *[1 mark]*, and protogyny *[1 mark]*, where the stigmas mature before the anthers *[1 mark]*. Some other plant species are dioecious *[1 mark]* and individual plants have either all male parts or all female parts *[1 mark]*.

c) Maximum of 3 marks available.
Self-pollination restricts the amount of genetic variation arising in a population *[1 mark]*. This can lead to a whole population or species being wiped out by a particular pathogen or set of environmental conditions *[1 mark]*.

Page 151 — Sexual Reproduction in Plants

1 a) Maximum of 4 marks available.
A = Pollen tube *[1 mark]*.
B = Embryo sac *[1 mark]*.
C = Tube nucleus *[1 mark]*.
D = Micropyle *[1 mark]*.

b) Maximum of 2 marks available.
The enzymes digest surrounding cells *[1 mark]*.
This makes a path through to the ovary *[1 mark]*.

2 Maximum of 3 marks available.
a) testa / seed coat *[1 mark]*.
b) seed *[1 mark]*.
c) fruit *[1 mark]*.

Page 153 — Human Reproductive Organs

1 Maximum of 10 marks available, for any 10 of the following:
Gametogenesis is the production of gametes (ova and sperm) *[1 mark]*. In the human male sperm are produced in the testes *[1 mark]*, in the walls of the seminiferous tubules *[1 mark]*. This is known as spermatogenesis *[1 mark]*. Diploid cells in the germinal epithelium *[1 mark]* divide constantly by mitosis *[1 mark]* to give spermatogonia *[1 mark]*, and these then grow into primary spermatocytes *[1 mark]*. The primary spermatocytes divide by meiosis *[1 mark]*, producing secondary spermatocytes *[1 mark]*. Secondary spermatocytes divide once more to produce spermatids *[1 mark]*, which mature into sperm *[1 mark]*.

Page 155 — Human Reproduction

1 Maximum of 5 marks available.
ii) vas deferens *[1 mark]*
iii) urethra *[1 mark]*
iv) vagina *[1 mark]*
v) cervix *[1 mark]*
vi) uterus *[1 mark]*

Answers

2   Maximum of 10 marks available.
1 mark for a definition of abortion, e.g. Abortion is the surgical removal of a foetus or the use of drugs to kill it.
Up to 3 marks for a statement similar to: The ethical arguments surrounding abortion are complex *[1 mark]*. Those in favour of allowing abortion tend to be concerned with the rights of the pregnant woman *[1 mark]*, while those who would like to see it made illegal are concerned with the rights of the foetus *[1 mark]*.
Up to 5 marks for any 5 of the following points:
Some women might need an abortion if a pregnancy posed a serious threat to their health *[1 mark]*.
Women have the right to end an unwanted pregnancy — it's their body that has to support the foetus, so they shouldn't be forced to keep it if they don't want to *[1 mark]*.
The existing family should be considered first, because the foetus is not yet a real person — only a potential one *[1 mark]*.
It might be better to abort a foetus with serious genetic defects, as the quality of life of the child might be seriously reduced *[1 mark]* and caring for the child might put a lot of strain on the rest of the family *[1 mark]*.
The mother might be very young and unable to cope with a baby *[1 mark]*.
There is no excuse for destroying a life *[1 mark]*.
If abortion is allowed, women may begin to use it as form of contraception, rather than it being reserved for cases where there are serious reasons for ending the pregnancy *[1 mark]*.
Women might also choose only to keep babies with certain features, e.g. they might only want a girl *[1 mark]*.
The father of the child may want to keep it and might be able to support it even if the mother can't, but his rights and opinions are often ignored *[1 mark]*.
Women can often be pressured into an abortion or feel it is their only option *[1 mark]*.
Some women experience emotional trauma and feel guilt and regret after having an abortion *[1 mark]*.
There are other options available, such as adoption — there is no need to end a potential life if a woman can't cope *[1 mark]*.
Other sensible arguments can also be accepted.
1 mark available for a well-structured answer that considers both sides of the argument.

Page 157 — Embryo Development

1 a) Maximum of 12 marks available, for any 12 of the following:
Oxygen passes from mother to foetus *[1 mark]* by diffusion down a concentration gradient *[1 mark]*. This movement is helped by the higher affinity of foetal haemoglobin for oxygen *[1 mark]*. Carbon dioxide passes from foetus to mother *[1 mark]* by diffusion down a concentration gradient *[1 mark]*. Nutrients move from mother to foetus *[1 mark]*. Glucose moves across via facilitated diffusion *[1 mark]*, amino acids by active transport *[1 mark]*, and vitamins and minerals by a combination of diffusion and active transport *[1 mark]*. Water moves from mother to foetus *[1 mark]* by osmosis *[1 mark]*. Urea moves from foetus to mother *[1 mark]* by diffusion *[1 mark]*. Some antibodies can pass from mother to foetus across the placenta *[1 mark]*, as can nicotine, alcohol and other drugs *[1 mark]*.

b) Maximum of 2 marks available.
The placenta acts as an endocrine organ / produces hormones / secretes oestrogen, progesterone and HCG during pregnancy *[1 mark]*. The placenta acts as a barrier to bacteria and some viruses *[1 mark]*.

Page 159 — Controlling Growth in Plants

1   Maximum of 6 marks available.
The increase in temperature means that the enzymes in the seeds have more energy *[1 mark]*, and so they move about more rapidly and collide with substrate molecules more often *[1 mark]*. This means that the enzymes are able to break down the food stores needed for germination more quickly *[1 mark]*. But above 40 °C, enzymes may start to denature *[1 mark]*. They have too much energy and vibrate so strongly that bonds holding the molecule in its 3D shape are broken *[1 mark]*, and the shape of the enzyme no longer corresponds to that of its substrate *[1 mark]*. If the enzymes are not working, germination is prevented *[1 mark]*.

2   Maximum of 3 marks available.
Pfr inhibits flowering in SDPs *[1 mark]*. A short day length / long night length means that the Pfr is turned into Pr *[1 mark]*, which means the SDPs can flower *[1 mark]*.

Page 161 — Hormonal Control in Humans

1 a) Maximum of 3 marks available.
A = progesterone *[1 mark]*.
B = prolactin *[1 mark]*.
C = HCG *[1 mark]*.
b) Maximum of 2 marks available.
Prolactin is produced in the anterior pituitary gland *[1 mark]*, and stimulates milk production *[1 mark]*.

Page 163 — Hormonal Control in Humans

1   Maximum of 5 marks available.
Hypothalamus – TRH *[1 mark]*
Pituitary – TSH *[1 mark]*
Thyroid – Thyroxine *[1 mark]*
Ovaries – Oestrogen *[1 mark]*
Testes – Testosterone *[1 mark]*

Section 12 — Behaviour and Populations

Page 165 — Courtship and Territory

1   Maximum of 3 marks available.
Stereotyped courtship behaviour ensures that any potential mate is of the right species *[1 mark]*. It ensures that mating only occurs if the female is in a receptive condition / mating will produce offspring *[1 mark]*. It also allows the female to select a mate with features that will give her young a good chance of survival if inherited (this could be explained in a variety of ways e.g. strongest, best display etc. — the key idea is that the female has a basis for choice) *[1 mark]*.

Answers

2 a) *Maximum of 2 marks available. Possible answers include:*
Pheromones travel long distances [1 mark], so the female can attract males over a wide area with little effort [1 mark]. Pheromones travel over long distances [1 mark] so the female can attract several males and select the best of them for her mate [1 mark].

b) *Maximum of 1 mark available, for any one of the following: Females can attract several males by producing pheromones, and can then choose between them to select the strongest alleles for her offspring [1 mark]. / If lots of females were attracted to a single male, some of them might not reproduce at all and fewer offspring would be produced overall [1 mark]. / It's better for females not to travel too far and compete for males, but to conserve their energy for producing and raising young [1 mark]. / Females only produce pheromones when they're ready for breeding — it would be pointless for males to attract females that were not ready to mate [1 mark].*

When an exam question says "suggest", examiners can usually give credit for any reason that makes sense, even if it's not one of the ones given in the mark scheme. So think about it and have a go, even if you're not sure.

3 *Maximum of 5 marks available.*
Having a territory allows the breeding pair exclusive use of the resources in an area [1 mark]. This means that their young are more likely to survive [1 mark]. Certain features (usually in the male) lead to successfully acquiring / defending a territory [1 mark]. Better survival of the young means that genes for successful territorial behaviour continue in the offspring [1 mark]. Over several generations, territorial behaviour will become widespread in the population because of its selective advantage [1 mark].

Page 167 — Infertility and Pregnancy

1 *Maximum of 4 marks available.*
Injection of extracted or synthetic gonadotrophin hormones [1 mark]. Gonadotrophins stimulate ovum / secondary oocyte development and release [1 mark]. Treatment using the drug clomiphene [1 mark]. Clomiphene stimulates natural gonadotrophin production [1 mark].

2 *Maximum of 6 marks available, for any 6 of the following:*
Pregnant women put on weight due to:
The weight of the baby [1 mark].
The weight of the placenta [1 mark].
An increased volume of blood / body fluids [1 mark] to provide extra energy for the demands of pregnancy and breast feeding [1 mark].
Breast development / growth [1 mark] ready for breast feeding [1 mark].
Increased fat storage [1 mark] to provide energy for the demands of pregnancy and breast feeding [1 mark].

3 *Maximum of 7 marks available.*
A drug is taken to stimulate superovulation / production of several ova [1 mark]. Sperm are collected from the male [1 mark]. Eggs are collected from the female's ovary using a needle and ultrasound [1 mark]. The eggs and sperm are incubated overnight [1 mark]. The fertilised eggs are selected and developed under laboratory conditions for a few days [1 mark]. The fertilised eggs are then transferred into the mother's uterus [1 mark]. The mother receives progesterone treatment throughout procedure, so that the uterus is suitable for implantation [1 mark].

Page 169 — Population Sizes

1 *Maximum of 5 marks available.*
The population would decrease in size [1 mark].
There are a large number of older people [1 mark] who are too old / unlikely to have children [1 mark].
There are not many young people [1 mark] to have children in the future [1 mark].
As an alternative to saying that there are a lot of old people / not many young people, it would be acceptable to say the pyramid is wide at the top / narrow at the base.

2 *Maximum of 4 marks available.*
In the low density population, deer tend to live longer / death rate increases sharply when the deer reach old age / deer tend to die from natural causes as they near maximum life expectancy [1 mark].
In the high density population, deer have a more constant death rate / die at a younger age more often [1 mark].
Deer tend to die at a younger age in the high density population due to greater competition for food [1 mark] and increased likelihood of disease [1 mark].

The use of the word "reasons" in this question should clue you in to the fact that more than one is required.

Page 171 — Infectious Disease and Immunity

1 a) *One from:*
More people were getting vaccinated / The vaccine may take a while to become effective [1 mark].
(Or other sensible answer.)

b) *One from:*
The polio virus is still present / there are still a small number of cases / the disease could be re-introduced from outside / the virus might be 'dormant' in the population [1 mark].

Page 173 — Effect of Lifestyle on Health

1 *Maximum of 6 marks available.*
The fats are converted into cholesterol [1 mark]. Cholesterol is deposited in the artery walls and causes atheroma / atherosclerosis [1 mark]. Blood flow through the arteries is reduced [1 mark]. This can put a strain on the heart [1 mark] and / or can cut off the blood supply to an area of the heart causing a heart attack [1 mark]. Reduced blood flow to the brain can cause localised brain damage resulting in a stroke / cerebrovascular accident [1 mark].

Answers

2   Maximum of 10 marks available, for any 10 of the following:
Sunbathing exposes the skin to UV light *[1 mark]*.
This penetrates the skin *[1 mark]* and can damage the DNA
in a cell *[1 mark]*. Sometimes the genes that control cell
division are damaged *[1 mark]*, and then the cell undergoes
uncontrolled mitosis and tumours develop *[1 mark]*. The cells
are said to be cancerous / malignant *[1 mark]* leading to skin
cancer *[1 mark]*.
The danger can be reduced by:
Avoiding exposure to the sun for long periods / covering up
the skin with clothing and / or hats *[1 mark]*.
Avoiding the midday sun *[1 mark]*.
Not using sunbeds *[1 mark]*.
Using sunscreen of at least factor 15 *[1 mark]*.

Page 175 — Screening Programmes

1   Maximum of 5 marks available, for any 5 of the following:
X-rays carry a risk of damaging cells / causing cancer /
causing mutations *[1 mark]*, so patients shouldn't be exposed
to them too often *[1 mark]*. Women under 50 are much less
likely to develop breast cancer, so the small risk involved in
having a mammogram isn't as worthwhile *[1 mark]*. Women
under 50 have denser breast tissue *[1 mark]* and so x-rays
don't work as well for them *[1 mark]*. Screening programmes
can be expensive, so only the women most at risk are tested
[1 mark].

2   Maximum of 3 marks available.
Ultrasound is harmless *[1 mark]*. X-rays can cause mutation
of cells *[1 mark]*, which would be particularly dangerous in a
developing foetus as it has relatively few cells, and they're
dividing to form important structures *[1 mark]*.

3   Maximum of 4 marks available.
Advantages — Patients and their families would be prepared
for the onset of the disease and could plan how best to cope
[1 mark]. They could avoid passing the disease on to any
children *[1 mark]*.
Disadvantages — The person would know they had a fatal
disease, perhaps many years before they died, which would
cause a lot of stress and anxiety *[1 mark]*.
The person could be refused life insurance *[1 mark]*.
The person might be refused employment *[1 mark]*.
The person's family life and relationships could be badly
affected *[1 mark]*.
*(Any 2 of the disadvantages listed, for a maximum of 2
marks.)*

Section 13 — Microbiology and Biotechnology

Page 177 — Bacteria, Viruses and Fungi

1   Maximum of 3 marks available.
Capsids and envelopes may protect the virus from chemicals
whilst outside a host *[1 mark]*, let the virus bind to host cell
membranes *[1 mark]* or assist in penetration of a host cell
[1 mark].

2 a) Maximum of 2 marks available.
The layer of polysaccharide might prevent desiccation /
drying out *[1 mark]* and attack from antibodies / phagocytes
/ white blood cells *[1 mark]*.

b) Maximum of 3 marks available.
The diagram should include a labelled cell membrane
overlying the cytoplasm *[1 mark]*, a thin cell wall overlying
this cell membrane *[1 mark]*, and another labelled membrane
overlying the cell wall *[1 mark]*. As shown:

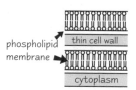

Page 179 — Cell Culture

1 a) Maximum of 5 marks available.
Hold a wire loop in a flame until red hot and then cool
[1 mark]. Ensure that there is minimal lifting of the lids from
the monoculture and sterile plate *[1 mark]*. Dip the loop into
the monoculture *[1 mark]*, then streak the surface of the
medium *[1 mark]*. One mark for any other precaution, e.g.
work close to bunsen flame, wear protective clothing, swab
bench with disinfectant.

b) Maximum of 2 marks available.
Heat in an autoclave *[1 mark]* to 121°C for 15 minutes
[1 mark].
Also accept:
The agar plates would be irradiated using gamma radiation.

2 a) Maximum of 3 marks available.
The bacteria respired aerobically *[1 mark]*, which released
more energy / made more ATP *[1 mark]* for cell growth /
division *[1 mark]*.

b) Maximum of 2 marks available.
There would be no population growth / cells would die
[1 mark] because oxygen is poisonous to them *[1 mark]*.

Page 181 — Measuring Bacterial Growth

1 a) Maximum of 3 marks available.
8.6 – 3.4 = 5.2 *[1 mark]*, 10 × 0.301 = 3.01 *[1 mark]*,
5.2 / 3.01 = 1.73 (generations per hour) *[1 mark]*.

b) Maximum of 8 marks available.
Take a known / fixed volume of culture at the start of 10
hours / at time = 0 *[1 mark]*. Dilute it by adding water of a
known volume / dilute by a known amount *[1 mark]*. Repeat
the dilution a fixed number of times *[1 mark]*. Spread a
known volume of the final dilution on an agar plate *[1 mark]*.
Incubate it *[1 mark]*, then count the number of colonies
formed *[1 mark]*. Multiply this number by each dilution
factor *[1 mark]*. Repeat for the sample taken at the end of 10
hours / at time = 1 *[1 mark]*.

2 a) Maximum of 2 marks available from any of the following.
Diauxic growth is when microorganisms are grown on a
medium with two different carbon sources *[1 mark]*, so that
one carbon source is used before the other *[1 mark]*. The
graph of their growth shows two separate exponential phases
[1 mark].

b) Maximum of 3 marks available.
1 mark for labelling the axes correctly. 1 mark for showing
two exponential peaks. 1 mark for correctly labelling lag and
exponential phases. As shown:

Answers

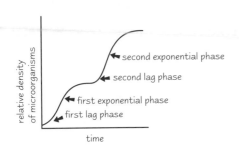

Page 183 — Industrial Growth of Microorganisms

1 a) Maximum of 2 marks available for any of the following.
Batch culture occurs in fixed volume, continuous culture
doesn't *[1 mark]*. Medium is added to the continuous culture
vessel at a constant rate, but none is added to a batch culture
after its start *[1 mark]*. Culture reaches the end of the
stationary phase in batch culture, but it's kept at the
exponential phase in continuous culture *[1 mark]*. Batch
cultures have to be started over again regularly, unlike
continuous cultures which last much longer *[1 mark]*.
 b) Maximum of 2 marks available.
Advantages of batch culture: easier to control conditions /
less costly to start again in event of contamination *[1 mark]*.
Advantages of continuous culture: method is more productive
/ culture lasts for longer / a smaller vessel can be used, linked
with stated advantage (e.g. easier to sterilise) *[1 mark]*.
 c) Maximum of 3 marks available. Possible answers include:
Penicillium would get less oxygen *[1 mark]*, so the culture
would respire less *[1 mark]*, so less energy / less growth
would occur *[1 mark]*, and less fungus / mycelium produces
less penicillin *[1 mark]*.

Page 185 — Biotechnology in Food Production

1 a) Maximum of 3 marks available.
Barley seeds contain starch *[1 mark]* and amylase *[1 mark]*.
Amylase hydrolyses starch into sugar / maltose *[1 mark]*.
 b) Maximum of 2 marks available.
Yeast grows / reproduces *[1 mark]* and uses up the oxygen
(in respiration) *[1 mark]*.
 c) Maximum of 2 marks available.
There would be a higher concentration of sugar / lower
concentration of ethanol (alcohol) *[1 mark]* because the rate
of fermentation / conversion of sugar to ethanol (alcohol) is
less *[1 mark]*.

2 a) Maximum of 4 marks available.
Lactobacillus produces lactic acid *[1 mark]* which gives a sour
taste *[1 mark]*. Both produce ethanal *[1 mark]*, which adds
to the flavour *[1 mark]*.
 b) Maximum of 4 marks available.
Lactobacillus produces peptides (from protein) *[1 mark]*
which are used by Streptococcus *[1 mark]*. Streptococcus
produces methanoic acid *[1 mark]* which is used by the
Lactobacillus *[1 mark]*.
 c) Maximum of 2 marks available from any of the following.
Sterilising prevents the addition of other kinds of bacteria
[1 mark] which could be pathogenic / disease-causing /
harmful *[1 mark]* and could spoil the product / flavour
[1 mark].

Page 187 — Biotechnology in Medicine

1 a) Maximum of 2 marks available.
Monoclonal antibodies are antibodies produced by the
descendents *[1 mark]* of a single B-lymphocyte *[1 mark]*.
 b) HCG / human chorionic gonadotropin *[1 mark]*.
 c) Maximum of 3 marks available.
Antigen binds to receptors / cells of B-lymphocytes *[1 mark]*.
These lymphocytes divide / undergo mitosis to form plasma
cells *[1 mark]*. These secrete / release antibodies specific to
the antigen *[1 mark]*.
 d) Maximum of 2 marks available.
Polyethylene glycol enables myeloma and lymphocyte cells to
fuse *[1 mark]* by fusing their cell membranes *[1 mark]*.

Page 189 — Biotechnology in Industry and Public Health

1 a) Maximum of 2 marks available. Possible answers include:
Attach amylase to fibres *[1 mark]* of collagen / cellulose
[1 mark]. OR Encapsulate / enclose it in a latticework
[1 mark] of silica gel / alginate *[1 mark]*.
 b) Maximum of 2 marks available.
Immobilised enzyme molecules move less / have lower
kinetic energy *[1 mark]*, so are less likely to collide with starch
/ substrate *[1 mark]*.
 c) Maximum of 3 marks available.
Enzyme can be reused *[1 mark]*. Enzyme does not
contaminate product *[1 mark]*. Enzyme is more stable / less
likely to be affected / denatured by high temperature or
extremes of pH *[1 mark]*.

2 a) Maximum of 4 marks available.
A source of carbohydrate / named example (e.g. plant
matter) is used *[1 mark]*. Yeast is used to ferment it *[1 mark]*.
This produces ethanol *[1 mark]*. The ethanol is mixed with
petrol *[1 mark]*.
 b) Maximum of 2 marks available.
It uses renewable fuel / biofuel / plant matter *[1 mark]*.
Burning / combustion of ethanol releases fewer polluting
gases / examples (e.g. sulphur dioxide) *[1 mark]*.

Page 191 — Bacterial Disease

1 Maximum of 4 marks available. Possible answers include:
Salmonella bacteria bind to the (epithelium of) the gut wall
[1 mark], where they are taken up by phagocytosis *[1 mark]*.
This causes inflammation *[1 mark]*. Salmonella bacteria
contain endotoxins *[1 mark]*, and their release prevents
absorption of water in the large intestine, leading to diarrhoea
[1 mark].

2 Maximum of 3 marks available.
Staphylococcus bacteria remain in the gut cavity *[1 mark]*
where they release exotoxins *[1 mark]*. Their release prevents
absorption of water in the large intestine, resulting in
diarrhoea *[1 mark]*.

Page 193 — Viral Disease

1 Maximum of 2 marks available.
The latency period is the time between the initial invasion of a
cell and the replication of the virus *[1 mark]* when the virus is
dormant *[1 mark]*.

Answers

2   Maximum of 3 marks available.
     Without the enzyme reverse transcriptase, DNA cannot be made from RNA *[1 mark]*; so DNA cannot be inserted into the host chromosome *[1 mark]*; so the virus cannot reproduce *[1 mark]*.

Page 195 — Protection Against Disease

1   Maximum of 3 marks available.
     Lysosomes fuse with the phagocytic vesicles containing the engulfed bacteria or material, giving phagolysosomes *[1 mark]*. The lysosomes contain hydrolytic enzymes *[1 mark]*, hydrogen peroxide and free radicals to kill bacteria and break down organic material *[1 mark]*.

2   Maximum of 2 marks available.
     As capillaries become more leaky, phagocytes / neutrophils squeeze through pores in capillary wall *[1 mark]* to reach the site of infection *[1 mark]*.

3   Maximum of 3 marks available.
     An antigen is a foreign particle *[1 mark]* that sets up an immune response *[1 mark]*. It is recognised by the body as non-self *[1 mark]*.

Page 197 — Cell and Antibody Mediated Immunity

1 a) Maximum of 6 marks available.
     An antigen binds to receptors on the T-lymphocytes *[1 mark]* or to membrane-bound antibodies on the surface of B-lymphocytes *[1 mark]*. This stimulates mitosis / cell division of the lymphocytes to produce clones *[1 mark]*. These lymphocytes then produce and secrete / release antibodies *[1 mark]* that can bind to / are complementary to the antigen *[1 mark]* making the antigen harmless *[1 mark]*.
  b) Maximum of 2 marks available.
     In the bone marrow and thymus when B- and T-lymphocytes are maturing, any that have receptors that fit self antigens are destroyed *[1 mark]*.

Page 199 — Use of Antibiotics

1   Maximum of 4 marks available. Possible answers include:
     A bacterium in the population has a random mutation *[1 mark]* that causes it to produce the enzyme penicillinase *[1 mark]*. This blocks the action of penicillin *[1 mark]*. The mutant bacteria are more likely to survive when exposed to penicillin *[1 mark]*. The gene for resistance is passed to their offspring and also transferred by conjugation *[1 mark]* via plasmids *[1 mark]*.

2   Maximum of 2 marks available.
     Microbicidal antibiotics kill microorganisms *[1 mark]*, while microbistatic ones inhibit (or reduce) their growth *[1 mark]*.

Section 14 — Human Health and Fitness

Page 201 — The Cardiovascular System

1   Maximum of 6 marks available.
     Cardiac muscle fibres are branched *[1 mark]*. Fibres are connected by intercalated discs in cardiac muscle *[1 mark]*. Striations occur *[1 mark]* caused by overlap of protein filaments *[1 mark]*. Many mitochondria are present *[1 mark]*. Many blood capillaries are present between the muscle fibres *[1 mark]*.

2 a) Maximum of 6 marks available from the following:
     Baroreceptors in the wall of the vena cava are stimulated by the increasing blood pressure *[1 mark]*. An action potential / nerve impulse is sent along the sensory nerve / neurone *[1 mark]* to the cardioaccelerator centre *[1 mark]* in the medulla oblongata *[1 mark]*. An action potential / nerve impulse is sent along the motor nerve / neurone *[1 mark]* which is part of the sympathetic nervous system *[1 mark]* to the SAN, where noradrenaline is released which then increases the heartbeat rate *[1 mark]*.
  b) Maximum of 3 marks available.
     Heartbeat wouldn't be able to speed up or slow down *[1 mark]* because action potentials / nerve impulses from medulla / cardioaccelerator centre in the medulla would not reach SAN *[1 mark]*. In a normally functioning heart these impulses control heartbeat rate *[1 mark]*.

Page 203 — The Pulmonary and Lymphatic Systems

1   Maximum of 4 marks available.
     An action potential / nerve impulse is sent from stretch receptors along the sensory neurone / nerve when the lungs inflate *[1 mark]* to the inspiratory centre of medulla *[1 mark]*. The inspiratory centre is inhibited *[1 mark]* so inspiration stops / expiration begins *[1 mark]*.

2   Maximum of 3 marks available.
     Thin / squamous epithelium / wall of the alveoli minimises the distance that substances have to diffuse *[1 mark]*. The large number of alveoli increases the surface area of the lung for diffusion *[1 mark]*. The presence of many blood capillaries maximises the concentration gradient of respiratory gasses *[1 mark]*.

3   Maximum of 3 marks available from the following:
     The lymphatic system has many lymph nodes *[1 mark]* which contain lymphocytes and neutrophils *[1 mark]*. The presence of antigens in the lymph triggers an immune response *[1 mark]*. Neutrophils are phagocytic *[1 mark]*.

Page 205 — Exercise and the Cardiovascular System

1 a) Maximum of 2 marks available.
     100×48 *[1 mark]* = 4800 cm^3 per minute *[1 mark]*.

  b) Maximum of 6 marks available.
     Exercise involves muscle contraction *[1 mark]* which uses up energy *[1 mark]* and produces CO_2 *[1 mark]*. The CO_2 brings about a decrease in the pH of the blood which is detected by the medulla *[1 mark]*. The medulla sends action potentials which increase the frequency and strength of heartbeats *[1 mark]*. Increasing cardiac output delivers more blood to the muscles per minute / unit time *[1 mark]* which increases the amount of oxygen available for respiration *[1 mark]*.

Answers

2   Maximum of 3 marks available.
Myoglobin acts as oxygen 'store' **[1 mark]** the oxygen in myoglobin is only offloaded when the oxygen concentration in the muscle gets very low **[1 mark]** such as during vigorous exercise **[1 mark]**.

Page 207 — Exercise and the Pulmonary System

1 a) (i) Exercise increases ventilation rate **[1 mark]**.
     (ii) Exercise increases the minute volume **[1 mark]**.

b)   Maximum of 2 marks available from the following:
The inspiratory reserve volume is the maximum volume of air that can be breathed in over and above the tidal volume **[1 mark]**, the value decreases as exercise increases **[1 mark]**.

2   Maximum of 6 marks available.
More capillaries develop around the alveoli **[1 mark]** which improves blood flow to the lungs **[1 mark]**. This increases the concentration difference of oxygen / carbon dioxide / respiratory gases / makes concentration gradients steeper **[1 mark]** which increases rate of diffusion **[1 mark]** of oxygen into the blood **[1 mark]** and carbon dioxide out of the blood **[1 mark]**. This means that more oxygen is carried to the muscles / more carbon dioxide removed from muscles **[1 mark]**.

Page 209 — Exercise and the Musculo-skeletal System

1 a) Maximum of 3 marks available.
Insufficient oxygen is delivered to the active muscles **[1 mark]** so anaerobic respiration begins to take place **[1 mark]** and lactate is a product of anaerobic respiration which cannot be removed from the blood quickly **[1 mark]**.

b) Maximum of 2 marks available.
Oxygen debt is the volume / amount of oxygen needed to get rid of the lactic acid **[1 mark]** that has accumulated in active muscles. The lactic acid is removed by oxidation **[1 mark]**.

2   Maximum of 3 marks available.
Taking creatine as a dietary supplement would mean that more phosphocreatine could be made **[1 mark]**. The phosphocreatine donates phosphate to ADP to make ATP **[1 mark]**. This extra ATP would provide more energy for muscle contraction **[1 mark]**.

3   Maximum of 6 marks available from the following.
Initially ATP from phosphocreatine is used **[1 mark]** — it is broken down to release phosphate for converting ADP to ATP **[1 mark]**. After that ATP from anaerobic respiration is used **[1 mark]** resulting in production of lactic acid **[1 mark]**; then ATP from aerobic respiration **[1 mark]**; from oxidative phosphorylation **[1 mark]**.

Page 211 — Training

1 a) Maximum of 5 marks available.
Anaerobic conditioning is needed in preparation for a sprint **[1 mark]**. This anaerobic training enables the muscles to get more ATP from the two anaerobic systems: the phosphocreatine / ATP-PCr system **[1 mark]** and the lactic acid system **[1 mark]**. Anaerobic training involves brief periods of rest **[1 mark]** to prevent lactate from building up too much **[1 mark]**.

b) Maximum of 6 marks available.
Glycogen loading is a method of increasing the glycogen available for muscle contraction in preparation for a specific event e.g. a marathon **[1 mark]**. A period of carbohydrate starvation takes place for a few days ahead of the event **[1 mark]**. This is followed by a diet high in carbohydrate **[1 mark]**. The body then overcompensates and large amounts of glycogen are stored **[1 mark]**. This means that there is a large amount of glycogen in the muscles during the event **[1 mark]** which helps to delay fatigue **[1 mark]**.

2   Maximum of 3 marks available from the following:
Trained muscle has more capillaries / a better blood supply **[1 mark]**. Trained muscle has more mitochondria / better ATP production **[1 mark]**. Trained muscle has a larger cross-sectional area **[1 mark]**. Trained muscle has thicker fibres **[1 mark]**. Trained muscle contains more myglobin **[1 mark]**. Trained muscle has more respiratory enzymes **[1 mark]**.

Section 15 — Food Science

Page 213 — Balanced Diet

1   Maximum of 5 marks available.
a) Rate of metabolism / energy release when the body is at rest **[1 mark]**.
b) Metabolism involves respiration **[1 mark]** which uses up oxygen and releases heat energy **[1 mark]**.
c) More body mass means more respiring tissue / cells **[1 mark]** which releases more energy / uses up more oxygen **[1 mark]**.

2   Maximum of 2 marks available.
Either of the following pairs of points, with 1 mark for each point in a pair: child needs relatively more protein **[1 mark]** to grow **[1 mark]**, child needs more calcium **[1 mark]** for development of bones / teeth **[1 mark]**.

3   Maximum of 8 marks available.
(Dietary) protein is hydrolysed in the gut **[1 mark]** to release amino acids **[1 mark]**. Amino acids are absorbed into the blood **[1 mark]** (taken to cells) and used for protein synthesis in cells **[1 mark]**. Muscle (fibres) are composed of protein **[1 mark]**, so if more protein is synthesised, more can be used for making muscle fibres **[1 mark]**. Excess amino acids are deaminated **[1 mark]** in the liver **[1 mark]** to form urea **[1 mark]**, so more amino acids leads to greater urea production **[1 mark]**.

Page 215 — Under-Nutrition

1 a) Maximum of 4 marks available.
Vitamin C acts as a cofactor / coenzyme (for the enzyme needed to make collagen) **[1 mark]**. The enzyme converts proline into hydroxyproline **[1 mark]** so that hydrogen bonds can form (between hydroxyproline units / residues) **[1 mark]**. Collagen needs to be replaced for wounds to heal **[1 mark]**.
b) Maximum of 2 marks available.
Citrus fruits **[1 mark]**, green vegetables **[1 mark]**, potatoes **[1 mark]**. Also, allow separate marks for named examples of citrus fruits.

Answers

2   Maximum of 4 marks available.
    Low protein levels in the blood / plasma [1 mark] make water potential higher here / solute concentration lower here [1 mark], so less water reabsorbed back into blood / capillary (from tissue fluid) [1 mark] by osmosis [1 mark].

3   Maximum of 6 marks available.
    Less iron in a vegetarian diet [1 mark]. Iron in a vegetarian diet is non-haem [1 mark], which is absorbed less effectively than iron from meat / haem iron [1 mark]. This means less iron is absorbed into the blood [1 mark], so less haemoglobin is made [1 mark]. Less haemoglobin leads to anaemia [1 mark].

Page 217 — Over-Nutrition

1 a) Maximum of 8 marks available.
    Cholesterol and saturated fats in the blood (plasma) lead to fatty deposits in artery walls [1 mark], beneath the artery endothelium [1 mark], called atheromas [1 mark]. This is called atherosclerosis [1 mark] and it causes narrowing of the arteries [1 mark]. If this happens in coronary arteries [1 mark], it restricts blood flow / oxygen carriage to cardiac / heart muscle [1 mark], resulting in heart attack / myocardial infarction [1 mark].
  b) Maximum of 1 mark available for any of the following:
    Increased intake of polyunsaturates [1 mark], decreased intake of cholesterol / saturated fats [1 mark], increased intake of soluble fibre [1 mark], increased exercise [1 mark].

2   Maximum of 3 marks available.
    Food / faeces stay in the intestine for less time [1 mark], so food / faeces are in contact with the intestine wall for less time [1 mark]. This means that harmful substances have less time to affect the wall [1 mark].

Page 219 — Food Additives

1 a) Maximum of 3 marks available.
    Sufferers do not produce lactase (in the intestine) [1 mark], so cannot hydrolyse lactose [1 mark]. Lactose gets fermented by bacteria in the intestine / gut, causing indigestion [1 mark].
  b) Maximum of 2 marks available.
    Lactase is added to the food [1 mark] to hydrolyse the lactose [1 mark] and make the food digestible.

2 a) Maximum of 3 marks available.
    Fructose is the sweetest [1 mark], so less fructose is needed to make food as sweet as sucrose would [1 mark]. This means fewer calories / less energy is taken in [1 mark].
  b) Maximum of 4 marks available.
    Immobilised enzymes can be reused [1 mark] and are more stable [1 mark]. The rate of reaction is higher at high temperatures [1 mark] and higher temperatures also reduce growth of microorganisms [1 mark].

Page 221 — Other Additives

1 a) Maximum of 4 marks available.
    Flavouring can be tasted [1 mark], examples: spices / herbs / named example, e.g. vanilla [1 mark]. Flavour enhancers increase the flavour of other substances [1 mark], examples: monosodium glutamate / sodium chloride [1 mark].

  b) Maximum of 6 marks available.
    *i) High concentration of salt causes microorganisms / cells to lose water by osmosis [1 mark] so that they can't grow / multiply / carry out metabolism properly [1 mark].
    ii) Vinegar has a low pH / is acidic [1 mark] which denatures enzymes / proteins in microorganisms so that they can't metabolise properly [1 mark].
    iii) Sulphites lower the oxygen concentration [1 mark] so that microorganisms are unable to respire aerobically [1 mark].*

Page 223 — Food Storage

1 a) Maximum of 3 marks available.
    Pectin makes up middle lamellae (between cells) [1 mark], so pectinase breaks down the middle lamellae [1 mark]. This means cells lose their cell sap / juice [1 mark].
  b) Maximum of 4 marks available.
    Respiration uses up sugar [1 mark] so reduces sweetness [1 mark]. Ripening converts malic acid to sugar [1 mark], so increases sweetness [1 mark].

2   Maximum of 8 marks available.
    Pasteurisation involves heating (milk) to 65°C [1 mark], sterilisation to at least 100 °C [1 mark]. Pasteurisation kills off harmful bacteria / does not kill off all bacteria [1 mark], sterilisation kills off more bacteria, but not spores [1 mark]. Pasteurised food must be refrigerated [1 mark] to reduce the growth / multiplication of bacteria [1 mark], sterilised food does not need to be refrigerated until it is opened, because there will be no viable bacteria [1 mark]. Shelf life of sterilised food is longer than that of pasteurised foods [1 mark]. Taste of sterilised food may be affected (more than that of pasteurised food) [1 mark].

Page 225 — Microorganisms in Food

1 a) Maximum of 2 marks available.
    Salt makes the sauerkraut lose water by osmosis [1 mark]. Anaerobic conditions are needed for the bacteria that produce lactic acid and other flavours [1 mark].
  b) Maximum of 2 marks available.
    In sauerkraut production lactic acid is produced, in wine production ethanol is produced [1 mark]. In sauerkraut production bacteria carry out the fermentation, in wine production yeast (a fungus) carries out the fermentation [1 mark].

2   Maximum of 6 marks available.
    Fermentation is done by microorganisms acting on starch / flour and boiled soy bean mixture [1 mark]. The mould Aspergillus oryzae [1 mark] produces amylases / proteases / enzymes [1 mark] to release sugars / amino acids [1 mark: product given must be appropriate for stated enzyme]. Bacillus / Lactobacillus [1 mark] produce(s) lactic acid [1 mark]. Yeast / Saccharomyces rouxii [1 mark] makes alcohol [1 mark].

Index

Index

Index

Index

Index